Severe Plastic Deformation and Thermomechanical Processing Nanostructuring and Properties

Severe Plastic Deformation and Thermomechanical Processing Nanostructuring and Properties

Editors

Andrea Bachmaier
Thierry Grosdidier
Yulia Ivanisenko

MDPI • Basel • Beijing • Wuhan • Barcelona • Belgrade • Manchester • Tokyo • Cluj • Tianjin

Editors
Andrea Bachmaier
Erich Schmid Institute of
Materials Science of the Austrian
Academy of Sciences
Austria

Thierry Grosdidier
Université de Lorraine
France

Yulia Ivanisenko
Karlsruhe Institute of Technology
Germany

Editorial Office
MDPI
St. Alban-Anlage 66
4052 Basel, Switzerland

This is a reprint of articles from the Special Issue published online in the open access journal *Metals* (ISSN 2075-4701) (available at: https://www.mdpi.com/journal/metals/special_issues/severe_plastic_deformation).

For citation purposes, cite each article independently as indicated on the article page online and as indicated below:

LastName, A.A.; LastName, B.B.; LastName, C.C. Article Title. *Journal Name* **Year**, *Volume Number*, Page Range.

ISBN 978-3-03943-687-3 (Hbk)
ISBN 978-3-03943-688-0 (PDF)

Contents

About the Editors

Andrea Bachmaier (Dr.) studied Materials Science at the University of Leoben and received her Ph.D. degree in 2011. In 2013, she was awarded an Erwin-Schrödinger scholarship that allowed her to work on the decomposition process of supersaturated solid solutions and its influence on thermal stability. In 2017, she received the noteworthy ERC Starting Grant. Currently, she is leading the research group "Novel bulk nanomaterials by severe plastic deformation" at the Erich Schmid Institute of Materials Science, Austria. Her research activities focus on the generation of metastable materials, novel nanocomposites and nanocrystalline metal matrix composites by SPD and the investigation of their functional and mechanical properties.

Thierry Grosdidier (Prof.) is a metallurgist, working mostly in microstructure–properties relationships, taking into account texture and processing conditions. He received his Ph.D. degree in Material Science from Institut National Polytechnique de Lorraine—Ecole des Mines de Nancy (France) in 1993 and carried on his research career at the University of Surrey (UK) and at the Université de Technologie Belfort Montebeliard (France). He is now a full time Professor at the University of Lorraine, France. He was Deputy Director of the Scientific Federation Genie Industriel Mecanique Materiaux (5 laboratories, 250 employees) in Metz (France) from 2006 to 2012 and Scientific Expert at the "Advanced Light Metal Innovation Center"—Shanghai Jiao Tong University (China) from 2015 to 2019. He is now co-responsible for the "Impact" department at LEM3 laboratory. Co-author of more than 150 scientific publications, his current topic of interest are (i) powder metallurgy, (ii) surface modifications, (iii) severe plastic deformation and (iv) improvement of H-storage in metal hydrides.

Yulia Ivanisenko (Dr.) graduated with a Diploma in Mechanical Engineering at Ufa State Aviation Technical University and received her Ph.D. in applied physics from the Institute for Metals Superplasticity Problems of Russian Academy of Sciences in Ufa, Russia in 1985 and 1997, respectively. She worked as a Research Scientist at the Institute for Nanotechnology, Karlsruhe Institute for Technology since September 2003, where she leads a research group now. Dr. Ivanisenko's research is mostly in the areas of severe plastic deformation, mechanical properties of nanostructured materials and mechanically driven phase transformations.

Editorial

Severe Plastic Deformation and Thermomechanical Processing: Nanostructuring and Properties

Andrea Bachmaier [1], Thierry Grosdidier [2] and Yulia Ivanisenko [3,*]

[1] Erich Schmid Institute of Materials Science of the Austrian Academy of Sciences, Jahnstrasse 12, 8700 Leoben, Austria; andrea.bachmaier@oeaw.ac.at

[2] Laboratoired'Etude des Microstructures et de Mécanique des Matériaux (LEM3), Université de Lorraine, CNRS UMR 7239, 7 rue Félix Savart, 57073 Metz, France; thierry.grosdidier@univ-lorraine.fr

[3] Karlsruhe Institute of Technology, Campus North, Institute of Nanotechnology, 76344 Eggenstein-Leopoldshafen, Germany

* Correspondence: julia.ivanisenko@kit.edu; Tel.: +49-721-60826961

Received: 21 September 2020; Accepted: 28 September 2020; Published: 29 September 2020

1. Introduction and Scope

The research field of severe plastic deformation (SPD) offers innovative potential for manufacturing bulk metallic materials as well as for modifying their surfaces. Significant grain refinement can be obtained using hot, warm, cold, and even cryogenic deformations by SPD and Thermo-Mechanical Processing (TMP), or a combination thereof. In addition to grain refinement in the respective phases of the metallic materials, microstructural design on different hierarchical levels and even alloy design by phase formations and transformations during processing is possible by SPD and TMP. Examples include mechanically driven phase transformations, the formation of metastable phases, grain boundary engineering, and the formation of desirable textures.

This Special Issue contains the selected papers presented at the 3rd Symposium, "Severe Plastic Deformation and Thermomechanical Processing: Nanostructuring and Properties", which was organised as a part of EUROMAT congress in Stockholm, Sweden, 1–5 September, 2019.

The published papers report the use and recent advancements of different established as well as novel SPD processes like high-pressure torsion (HPT), Equal-Channel Angular Pressing (ECAP), High-Pressure Torsion Extrusion (HPTE) and Surface Mechanical Attrition Treatment (SMAT). In the scope of these studies are microstructural evolution, phase formations and grain refinement in single- and multi-phase alloys, strategies to enhance the microstructural stability at elevated temperatures as well as during thermal cycling and the improvement of mechanical and physical properties by SPD processing. Additionally, the biomedical properties of SPD-processed materials are investigated.

2. Contributions

Thirteen research papers have been published in this Special Issue of *Metals*. The papers cover a wide spectrum of topics, including (i) deformation mechanisms, the mechanical properties and fatigue behaviour of SPD-processed materials [1–7], (ii) microstructural and mechanical stability at elevated temperatures [8,9], phase formations in alloys [10], the improvement of physical properties (electrical conductivity [11], magneto-resistance [12]) and biomedical properties (biocorrosion resistance [6], biocompatibility [13]) by SPD.

In the thematic area (i), Klu et al. [1] investigated a combination of multi-pass warm ECAP followed by rolling at room temperature for the development of a high-strength Mg-9Li duplex alloy and carried out a systematic investigation of the microstructure–mechanical properties relationship. They found that grain boundary strengthening and dislocation strengthening are the main factors determining the strength of this alloy. Additionally, the basal texture of the α-Mg phase, which was induced by the rolling process, contributed to the strength. Nugmanov et al. [2] examined the structure and tensile strength of pure Cu after HPTE. A gradient structure developed due to the strain varying across the sample from the central area to the edge. Nevertheless, similar strength values as in pure Cu with a homogeneous microstructure after SPD were reached, which demonstrates the potential of HPTE for future industrial use. Stráská et al. [3] processed a low alloyed Mg-Zn-Nd alloy by hot extrusion and subsequent ECAP deformation. The influence of texture and microstructure on the mechanical properties and deformation mechanisms was thoroughly studied and a compressive yield strength more than twice as high as that of the undeformed alloy was reached. Dureau et al. [4] investigated the influence of ultrasonic SMAT treatment at room and cryogenic temperatures on the fatigue behaviour of a 304L austenitic stainless steel with a focus on the nature of the cyclic loading conditions. In the case of this steel, the higher fraction of martensite induced by the cryogenic SMAT did not provide an enhancement of fatigue performance compared with the room temperature treatment. However, the fatigue limit was increased by approximately 30% for both peening temperatures in comparison with the untreated samples. Industrial scale multi-pass rotary-die ECAP processing was used in [5] to tailor the morphology and distribution of Al_2Ca particles in a Mg-Al-Ca-based alloy to improve its strength and ductility. The alloy with the finest and most homogenously dispersed Al_2Ca particles exhibited superior mechanical properties, which were also attributed to refined grains of the α-Mg phase and nanosized $Mg_{17}Al_{12}$ precipitates. The effects of thermomechanical processing (SPD by HPT and thermal treatment) on the mechanical properties of biodegradable Mg alloys were investigated by Ojdanic et al. in [6]. Thermomechanical processing lead to a strength increase of up to 250%, whereby about 1/3 of the increase could be related to the thermal treatment. Differential scanning calorimetry and X-ray line profile analysis proofed a significant contribution of the high vacancy concentration to the extensive hardening of the investigated alloys. Furthermore, intermetallic precipitates contributed to the strength. Veverková et al. [7] evaluated the mechanical properties of a metastable β-Ti alloy Ti-15Mo, which was prepared by cryogenic milling and spark plasma sintering. By using this process, a refined microstructure with very high strength levels could be obtained.

In the framework of topic (ii), the microstructural and mechanical stability during elevated temperatures is discussed in two papers. Kriegel et al. [8] investigated the formation and thermal stability of the ω-Ti(Fe) phase in α-phase-based Ti(Fe) alloys, which were processed by HPT. The formation of the ω-Ti(Fe) phase was mainly at the expense of α-Ti. The thermal stability of the studied alloys was lower than that of samples, annealed above the eutectoid reaction. However, a similar decomposition pathway was found. Churakova and Gunderov [9] analysed the influence of thermal cycling on the microstructural and mechanical stability of a Ti-Ni shape memory alloy, which was processed by ECAP, and compared it to a coarse-grained, undeformed counterpart. It was found that the ECAP-processed alloy is more attractive for applications, due to its higher level of properties compared to the coarse-grained state, and its higher stability during thermal cycling.

The effect of HPT and subsequent isothermal annealing on phase evolution and transformation in a Ti15Mo alloy was studied by Bartha et al. in [10]. They showed that, thanks to the increase in nucleation sites due to HPT-induced lattice defects, the α-phase formation is enhanced. Furthermore, it was found that the α-precipitates are small and equiaxed. Additionally, the microhardness is increased, which is mainly attributed to the microstructural refinement in combination with the formation of the ω phase.

Finally, the improvement of physical and biomedical properties by SPD is investigated in three papers. Lapovok et al. [11] studied the influence of ECAP on the electrical conductivity and strength of Cu-clad Al conductors with different sheet thicknesses, which were subsequently annealed. Although the conductivity decreased after deformation, annealing led to conductivity values exceeding the predicted, theoretical ones. It could be shown that ECAP in combination with short annealing can be used to produce conductors with high conductivity and strength. Wurster et al. [12] investigated the influence of HPT such as deformation temperature and strain rate on the granular magnetoresistance of different materials consisting of ferromagnetic and diamagnetic elements and related it to the microstructure of SPD-processed materials. It is shown that the magnitude of the granular magnetoresistance can be tuned by changing the HPT process parameters. In [13], the effect of SMAT was compared for two binary functionally graded materials (Ti-Nb and Ti6Al4V-Mo) to study the effect of chemistry, roughness and SPD microstructure on mesenchymal stem cell adhesion and proliferation. The increased roughness introduced by SMAT improved the cellular adhesion, but did not influence their proliferation capability. It was further found that the SPD treatment has an effect on cell distribution during the first stages of proliferation due to the induced microstructural refinement and structural defects.

3. Conclusion and Outlook

The current Special Issue of *Metals* provides a comprehensive insight into current research in the field of SPD. The papers cover several research topics and we hope that this Special Issue will be a starting point for future scientific discussions. As Guest Editors of this Special Issue, we hope that the papers will catch the interest of many scientists and will be useful for their future work.

Acknowledgments: We would like to express our deep gratitude to all the authors for their contributions. We further thank the anonymous reviewers for their efforts to ensure high-quality publications were accepted. Sincere thanks are also due to the editors and editorial assistants of *Metals* for their help and support during the preparation of this Special Issue.

Conflicts of Interest: The authors declare no conflict of interest.

References

1. Klu, E.E.; Song, D.; Li, C.; Wang, G.; Zhou, Z.; Gao, B.; Sun, J.; Ma, A.; Jiang, J. Development of a High Strength Mg-9Li Alloy via Multi-Pass ECAP and Post-Rolling. *Metals* **2019**, *9*, 1008. [CrossRef]

2. Nugmanov, D.; Mazilkin, A.; Hahn, H.; Ivanisenko, Y. Structure and Tensile Strength of Pure Cu after High Pressure Torsion Extrusion. *Metals* **2019**, *9*, 1081. [CrossRef]

3. Stráská, J.; Minárik, P.; Šašek, S.; Veselý, J.; Bohlen, J.; Král, R.; Kubásek, J. Texture Hardening Observed in Mg–Zn–Nd Alloy Processed by Equal-Channel Angular Pressing (ECAP). *Metals* **2019**, *10*, 35. [CrossRef]

4. Dureau, C.; Novelli, M.; Arzaghi, M.; Massion, R.; Bocher, P.; Nadot, Y.; Grosdidier, T. On the Influence of Ultrasonic Surface Mechanical Attrition Treatment (SMAT) on the Fatigue Behavior of the 304L Austenitic Stainless Steel. *Metals* **2020**, *10*, 100. [CrossRef]

5. Wang, C.; Ma, A.; Sun, J.; Zhuo, X.; Huang, H.; Liu, H.; Yang, Z.; Jiang, J. Improving Strength and Ductility of a Mg-3.7Al-1.8Ca-0.4Mn Alloy with Refined and Dispersed Al2Ca Particles by Industrial-Scale ECAP Processing. *Metals* **2019**, *9*, 767. [CrossRef]

6. Ojdanic, A.; Horky, J.; Mingler, B.; Fanetti, M.; Gardonio, S.; Valant, M.; Sulkowski, B.; Schafler, E.; Orlov, D.; Zehetbauer, M.J. The Effects of Severe Plastic Deformation and/or Thermal Treatment on the Mechanical Properties of Biodegradable Mg-Alloys. *Metals* **2020**, *10*, 1064. [CrossRef]

7. Veverková, A.; Kozlík, J.; Bartha, K.; Chráska, T.; Corrêa, C.A.; Stráský, J. Mechanical Properties of Ti-15Mo Alloy Prepared by Cryogenic Milling and Spark Plasma Sintering. *Metals* **2019**, *9*, 1280. [CrossRef]

8. Kriegel, M.J.; Rudolph, M.; Kilmametov, A.; Straumal, B.B.; Ivanisenko, J.; Fabrichnaya, O.; Hahn, H.; Rafaja, D. Formation and Thermal Stability of ω-Ti(Fe) in α-Phase-Based Ti(Fe) Alloys. *Metals* **2020**, *10*, 402. [CrossRef]

9. Churakova, A.; Gunderov, D. Microstructural and Mechanical Stability of a Ti-50.8 at.% Ni Shape Memory Alloy Achieved by Thermal Cycling with a Large Number of Cycles. *Metals* **2020**, *10*, 227. [CrossRef]

10. Bartha, K.; Stráský, J.; Veverková, A.; Barriobero-Vila, P.; Lukáč, F.; Doležal, P.; Sedlák, P.; Polyakova, V.; Semenova, I.; Janeček, M. Effect of the High-Pressure Torsion (HPT) and Subsequent Isothermal Annealing on the Phase Transformation in Biomedical Ti15Mo Alloy. *Metals* **2019**, *9*, 1194. [CrossRef]

11. Lapovok, R.; Dubrovsky, M.; Kosinova, A.; Raab, G. Effect of Severe Plastic Deformation on the Conductivity and Strength of Copper-Clad Aluminium Conductors. *Metals* **2019**, *9*, 960. [CrossRef]

12. Wurster, S.; Weissitsch, L.; Stückler, M.; Knoll, P.; Krenn, H.; Pippan, R.; Bachmaier, A. Tuneable Magneto-Resistance by Severe Plastic Deformation. *Metals* **2019**, *9*, 1188. [CrossRef]
13. Weiss, L.; Nessler, Y.; Novelli, M.; Laheurte, P.; Grosdidier, T. On the Use of Functionally Graded Materials to Differentiate the Effects of Surface Severe Plastic Deformation, Roughness and Chemical Composition on Cell Proliferation. *Metals* **2019**, *9*, 1344. [CrossRef]

Publisher's Note: MDPI stays neutral with regard to jurisdictional claims in published maps and institutional affiliations.

Article

Development of a High Strength Mg-9Li Alloy via Multi-Pass ECAP and Post-Rolling

Edwin Eyram Klu [1], Dan Song [1,2,3,*], Chen Li [1], Guowei Wang [1], Zhikai Zhou [1], Bo Gao [4], Jiapeng Sun [1], Aibin Ma [1,2] and Jinghua Jiang [1,*]

[1] College of Mechanics and Materials, Hohai University, Nanjing 210098, China;
    kluedwin@outlook.com (E.E.K.); muzi6318@163.com (C.L.); Guowei_wang9606@163.com (G.W.);
    18260056968@163.com (Z.Z.); sun.jiap@gmail.com (J.S.); aibin-ma@hhu.edu.cn (A.M.)
[2] Suqian Research Institute of Hohai University, Suqian 223800, China
[3] Nantong Research Institute of Materials Engineering, Nanjing University, Nantong 226000, China
[4] School of Material Science and Engineering, Nanjing University of Science and Technology, Nanjing 210094,
    China; gaobo@njust.edu.cn
*   Correspondence: songdancharls@hhu.edu.cn (D.S.); jinghua-jiang@hhu.edu.cn (J.J.);
    Tel.: +86-25-8378-7239 (D.S.); Fax: +86-25-8378-6046 (D.S.)

Received: 29 August 2019; Accepted: 11 September 2019; Published: 14 September 2019

Abstract: In this study, a high-strength Mg-9Li alloy was developed via multi-pass equal-channel-angular-pressing (ECAP) and post rolling, of which the yield tensile stress (YTS) and ultimate tensile stress (UTS) were 166 MPa and 174 MPa representing about 219% and 70% increase in YTS and UTS respectively, compared to the cast alloy. The cast alloy was ECAP processed at 200 °C for 4, 8, and 16 passes, followed by room-temperature rolling to a total thickness reduction of 50%. The 8-passes ECAPed (E8) alloy presented the best strength of all the ECAPed alloys, and the post rolling endowed the alloy (E8R) further strengthening and the best strength of all the alloys. Grain-boundary strengthening and dislocation strengthening were the two major factors for the high strength of the processed alloys. The α-Mg phase grains were greatly refined to about 2 μm after 8-passes ECAP, and was further refined to about 800 nm ~1.5 μm after rolling. Significant grain refinement endowed the alloy with sufficient grain-boundary strengthening. Profuse intragranular dislocation accumulated in the deformed matrix, leading to the significant dislocation hardening of the alloy. Rolling-induced strong basal texture of the α-Mg phase also enhanced the further strengthening of the E8R alloy.

Keywords: Mg-9Li duplex alloy; ECAP; rolling; high strength; microstructure

---

1. Introduction

Magnesium (Mg) alloys have high specific strength and ductility resulting in their use in various applications from automobile, aerospace, and orthopedic applications. However, its low cold workability near room temperature due to its hexagonal packed crystal structure limits its use [1–3]. Magnesium alloys have been classified by many researchers as a super lightweight structural metal, attracting a lot of attention in the automobile and aerospace industries where production of light, super-fast machines requiring less fuel consumption through weight reduction is of high priority [4,5]. Magnesium–Lithium (Mg–Li) alloys are gaining more and more interest in scientific research studies as well as in industrial applications owing to their super lightweight, high specific strength and good formability making it one of the most ideal structural materials for 3C intelligent electronics, medical devices etc. [6–8].

Many researchers have employed the use of alloying and various mechanical procedures to enhance the mechanical property of Mg-Li base alloys [9,10]. Notable amongst them is the ECAP (equal channel angular pressing) process for achieving ultrafine grains (UFG) by severe plastic deformation (SPD) [11,12]. ECAP has advantages of producing large bulk materials for various industrial applications, with improvement in strength and ductility properties [13–15]. Conventional ECAP process however has typical drawbacks of reinserting billets into the die for every pass, resulting in inconsistencies in temperature applied for each pass [16–18]. Some researchers have therefore resorted to the development of novel processes with higher production efficiency, such as repetitive upsetting (RU) [19–21] and rotary die equal channel angular pressing (RD-ECAP) [22–24]. Also, another effective plastic deformation in use is the rolling technique. Magnesium alloys have a dense hexagonal structure with few slip systems. This makes it easy to induce stresses when rolling at large strain rates which results in very low yield of magnesium alloy sheets [25]. However, the addition of lithium to the magnesium alloy forming Mg–Li alloy results in the formation of a β-rich Li phase which greatly increases the dislocation slip system and improves the plastic deformation ability of the Mg-Li alloys [26]. The results obtained from the reported research [27,28] show that, the use of rolling technique leads to an improvement in the strength properties of the Mg-Li alloy, activate more slip systems, enhance the ability of intergranular co-opening and improve the ability of plastic deformation at room temperature (RT). In the rolling process, strong basal texture will be formed [29]. Research done by the reference [28] employed unidirectional (transverse and longitudinal) and cross (combination of transverse and longitudinal) rolling routes to improve the mechanical properties of Mg-9Li-1Al alloy. The unidirectional rolling routes reached 170 MPa tensile strength whereas the cross rolling route reached 243 MPa respectively. During the rolling process, compression twins may be formed. Therefore, in the rolling process, the original structure can be refined by pretreatment and the morphology and texture of the material can be controlled during the rolling process. Different researchers [30,31] have employed the use of ECAP with further rolling, applying it in a wide variety of alloys and have achieved sufficiently good tensile strength and ductility.

This research therefore seeks to investigate the effect of employing a combination of RD-ECAP with further room-temperature rolling techniques on the microstructural evolution and mechanical property changes of Mg-9Li duplex alloy.

2. Experimental

2.1. Materials and Processing

The raw material Mg-9Li alloy ingots used in the experiment was purchased from Jiangsu Li Mg Aero Material Co., Ltd. The composition of the alloy was analyzed by GNR S3 spark direct reading spectrometer. The results are shown in Table 1. The raw material can be identified as Mg-9Li alloy according to the mass ratio of Mg to Li element. According to the binary phase diagram of Mg–Li alloy, the lithium content of the alloy in this paper falls in the range (between 5.7–10.3 wt. %) of the (α + β) duplex phase structure.

Table 1. Analysis of Mg-9Li alloy composition (wt. %).

Mg	Li	Fe	Mn	Zn	Cd
91.052	8.809	0.010	0.024	0.014	0.030

ECAP was conducted at 200 °C with the samples extruded for 4, 8, and 16 passes. The rotary Die equal channel angular pressing (RD-ECAP) set-up as diagrammatically illustrated in Figure 1a with detailed information obtained from our previous investigation [22,27] consists of a plunger for forcing the sample through the die orifice with all four sides of the die sealed with punches with the upper punches protruding out of the die to be pressed by the plunger. This is used for continuously processing the samples. This method is in tandem with the principles of the conventional ECAP with the extruded inner angle being 90°. The advantage over the conventional ECAP set-up is that unlike in the former, each pass needs to be rotated and then refilled with the mold and rotated clockwise at the end of each pass as shown in Figure 1a. The rotary die can save a lot of experiment time, shorten the holding time of each pass and effectively reduce the dynamic recovery behavior after long time heating [27]. The inner parts of the die is cleaned with graphite emulsion to reduce friction during the extrusion process. After each pass, the sample squeezes from the upper channel into the left channel. The mold is then rotated clockwise with the sample reverting to its initial upper channel position, which is ready for the next pass. The as-cast sample is heated to 200 °C for half an hour in an oven and then processed for 4, 8, and 16 passes respectively. The preparation of the 16-pass ECAP sample requires another half an hour reinsertion of the die into the furnace after 8 passes is completed before performing the further 8 passes. This is to ensure that the sample does not crack during the extrusion process.

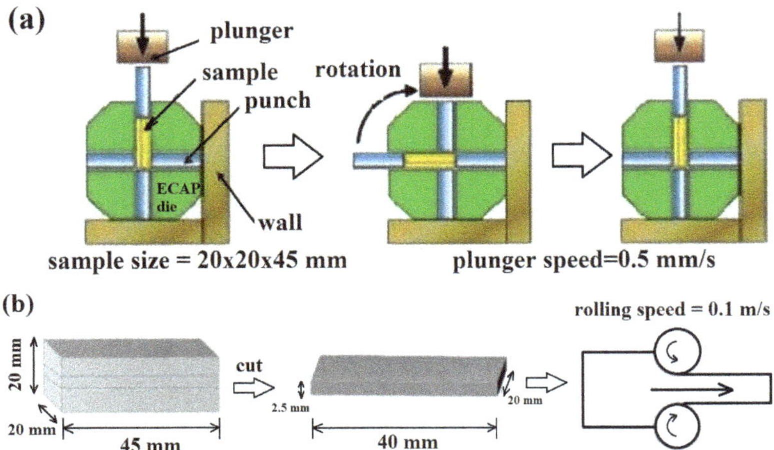

Figure 1. Schematic diagram of (**a**) rotary die equal channel angular pressing (RD-ECAP) process and (**b**) rolling technique [27].

The ECAPed specimens are further rolled at room temperature (RT) to obtain high strength and toughness. All the ECAP-processed specimens were further rolled along a $20 \times 40 \times 2.5$ mm^3 thin section perpendicular to the extruded surface using a rolling machine as diagrammatically illustrated in Figure 1b with detailed information obtained from references [27,31]. Multi-pass rolling at room temperature was used via the rolling speed of 0.1 mm·s^{-1} in this process. Each pass reduction was 10% until the total reduction reached 50%. This method can effectively prevent the sample from cracking during the rolling process. The height between the two rollers were adjusted with the speed controlled. After each pass, the sample is reversed for the next roll to ensure the uniformity of the rolling. Each sample is designated with a specific name. The as-cast alloy, ECAPed alloys with 4, 8, 16 passes, cast-rolled alloy, 4-passes ECAP plus rolling alloy, 8-passes ECAP plus rolling alloy and 16-passes ECAP plus rolling alloy are given designated names of C, E4, E8, E16, CR, E4R, E8R, and E16R respectively.

2.2. Microstructure Characterization

The microstructure of Mg-9Li dual phase alloy after ECAP and rolling was analyzed by Optical Microscopy (OM, Olympus BX51M, Tokyo, Japan). The samples were cut into 10×10 mm small squares, then grounded using sandpaper increasing the grit size after each grinding cycle. The samples were then polished using 0.5 μm Al$_2$O$_3$ suspension solution until a mirror-like specimen surface was observed. The polished samples were etched with a solution mixture of 4.3 mL picric acid, 95 mL ethanol, and 0.7 mL phosphoric acid.

Transmission electron microscopy (TEM, FEI Tecnai G2, Hillsboro, OR, USA) was applied to observe the microstructure and grain size of the Mg-9Li alloy after ECAP and post-rolling. The samples were cut into a $10 \times 10 \times 0.6$ mm^3 pieces. The cut piece was grounded using a 1000 and 2000 grit sandpaper to about 100 μm. The sample was then punched into circular disks of 3 mm diameter and a special auxiliary equipment was used to polish the disc to 20–30 μm on the 1500 grit sandpaper. The TEM foil was finally perforated by ion milling (Gatan 695C, Pleasanton, CA, USA).

The phase composition of Mg-9Li dual phase alloy in different states was qualitatively characterized by X-ray diffractometer with a Bruker D8 (XRD, Karlsruhe, Germany) with test conditions: Cu target Ka-ray, tube current 40 mA, scanning range 10–90°, scanning speed 5°/min. The texture of the samples in different states were observed by XRD. The samples were polished using a 1000 grit sandpaper to ensure a smooth surface and etched with 3% nitric acid to remove its surface stress with the direction of extrusion and rolling indicated on the sample surface.

2.3. Tensile Test

Tensile test of the Mg-9Li alloys were conducted using the uniaxial tensile test at room temperature via a UTM4204X electronic tensile machine (Suns, Shenzhen, China) with a strain rate of 1×10^{-3} s^{-1}. A dog bone shape of dimensions $6 \times 2 \times 2$ mm^3 was used with at least five samples tested for each process sample and the average value obtained. The mechanical properties were measured and the fracture morphologies of the different processed alloys were observed using a scanning electron microscope (FEI Quanta 3D FEG, Hillsboro, OR, USA).

3. Results

3.1. Microstructure of ECAPed and ECAP-Rolled Alloys

The optical microstructure of the cast alloy and ECAPed alloys with different processing passes are shown in Figure 2. As seen in Figure 2a, the grey phase of the cast alloy is indicative of the α-rich Mg phase and the darker phase is β-rich Li phase. Clearly, the α-Mg phase is large and irregularly distributed in the β-Li matrix. After ECAP processing, the microstructure of α-Mg phase is continuously deformed with increase in number of passes and the reduction in grain size accordingly. As shown in Figure 2b, after four passes, the α-Mg phase shows elongated phases distributed in a particular angle showing a distinct plastic flow pattern. In addition, compared with the as-cast microstructure, the β-Li phase also decreased after 4 passes ECAP. After ECAP for 8 passes, the α-Mg phase grains are further lengthened with both phases oriented in the direction of the extrusion flow lines, as shown in Figure 2c. After 16-passes ECAP, as shown in Figure 2d, the α-Mg phase is further refined, presenting the typical plastic-deformation flows.

Figure 3a shows the optical microstructure of cast-rolled Mg-9Li alloy in the rolling direction. Clearly, the α-Mg phase is elongated and arranged in order along the rolling direction. Figure 3b,c is the optical microstructure of the ECAP-rolled alloy observed from the rolling direction (RD) and normal to the rolling direction (ND) respectively. Note that, this alloy was firstly ECAP processed for 8 passes, and then subjected to rolling until a 50% reduction in thickness. Thus, the samples of Figure 3b,c are named as E8R-RD and E8R-ND, respectively. Compared to the ECAPed alloys, the duplex phases of the alloy was further elongated and arranged in order along the rolling direction, presenting the typical plastic deformation flow. Meanwhile, the alloy has a more severe plastic-deformation flow in the direction normal to the rolling direction, presenting the typical fibrous microstructure.

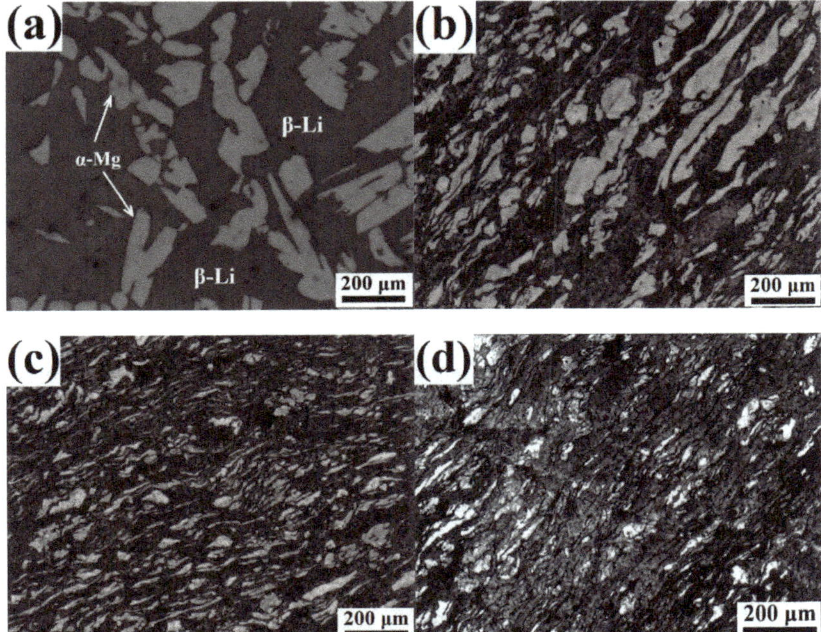

Figure 2. Optical Micrographs of as-cast and equal-channel-angular-pressing (ECAP)ed Mg-9Li alloys: (**a**) cast; (**b**) E4; (**c**) E8; (**d**) E16.

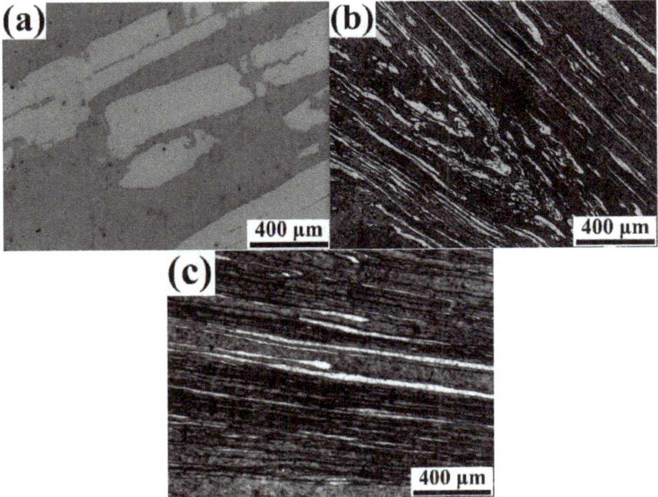

Figure 3. Optical Micrographs of Rolled Mg-9Li alloys: (**a**) CR along the rolling direction (CR-RD); (**b**) E8R along the rolling direction (E8R-RD); (**c**) E8R normal to the rolling direction ND (E8R-ND).

Figure 4 presents the XRD spectra of the Mg-9Li alloys after ECAP and post rolling. As shown in Figure 4a with the cast and ECAPed alloys with different passes, it is clear that the peaks, as well as their intensities, of the two-phase structure changes with the number of ECAP passes in the alloy. Compared with the as-cast alloy, the peaks of the α-Mg and β-Li phases have many different crystal faces. For example, (200), (220), (310) peaks are formed in the Li-rich phase whiles ($10\bar{1}2$), ($11\bar{2}0$) peaks are formed in the α-Mg phase. Comparatively, after 8 passes, the α-Mg shows peaks of (10$\bar{1}$0), (10$\bar{1}$1), (11$\bar{2}$0) whereas the peak increased obviously and the diffraction peaks of (200) and (211) crystal planes are enhanced in the β-Li phase. The XRD spectra of the ECAP-rolled alloys are shown in Figure 4b. The diffraction peaks of the β-Li phase (110) crystal plane are weakened after rolling whereas the diffraction peaks are stronger on the basal (200) and cone (310) planes after rolling. Among all the diffraction peaks of the α-Mg phase, the diffraction peak of the basal plane (0002) is the strongest, which indicates the existence of strong basal texture in the ECAP-rolled alloys [32–34]. Meanwhile, the peaks of (200) crystal planes of the ECAPed alloys were further strengthened after rolling, and this phenomenon was more obvious in the alloys with larger ECAP passes.

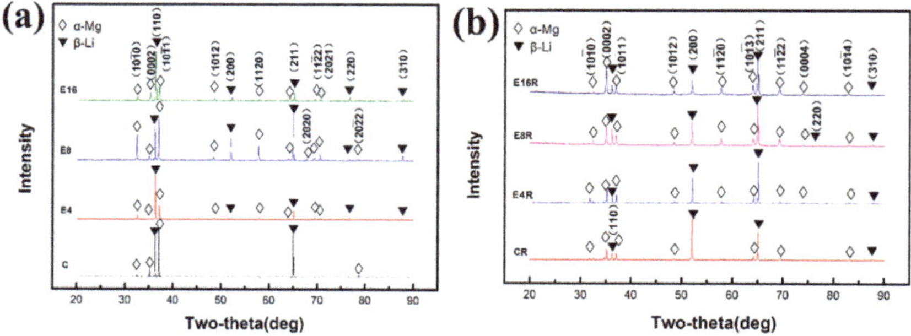

Figure 4. X-ray diffraction patterns of Mg-9Li alloys: (**a**) as-cast and ECAPed alloys; (**b**) rolled alloys.

3.2. Tensile Mechanical Properties of the ECAPed and ECAP-Rolled Alloy

The tensile engineering strain-stress curves of the alloy after different-passes ECAP process and post rolling were obtained and presented in Figure 5, and the related mechanical parameters summarized in Table 2. Due to the amount of soft β-Li phase of the duplex structure, the cast alloy has limited strength and excellent ductility, presenting the extremely low yield strength (YTS, about 52 MPa) and high elongation to failure (E_f, about 33%). One characteristic noteworthy is that, the cast alloy presents sufficient tension work-hardening ability, presenting the continuously strengthened tensile bearing capacity. Benefited from this advantage, the ultimate strength (UTS, about 102 MPa) of the cast alloy is nearly double of the YTS and also presents the excellent performance in the uniform elongation (E_u, about 15%).

ECAP process improved the strength but decreased the ductility of the Mg-9Li alloy, and the mechanical performance changed with different ECAP passes. After ECAP for 4 passes, the YTS and UTS of the E4 alloy were improved to 88 MPa and 106 MPa, and the E_u and E_f decreased to 5% and 25% respectively. With increased ECAP passes, the 8-passes ECAPed alloy (E8) was further greatly strengthened with better ductility compared to the E4 alloy, of which the YTS and UTS are about 110 MPa and 133 MPa, and the E_u and E_f are about 7% and 24%. However, remarkable softening of the alloy occurred when the ECAP process was further increased to 16 passes. Compared to the E8 alloy, the E16 alloy presented lower YTS (about 100 MPa) and UTS (about 116 MPa). However, the ductility of E16 alloy was improved, of which the E_f was about 31%, nearly the same as that of the cast alloy. However, it should be emphasized that the strength of the E16 alloy is still significantly higher than that of cast alloy, especially the YTS. Different to the sufficient work-hardening ability of the cast alloy during the tensile process, the ECAPed alloys presented a reduced difference between values of YTS and UTS, indicating their less tension work-hardening ability.

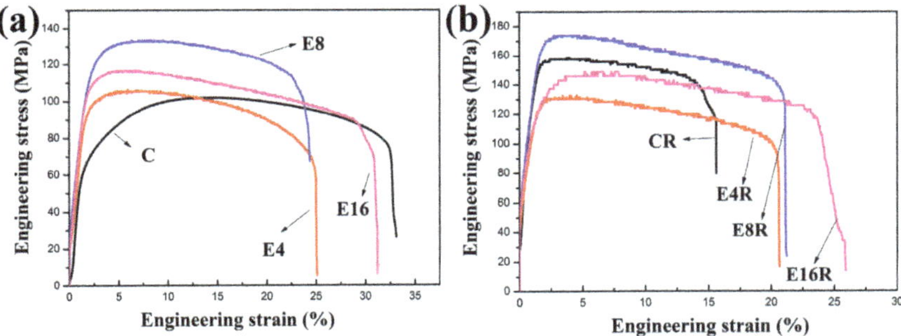

Figure 5. Tensile strain-stress curves of Mg-9Li alloys: (**a**) Cast and ECAPed alloys; (**b**) cast rolled and ECAP-rolled alloys.

Figure 5b presents the engineering strain-stress curves of the rolled alloys. Different to the tensile curves of the cast and ECAPed alloys, the tensile curves of the rolled sample showed the obvious dense serrated fluctuations after yielding which can be adjudged to be the C-type Portevin-Le Chatelier (PLC) effect [9]. The appearance of this phenomenon may be caused by the instability of the Mg atoms in the β-Li matrix after rolling. When plastic deformation occurs under the action of tensile stress, dislocation passes through these unstable solid-solution Mg atoms, which promotes the dissolution of these Mg atoms from the β-Li matrix. At the same time, the dislocation will be pinned at the grain boundary or the dislocation accumulation during the movement, so the interaction between the movable dislocation and the unstable dissolved Mg atom will cause the fluctuation of the tensile curve [9,35]. Significant strengthening and dramatic reduction in ductility occurred to the cast alloy after rolling process, of which the YTS and UTS increased to about 152 and 158 MPa, and the E_u and E_f decreased to about 3% and 16% respectively. Judging from the typical change in strength and ductility of the cast-rolled alloy, one can point out that strain-induced dislocation hardening is the major factor dominating the above phenomenon. During the rolling process, a large number of dislocations are induced within the grains. With further deformation, the edge dislocations and screw dislocations move along (0002) planes. Until the grain boundary is reached, a large number of dislocations in the grain boundary tangles further from the dislocation cell walls, further impeding the slip of dislocations.

Table 2. Tensile test properties of as-cast, ECAP and ECAP-rolled Mg-9Li alloy.

Mechanical Properties	C	E4P	E8P	E16P	CR	E4R	E8R	E16R
UTS (MPa)	102	106	133	116	158	133	174	149
YTS (MPa)	52	88	110	100	152	126	166	120
E_u (%)	15	5	7	5	3	3	2	7
E_f (%)	33	25	24	31	16	21	22	26

Post rolling also improved the strength and decreased the ductility of the ECAPed alloys. However, all the ECAP-rolled alloys had better ductility compared to the cast-rolled alloy. Among them, the E8R alloy had the best strength with satisfactory ductility, of which the YTS, UTS, and E_f values reached 166 MPa, 174 MPa and 22% respectively. Compared to the cast alloy, the E8R alloy dramatically enhanced the YTS and UTS, which increased by 219% and 70% respectively. Compared to the cast-rolled alloy, the YTS and UTS increased by 10% and 9%. The most important thing is that, the E8R still kept satisfactory ductility after rolling while the cast-rolled alloy suffered dramatic reduction in ductility. Different to the E8 alloy, the rolling process endowed a quite limited strengthening to the E4 and E16 alloys, of which the YTS and UTS were all less than that of the CR alloy. This phenomenon may have a close relationship to the incompletely refined microstructure of the E4 alloy and the softening process of the E16 alloy.

Figure 6 presents the SEM fracture morphology of Mg-9Li alloys. Generally, as seen in Figure 6a, the cast alloy has a typical tearing fracture, indicating the excellent ductility. The fracture of the as-cast alloy shows major deep dimples in the β-Li phase induced by further growth of the voids due to the accumulation of dislocations and the few quasi-cleavage fracture along the α-Mg phase. In the E8 alloy (Figure 6b), the fracture surface is relatively flat with fewer marks of severe tearing compared to the cast alloy, indicating less plasticity. In addition, the fracture is majorly occupied by a mass of small shallow dimples while a few cleavage steps can be also observed. The great decrease in dimple size should be closely related to the grain refinement. Post rolling created more obvious flat fracture to both the cast and ECAPed alloy, indicating remarkable reduced ductility. As seen in Figure 6c, the CR alloy has an equal amount of dimples, mixed with large cleavage steps. Meanwhile, its dimples are much shallower than that of the cast alloy. As seen in Figure 6d, the E8R alloy also presents increased cleavages compared to the E8 alloy. However, the cleavage of this alloy was obviously less than that of the CR alloy, indicating the satisfactory and better ductility. From the above analysis, it can be concluded that the SEM fracture morphologies properly reflect the alternating ductility of the alloys after ECAP and post rolling.

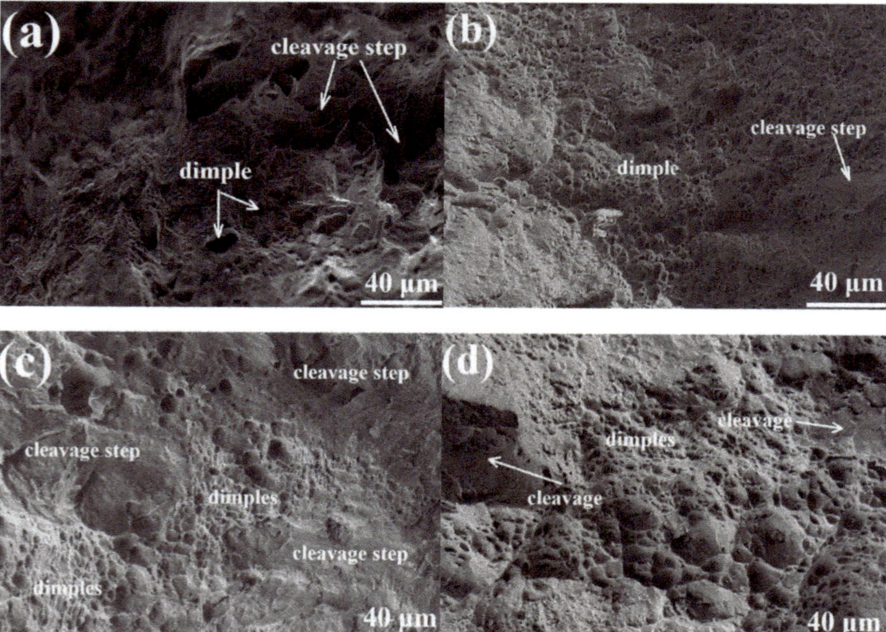

Figure 6. SEM fracture morphology of Mg-9Li alloys: (**a**) Cast alloy; (**b**) E8 alloy; (**c**) CR alloy; and (**d**) E8R alloy.

4. Discussion

4.1. Grain-Boundary Strengthening and Dislocation Strengthening

Generally, thermoplastic processing greatly influences the mechanical properties of a metal from three aspects [36–38]. Firstly, the process eliminates the casting defects of the raw materials. Secondly, the coarse grains will be greatly refined due to the strain-induced dislocation evolution, resulting in the significant grain refinement. Thirdly, the coupling effect of heat and strain-induced dislocation entanglement will stimulate the dynamic recrystallization of the deformed matrix. For the Mg-9Li alloy, due to the bcc structure and its softer essence of β-Li phase, it will be firstly deformed with lots of dislocations accumulated in the α/β phase interface, which inhibits the recovery of the β-Li phase but stores up enormous distortion energy [39]. With increase in ECAP passes and induced strain, a large number of dislocations will concentrate inside the deformed β-Li phase grains [26,40]. The increase in dislocation density leads to the sub-grain boundary becoming the nucleation point of recrystallization, eventually growing up to from new grains [18,41]. Therefore, dynamic recrystallization takes place preferentially in β-Li phase, which may lead to a limited strengthening and the potential softening of the alloy [42].

As revealed in the tensile curves, the E4 alloy has an improved YTS compared to the cast alloy with limited UTS and decreased ductility. The dynamic-recrystallization of β-Li phase and the limited grain refinement of the α-Mg phase is the major contributing factor to this phenomenon as well as the extremely limited further strengthening to the E4 alloy after further rolling. As revealed in the tensile curves, both E8 and E8R alloys presented the best strength, due to the combination of grain-boundary strengthening and dislocation strengthening from both α-Mg and β-Li phases. Generally believed, the finer the grain size during thermal plastic deformation, the more easily dynamic recrystallization occurs [40,43]. Due to the re-heating of the E16 alloy after 8-passes followed by an additional 8 passes, a high possibility of intensive dynamic recrystallization occurring to both α-Mg phase and β-Li phase due to the lattice distortion energy obtained from the further ECAP-induced plastic deformation results in the typical softening phenomenon of the E16 alloy.

TEM images of the E8 and E8R alloys are presented in Figure 7. Figure 7a presents the typical grain morphologies of deformed α-Mg phase. The selected electron diffraction pattern (SAED) in the bottom right corner of Figure 7a was observed in the ($[41\bar{5}3]$) zone axis. As marked, the crystal planes of ($1\bar{1}10$), (1102), and ($\bar{1}013$) clearly shows the typical lattice feature of the Mg phase. After continuous ECAP process for 8 passes, obvious grain refinement of approximately 2 μm has been achieved in the deformed α-Mg phase. As shown in Figure 7b, the deformed Mg grains has been further refined to about 800 nm to 1.5 μm with improved grain boundary strengthening, satisfying the Hall-petch equation, stated as $\sigma_y = \sigma_0 + kd^{-1/2}$ [44]. Figure 7c presents the typical intragranular dislocations of the E8R alloy. Based on the TEM microstructure analysis, it can be deduced that the strength enhancement of E8 and E8R alloys was dominated by the combined effect of sufficient grain-boundary strengthening and dislocation strengthening with additional induced straining at room temperature and the obtained finer grains. A number of researches have shown that the intensity and plasticity of the α-Mg phase in the hcp structure can be improved synchronously after severe grain refinement via severe plastic deformation (SPD) process [45]. Also, according to the reported results of the SPDed UFG Mg alloys [46,47], the great refinement of the Mg grains will benefit both high strength and high ductility of the α-Mg phase of the alloy. However, the large plastic deformation also leads to the grain refinement, as well as the strain-induced work hardening of the β-Li phase, leading to the observed decrease in ductility of the E8 alloy.

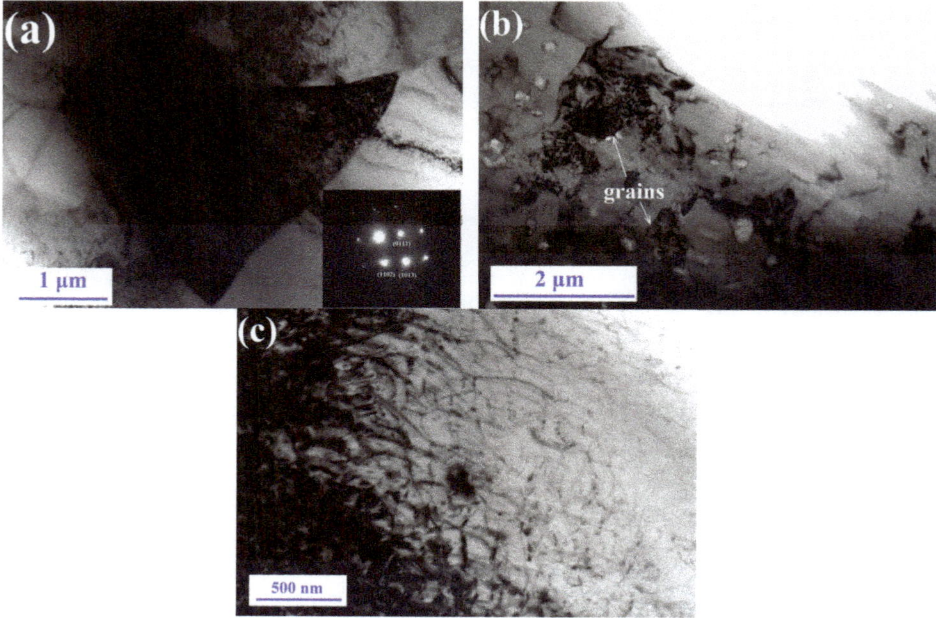

Figure 7. TEM microstructure of the Mg-9Li alloys: (**a**) 8-passes ECAPed (E8) alloy grain morphology; (**b**) E8R alloy grain morphologies; (**c**) intragranular dislocations of the E8R alloy.

4.2. Texture Strengthening

As reported by many researches [26,48], texture also greatly influences the strength and ductility of the Mg alloys. As revealed in the XRD patterns of the alloys, the $(10\bar{1}0)$, $(10\bar{1}1)$, and $(11\bar{2}0)$ peaks of the ECAPed alloys were strengthened. Meanwhile, the strongest diffraction peak of the basal plane (0002) was detected. Thus, one can infer that the texture may play an important role in the mechanical property of the processed Mg-9Li alloy. The XRD texture of the E8 and E8R alloys are presented in Figure 8. Due to the cumulative effect of shear stress during the ECAP deformation, the basal texture of the α-rich Mg phase is 40–60° with the ECAP-extrusion direction, and the maximum texture density is up to 6.1. Prismatic texture $(10\bar{1}0)$ and conical texture $(10\bar{1}1)$ and $(11\bar{2}0)$ are found with weak texture density. Generally, during plastic deformation of magnesium alloys, the slip in the direction of the basal plane is the first thing to occur [16]. Due to the addition of Li, the ratio of c/a crystal axis is decreased, which is favorable to the dislocation slip in the cylindrical and conical directions. In conjunction with the orientation distribution function (ODF) diagram in Figure 8b, it is found that it is not only the basal plane, prismatic, and cone textures that appear in the α-Mg phase but also there exists $(11\bar{2}0)$ <$1\bar{1}00$> textures. The texture shows that the local grains were ordered after ECAP processing since the grains in different directions are arranged differently by shear force after multi-pass ECAP processing. These multi-directionally ordered textures block the slip of the grains, leading to an increase in the strength. Figure 8c depicts the pole figures for the {0002} plane of the α-Mg phase, a strong basal texture along the rolling direction can be seen, and it is the appearance of this texture that leads to the significant increase in strength of the E8R alloy. The distribution of the $\{10\bar{1}0\}$ prismatic weak structure in the rolling direction is 15°. The $\{10\bar{1}1\}$, $\{11\bar{2}0\}$ conical planes have no particularly apparent texture.

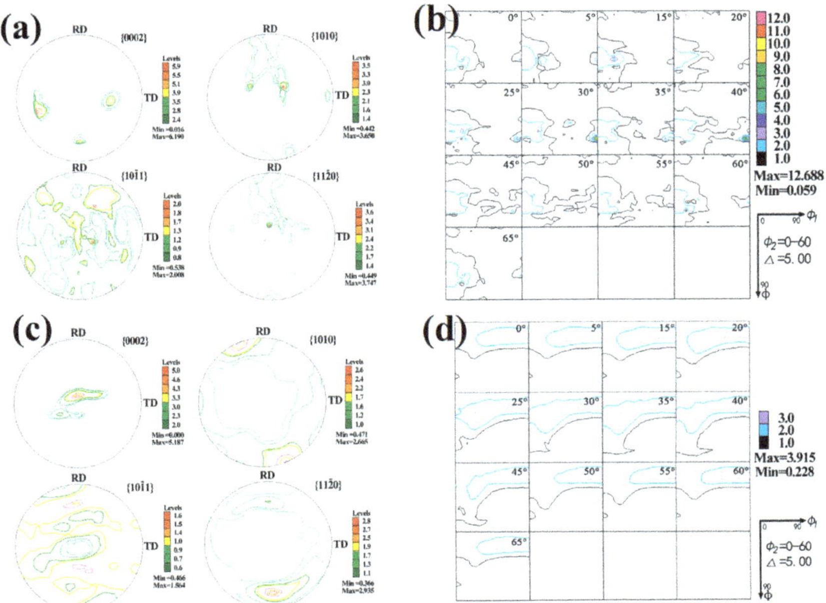

Figure 8. XRD texture analysis of the α-Mg phase of Mg-9Li alloys. (**a**,**b**) are polar diagram and ODF diagram of the E8 alloy, respectively; (**c**,**d**) are polar diagram and ODF diagram of the E8R alloy, respectively.

5. Conclusions

In summary, a combined process of multi-pass ECAP and post rolling technologies were employed in the development of a high-strength Mg-9Li duplex alloy. The microstructure and their influence on the strength and ductility of the processed Mg-9Li alloys were systematically investigated. The main conclusions are:

(1) Cast Mg-9Li alloy was firstly processed via multi-pass ECAP at 200 °C for 4, 8, and 16 passes to achieve grain refinement in both α-Mg and β-Li phases of the ECAPed alloys. Post rolling was conducted at room temperature to obtain further strengthening of the alloys. All the alloys after the combined process presented enhanced strength and decreased ductility compared to the cast alloy.

(2) Among all the ECAPed alloys, the E8 alloys presented the best strength, of which the YTS and UTS are 110 MPa and 133 MPa, respectively. Post rolling of the E8 alloy further strengthened the alloy and endowed it with the best strength of all the alloys in this research. The YTS and UTS of the E8R alloys reached 166 MPa and 174 MPa. Approximately 219% and 70% increase in YTS and UTS was achieved compared to the cast alloy, respectively.

(3) Grain-boundary strengthening and dislocation strengthening are the key factors to the greatly improved strength of the Mg-9Li alloys after the combined processing. Significant grain refinement of the α-Mg phase was achieved in the E8 alloy, of which the grain size was about 2 μm. Post rolling further reduced the grain size to between 800 nm and 1.5 μm. With the greatly refined grains, the grain-boundary strengthening of the E8 and E8R alloys was obtained. Profuse intragranular dislocation was accumulated in the deformed matrix of the E8 and E8R alloys, leading to the significant dislocation hardening of the alloy.

(4) Prismatic texture $\left(10\bar{1}0\right)$ and conical texture $\left(10\bar{1}1\right)$ and $\left(11\bar{2}0\right)$ were detected with weak texture density in the E8 alloy. The strong basal texture along {0002} plane of the α-Mg phase was

formed in the rolled Mg-9Li alloys along the rolling direction, which also contributed to the most improved strength of the E8R alloy.

Author Contributions: D.S. and J.J. conceived and designed the experiments; E.K., C.L., G.W., Z.Z., and B.G. contributed to the sample preparation, J.S. and A.M. contributed to the data analysis; D.S. wrote the paper.

Funding: This research was financial support from Funds for the Central Universities of Hohai University (2018B57714, 2018B48414 and 2019B76814), Natural Science Foundation of China (51878246, 51774109 and 51979099), Fundamental Research Six Talent Peaks Project in Jiangsu Province (2016-XCL-196), Science and Technology Support Program funded project of Suqian City (Industrial H201817), Applied Fundamental Research Foundation of Nantong City (JC2018110), Key Research and Development Project of Jiangsu Province of China (BE2017148).

Conflicts of Interest: The authors declare no conflict of interest.

References

1. Xu, W.; Birbilis, N.; Sha, G.; Wang, Y.; Daniels, J.E.; Xiao, Y.; Ferry, M. A high-specific-strength and corrosion-resistant magnesium alloy. *Nat. Mater.* **2015**, *14*, 1229–1235. [CrossRef] [PubMed]

2. Song, D.; Ma, A.; Jiang, J.; Lin, P.; Yang, D.; Fan, J. Corrosion behavior of equal-channel-angular-pressed pure magnesium in NaCl aqueous solution. *Corros. Sci.* **2010**, *52*, 481–490. [CrossRef]

3. Zhou, H.; Wang, Q.; Ye, B.; Guo, W. Hot deformation and processing maps of as-extruded Mg–9.8Gd–2.7Y–0.4Zr Mg alloy. *Mater. Sci. Eng. A* **2013**, *576*, 101–107. [CrossRef]

4. Jiang, B.; Qiu, D.; Zhang, M.X.; Ding, P.; Gao, L. A new approach to grain refinement of an Mg–Li–Al cast alloy. *J. Alloys Compd.* **2010**, *492*, 95–98. [CrossRef]

5. Zhou, H.; Wang, Q.D.; Guo, W.; Ye, B.; Jian, W.W.; Xu, W.Z.; Ma, X.L.; Moering, J. Finite element simulation and experimental investigation on homogeneity of Mg-9.8Gd-2.7Y-0.4Zr magnesium alloy processed by repeated-up setting. *J. Mater. Process. Technol.* **2015**, *225*, 310–317. [CrossRef]

6. Wu, R.; Yan, Y.; Wang, G.; Murr, L.E.; Han, W.; Zhang, Z.; Zhang, M. Recent progress in magnesium–lithium alloys. *Int. Mater. Rev.* **2014**, *60*, 65–100. [CrossRef]

7. Haferkamp, H.; Niemeyer, M.; Boehm, R.; Holzkamp, U.; Jaschik, C.; Kaese, V. Development, Processing and Applications Range of Magnesium Lithium Alloys. *Mater. Sci. Forum* **2000**, *350*, 31–42. [CrossRef]

8. Bronfin, B.; Aghion, E. Magnesium Alloys Development towards the 21st Century. *Mater. Sci. Forum* **2000**, *350*, 19–30.

9. Hanwu, D.; Limin, W.; Ke, L.; Lidong, W.; Bin, J.; Fusheng, P. Microstructure and deformation behaviors of two Mg–Li dual-phase alloys with an increasing tensile speed. *Mater. Des.* **2016**, *90*, 157–164. [CrossRef]

10. Valiev, R. Nanostructuring of metals by severe plastic deformation for advanced properties. *Nat. Mater.* **2004**, *3*, 511.

11. Djavanroodi, F.; Ebrahimi, M. Effect of die channel angle, friction and back pressure in the equal channel angular pressing using 3D finite element simulation. *Mater. Sci. Eng. A* **2010**, *527*, 1230–1235. [CrossRef]

12. Jiang, J.; Yuan, T.; Zhang, W.; Ma, A.; Song, D.; Wu, Y. Effect of equal-channel angular pressing and post-aging on impact toughness of Al-Li alloys. *Mater. Sci. Eng. A* **2018**, *733*, 385–392. [CrossRef]

13. Langdon, T.G. The principles of grain refinement in equal-channel angular pressing. *Mater. Sci. Eng. A* **2007**, *462*, 3–11. [CrossRef]

14. Sun, J.; Yang, Z.; Han, J.; Liu, H.; Song, D.; Jiang, J.; Ma, A. High strength and ductility AZ91 magnesium alloy with multi-heterogenous microstructures prepared by high-temperature ECAP and short-time aging. *Mater. Sci. Eng. A* **2018**, *734*, 485–490. [CrossRef]

15. Ma, A.; Jiang, J.; Saito, N.; Shigematsu, I.; Yuan, Y.; Yang, D.; Nishida, Y. Improving both strength and ductility of a Mg alloy through a large number of ECAP passes. *Mater. Sci. Eng. A* **2009**, *513*, 122–127. [CrossRef]

16. Sun, J.; Yang, Z.; Han, J.; Yuan, T.; Song, D.; Wu, Y.; Yuan, Y.; Zhuo, X.; Liu, H.; Ma, A. Enhanced quasi-isotropic ductility in bi-textured AZ91 Mg alloy processed by up-scaled RD-ECAP processing. *J. Alloys Compd.* **2018**, *780*, 12. [CrossRef]

17. Min, X.; Xinying, T.; Chaoping, Y.; Jinyang, Z. Effect of Heat Treatment on Microstructure and Mechanical Properties of Mg 94 Zn 2 Y 4 Alloy. *Rare Met. Mater. Eng.* **2016**, *45*, 2804–2808. [CrossRef]

18. Chang, S.Y.; Lee, S.W.; Kang, K.M.; Kamado, S.; Kojima, Y. Improvement of Mechanical Characteristics in Severely Plastic-deformed Mg Alloys. *Mater. Trans.* **2004**, *45*, 488–492. [CrossRef]

19. Zhou, H.; Wang, Q.; Chen, J.; Ye, B.; Guo, W. Microstructure and mechanical properties of extruded Mg-8.5Gd-2.3Y-0.8Ag-0.4Zr alloy. *Trans. Nonferr. Met. Soc. China* **2012**, *22*, 1891–1895. [CrossRef]

20. Zhou, H.; Ning, H.; Ma, X.; Yin, D.; Xiao, L.; Sha, X.; Yu, Y.; Wang, Q.; Li, Y. Microstructural evolution and mechanical properties of Mg-9.8Gd-2.7Y-0.4Zr alloy produced by repetitive upsetting. *J. Mater. Sci. Technol.* **2018**, *34*, 1067–1075. [CrossRef]

21. Zhou, H.; Xu, W.Z.; Jian, W.W.; Cheng, G.M.; Ma, X.L.; Guo, W.; Mathaudhu, S.N.; Wang, Q.D.; Zhu, Y.T. A new metastable precipitate phase in Mg-Gd-Y-Zr alloy. *Philos. Mag.* **2014**, *94*, 2403–2409. [CrossRef]

22. Song, D.; Li, C.; Liang, N.; Yang, F.; Jiang, J.; Sun, J.; Wu, G.; Ma, A.; Ma, X. Simultaneously improving corrosion resistance and mechanical properties of a magnesium alloy via equal-channel angular pressing and post water annealing. *Mater. Des.* **2019**, *166*, 107621. [CrossRef]

23. Nishida, Y.; Arima, H.; Kim, J.C.; Ando, T. Rotary-die equal-channel angular pressing of an Al—7 mass% Si—0.35 mass% Mg alloy. *Scr. Mater.* **2001**, *45*, 261–266. [CrossRef]

24. Wei, J.; Huang, G.; Yin, D.; Li, K.; Wang, Q.; Zhou, H. Effects of ECAP and Annealing Treatment on the Microstructure and Mechanical Properties of Mg-1Y (wt. %) Binary Alloy. *Metals* **2017**, *7*, 119. [CrossRef]

25. Liu, Y.; Chen, X.; Wei, K.; Xiao, L.; Chen, B.; Long, H.; Yu, Y.; Hu, Z.; Zhou, H. Effect of Micro-Steps on Twinning and Interfacial Segregation in Mg-Ag Alloy. *Materials* **2019**, *12*, 1307. [CrossRef] [PubMed]

26. Zou, Y.; Zhang, L.; Li, Y.; Wang, H.; Liu, J.; Liaw, P.K. Improvement of mechanical behaviors of a superlight Mg–Li base alloy by duplex phases and fine precipitates. *J. Alloys Compd.* **2017**, *735*, 2625–2633. [CrossRef]

27. Wu, H.; Jiang, J.; Liu, H.; Sun, J.; Gu, Y.; Tang, R.; Zhao, X.; Ma, A. Fabrication of an Ultra-Fine Grained Pure Titanium with High Strength and Good Ductility via ECAP plus Cold Rolling. *Metals* **2017**, *7*, 563. [CrossRef]

28. Jiang, B.; Yang, Q.S.; Gao, L.; Pan, F.S. Effect of the Rolling Process on Microstructures and Mechanical Properties of the Extruded LA91 Alloy Sheet. *Mater. Sci. Forum* **2011**, *686*, 90–95. [CrossRef]

29. Chih-Te, C.; Shyong, L.; Chun-lin, C. Rolling route for refining grains of super light Mg–Li alloys containing Sc and Be. *Trans. Nonferrous Met. Soc. China* **2009**, *20*, 1374–1379.

30. Hajizadeh, K.; Eghbali, B. Effect of two-step severe plastic deformation on the microstructure and mechanical properties of commercial purity titanium. *Met. Mater. Int.* **2014**, *20*, 343–350. [CrossRef]

31. Yu, H.L.; Lu, C.; Tieu, A.K.; Li, H.J.; Godbole, A.R. Special rolling techniques for improvement of mechanical properties of ultrafine-grained metal sheets: A review. *Adv. Eng. Mater.* **2016**, *18*, 754–769. [CrossRef]

32. Xiao, L.; Cao, Y.; Li, S.; Zhou, H.; Ma, X.; Mao, L.; Sha, X.; Wang, Q.; Zhu, Y.; Han, X. The formation mechanism of a novel interfacial phase with high thermal stability in a Mg-Gd-Y-Ag-Zr alloy. *Acta Mater.* **2019**, *162*, 214–225. [CrossRef]

33. Han, B.; Dunand, D. Microstructure and mechanical properties of magnesium containing high volume fractions of yttria dispersoids. *Mater. Sci. Eng. A* **2000**, *277*, 297–304. [CrossRef]

34. Xu, C.; Horita, Z.; Furukawa, M.; Langdon, T.G. Using Equal-Channel Angular Pressing for the Production of Superplastic Aluminum and Magnesium Alloys. *J. Mater. Eng. Perform.* **2004**, *13*, 683–690. [CrossRef]

35. Wei, G.B.; Peng, X.D.; Hu, F.P.; Hadadzadeh, A.; Yang, Y.; Xie, W.D.; Wells, M.A. Deformation behavior and constitutive model for dual-phase Mg–Li alloy at elevated temperatures. *Trans. Nonferr. Met. Soc. China* **2016**, *26*, 508–518. [CrossRef]

36. Chang, L.; Wang, Y.; Zhao, X.; Huang, J. Microstructure and mechanical properties in an AZ31 magnesium alloy sheet fabricated by asymmetric hot extrusion. *Mater. Sci. Eng. A* **2008**, *496*, 512–516. [CrossRef]

37. Hsiang, S.; Lin, Y. Investigation of the influence of process parameters on hot extrusion of magnesium alloy tubes. *J. Mater. Process. Technol.* **2007**, *192*, 292–299. [CrossRef]

38. Liu, Y.; Liu, M.; Chen, X.; Cao, Y.; Roven, H.J.; Murashkin, M.; Valiev, R.Z.; Zhou, H. Effect of Mg on microstructure and mechanical properties of Al-Mg alloys produced by high pressure torsion. *Scr. Mater.* **2019**, *159*, 137–141. [CrossRef]

39. Zhou, H.; Huang, C.H.; Sha, X.C.; Xiao, L.R.; Ma, X.L.; Höppel, H.W.; Göken, M.; Wu, X.L.; Ameyama, K.; Han, X.D.; et al. In-situ observation of dislocation dynamics near hterostructured interface. *Mater. Res. Lett.* **2019**, *7*, 376–382. [CrossRef]

40. Hu, H.; Zhen, L.; Zhang, B.; Yang, L.; Chen, J. Microstructure characterization of 7050 aluminum alloy during dynamic recrystallization and dynamic recovery. *Mater. Charact.* **2008**, *59*, 1185–1189. [CrossRef]

41. Zou, Y.; Zhang, L.; Wang, H.; Tong, X.; Zhang, M.; Zhang, Z. Texture evolution and their effects on the mechanical properties of duplex Mg–Li alloy. *J. Alloys Compd.* **2016**, *669*, 72–78. [CrossRef]

42. Zeng, Y.; Jiang, B.; Li, R.; Yin, H.; Al-Ezzi, S. Grain Refinement Mechanism of the As-Cast and As-Extruded Mg–14Li Alloys with Al or Sn Addition. *Metals* **2017**, *7*, 172. [CrossRef]

43. Kugler, G.; Turk, R. Modeling the dynamic recrystallization under multi-stage hot deformation. *Acta Mater.* **2004**, *52*, 4659–4668. [CrossRef]

44. Hansen, N. Hall–Petch relation and boundary strengthening. *Scr. Mater.* **2004**, *51*, 801–806. [CrossRef]

45. Valiev, R.Z.; Langdon, T.G. Principles of equal-channel angular pressing as a processing tool for grain refinement. *Prog. Mater. Sci.* **2006**, *51*, 881–981. [CrossRef]

46. Roodposhti, P.S.; Farahbakhsh, N.; Sarkar, A.; Murty, K.L. Microstructural approach to equal channel angular processing of commercially pure titanium—A review. *Trans. Nonferr. Met. Soc. China* **2015**, *25*, 1353–1366. [CrossRef]

47. Cao, Y.; Ni, S.; Liao, X.; Song, M.; Zhu, Y. Structural evolutions of metallic materials processed by severe plastic deformation. *Mater. Sci. Eng. R: Rep.* **2018**, *133*, 1–59. [CrossRef]

48. Lentz, M.; Klaus, M.; Beyerlein, I.J.; Zecevic, M.; Reimers, W.; Knezevic, M. In situ X-ray diffraction and crystal plasticity modeling of the deformation behavior of extruded Mg–Li–(Al) alloys: An uncommon tension–compression asymmetry. *Acta Mater.* **2015**, *86*, 254–268. [CrossRef]

Article

Structure and Tensile Strength of Pure Cu after High Pressure Torsion Extrusion

Dayan Nugmanov [1,*], Andrey Mazilkin [1,2], Horst Hahn [1] and Yulia Ivanisenko [1]

[1] Institute of Nanotechnology (INT), Karlsruhe Institute of Technology (KIT), 76021 Karlsruhe, Germany; andrey.mazilkin@kit.edu (A.M.); horst.hahn@kit.edu (H.H.); julia.ivanisenko@kit.edu (Y.I.)
[2] Institute of Solid State Physics, Russian Academy of Sciences, 142432 Chernogolovka, Russia
* Correspondence: dayan.nugmanov@kit.edu; Tel.: +49-(0)721-608-28902

Received: 9 September 2019; Accepted: 25 September 2019; Published: 7 October 2019

Abstract: The microstructure and mechanical properties of rod-shaped samples (measuring 11.8 mm in diameter and 35 mm in length) of commercially pure (CP) copper were characterized after they were processed by high pressure torsion extrusion (HPTE). During HPTE, CP copper was subjected to extremely high strains, ranging from 5.2 at central area of the sample to 22.4 at its edge. This high but varying strain across the sample section resulted in HPTE copper displaying a gradient structure, consisting of fine grains in the central area and of ultrafine grains both in the middle-radius area and at the sample edge. A detailed analysis of the tensile characteristics showed that the strength of HPTE copper with its gradient structure is similar to that of copper after severe plastic deformation (SPD) techniques, typically displaying a homogeneous structure. Detailed analysis of the contributions of various strengthening mechanisms to the overall strength of HPTE coper revealed the following: The main contribution comes from Hall–Petch strengthening due to the presence of high and low angle grain boundaries in gradient structure, which act as effective obstacles to dislocation motion. Therefore, both types of boundaries should be taken into account in the Hall–Petch equation. This study on CP copper demonstrated the potential of using the HPTE method for producing high-strength metallic materials in bulk form for industrial use.

Keywords: high pressure torsion extrusion; severe plastic deformation; gradient structure; microstructure; hardness distribution; tensile properties; copper

1. Introduction

During past three decades, numerous studies have been conducted on the structure and mechanical properties of bulk nanostructured (NS) and ultra-fine-grained (UFG) materials produced by severe plastic deformation (SPD) processing. In these studies, considerable effort has been devoted to high pressure torsion (HPT) [1,2] as well as equal channel angular pressing (ECAP) [3–5] since these two are the best-known SPD techniques for producing UFG metallic materials. UFG materials with unique combinations of mechanical properties such as high strength and plasticity [6] have generated intensive interest because the results of some studies have suggested the presence of new strengthening mechanisms.

NS metals, including those processed by HPT, generally exhibit very high strength but limited tensile ductility (with a uniform elongation, only reaching a few percent) with almost no work-hardening [7]. Low ductility is believed to be an intrinsic "Achilles heel" of NS metals because the conventional deformation mechanisms cease to operate at the nanoscale level such that: (i) the dislocation slip is substantially suppressed by the extremely small grains (which, however, account for the extreme strength values in NS metals); and (ii) grain boundary (GB) sliding or diffusional creep is not active enough to accommodate plastic straining at ambient temperature [8]. However, some experimental data have hinted to the possibility that the generally observed low plasticity in NS metals might be extrinsic rather than intrinsic to these materials. For instance, dimples have been observed on fracture surfaces of various NS metals, indicating substantial plastic deformation before failure [7]. In addition, large plastic strains were also obtained when using other deformation schemes such as compression and rolling [3,5]. Indeed, limited tensile ductility of NS metals has often been attributed to the absence of work hardening and to their nano-sized grains, so that strain localization and early necking occur immediately after the onset of yielding. On the other hand, intrinsic tensile plasticity of NS metals can be detected by effective suppression of strain localization.

There are several kinds of SPD techniques, which combine extrusion and shear, e.g., twist extrusion [9], extrusion compression [10] and cyclic expansion extrusion [11]. In this study, we used the high pressure torsion extrusion (HPTE) approach, with prevailing shear strain at the periphery of bulk rod samples and extrusion-like deformation near the center of the rods [12–14]. The strain conditions at the center and at the edge of the HPTE-processed rods depend on the processing parameters and they can vary by several orders of magnitude. Therefore, HPTE is the one of the SPD methods which is capable of producing rod samples with controlled gradient heterogeneous structures.

During the past decades, comprehensive analyses on the influence of the deformation mode on the structural evolution and the resulting physical properties of initially structurally homogeneous materials have been conducted. For example, refer to the study on the hardness and electrical conductivity of pure copper processed by ECAP and HPT [15]; the study on the structural evolution in a Siclanic alloy (Cu–Ni–Si) processed by ECAP and HPT [16]; and the studies on the effect of the SPD deformation mode on the microstructure of pure copper [17,18].

Recently, a few studies have been reported on the mechanical properties of structurally heterogeneous materials [7,19], consisting of hard and soft domains, which may demonstrate the combination of very attractive properties such as high strength and high tensile ductility values. Knowledge on the effects of fine–ultrafine and nano-grained structures on mechanical properties will make it possible to significantly extend our understanding of the deformation processes from the analysis of homogeneous materials to that of heterogeneous counterparts and, in particular, structurally gradient materials [20].

One of the clearest structural features of Cu after the HPTE processing is a strong grain refinement in the peripheral area, and a less pronounced grain refinement in the sample's core [12]. Such heterogeneous structure was the result of strain gradient along the sample radius. Tensile tests, carried out on the gradient Cu samples after a surface mechanical grinding treatment, have shown that the combination of NS to coarser grained microstructure provide an effective approach for enhancing the "strength–ductility" synergy of materials which offer a potential for using the gradient NS layers as an advanced coating of the bulk materials [7].

However, since HPTE it is a newly established SPD technique, data on the applicability of this method to produce structurally gradient materials are still not available. Consequently, results on the characterization of their mechanical properties are also limited. This study aims to bridge this gap by conducting a critical analysis of both the structure and the mechanical test results obtained on HPTE-processed Cu rod samples to reveal possible strengthening effects of varying grain refinement steps giving rise to gradient (heterogeneous) structures.

2. Materials and Methods

CP copper samples containing 99.0 wt% of Cu and 1.0 wt% of Al were first machined, annealed for two hours at 600 °C, and then water quenched to obtain a coarse-grained (CG) microstructure. A typical orientation imaging microscopy (OIM) map of the initial structure is shown in Figure 1a. The OIM map was obtained from the electron backscatter diffraction (EBSD) data and it demonstrates, random grain orientations. The initial structure is characterized by near-ellipsoid-shaped grains (the aspect ratio was ~1.3) with numerous annealing twins (Figure 1a) and an average grain size of ~30 μm. As shown by the back-scattered electron (BSE) image in Figure 1b, the microstructure also contains second-phase particles. The BSE image shows that the second phases are nearly spherical with an average diameter of ~1–5 μm. Usually, these particles are Cu-rich $AlCu_3$, Al_4Cu_9 and $AlCu_2$ intermetallic phases [21,22]. According to the technical specifications of CP copper [22], the particles have a metallurgic origin and they can also include oxides, hydrides, and flux-income impurities.

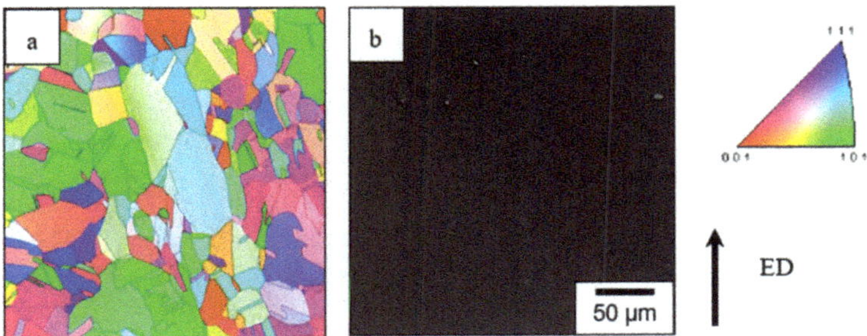

Figure 1. (**a**) Typical orientation imaging microscopy (OIM) map and (**b**) back-scattered electron (BSE) image of the annealed Cu rod (longitude section).

The HPTE apparatus (Karlsruhe Institute of Technology, Institute of Nanotechnology, Eggenstein-Leopoldshafen, Germany), with a specially designed die, is shown in Figure 2a. The material processing was performed at 100 °C with an extruding velocity (v) of 1 mm/min and a rotational velocity (ω) of 1 rpm. According to [12], HPTE-processed copper with v1w1 regime at room temperature led to the formation of a UFG structure. In our study, cylindrical shaped (rod) samples with an initial diameter of 11.8 mm and a length of 35 mm were processed using molybdenum disulfide (MoS_2) as a lubricant to facilitate the extrusion process. The HPTE apparatus was described in detail in [12]. During the processing, the Cu rod samples were exposed to expansion and extrusion as well as torsion.

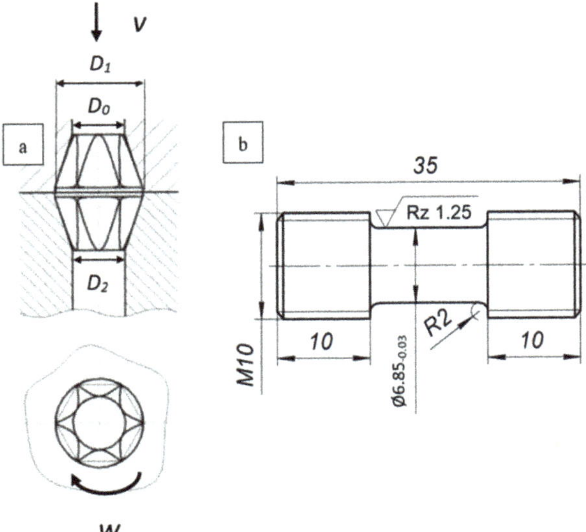

Figure 2. HPTE die with (**a**) holding elements and (**b**) a schematic drawing of the tensile test sample. Dimensions are given in millimeters.

Additional details regarding the development of the die design and the compressive strength the material samples are subjected to during the HPTE processing have been described in an earlier study [23]. The resulting equivalent strain depends on the ratio of the channel diameters (D1/D0 and D1/D2), the extruding velocity (v), and the angular velocity of the die (ω), and it can be calculated as follows [12]:

$$e = 2\ln D1/D0 + 2\ln D1/D2 + (\omega^*R^*D1)/(\sqrt{3}^*v^*D2) \tag{1}$$

where D0, D1, and D2 are fixed and equal to 12, 14, and 10.6 mm, respectively, and R is the distance from the center in the sample's cross section.

Under the above-mentioned parameters, Table 1 contains the strain values calculated using Equation (1). The equivalent strain is in the range between 5.2 and 22.4, depending on the distance from the center, which ensures the formation of a gradient structure.

Table 1. Structure parameters of CP Cu rod after the HPTE treatment, based on the SEM–EBSD and BSE imaging *.

Process ID		E	D_{15}, µm	D_3, µm	N_X/N_Y	V_{HAB}, %	Θ_{av}, deg	V_{twin}, %	d_p, µm	f, %
	initial	0	30.0	-	1.3	90	47	56.4	0.61	0.21
	center	5.2	(16.7,1.2) **/1.4	0.8/0.8	0.9/0.9	25/41	14/19	3.8		
HPTE	mid-rad	13.8	0.9/0.8	0.4/0.6	1.0/0.9	33/49	16/26	9.0	1.22	0.17
	edge	22.4	0.7/0.7	0.4/0.6	1.1/1.0	61/79	31/35	9.1		

* All microstructure parameters are given for the transverse/longitudinal sections. ** Here the grain size values for two maxima of the grain size distribution by the specific area are given.

The HPTE-processed samples were afterwards sectioned both in the transverse (normal) and longitude directions, thinned to the foil thickness, grinded, and mechanically polished. The polished samples were subsequently examined by EBSD technique, using a ZeissAuriga60 scanning electron microscope (SEM), equipped with an EDAX-TSL EBSD system, and operated at 20 kV. The samples for SEM observations were prepared by conventional electro-polishing procedure with a Tenupol-5 twinjet polisher, using standard Struers solution for copper. Final polishing on GATAN PIPS system was performed to remove the surface oxide layer.

TSL OIM EDAX v.7 (EDAX Inc., Draper, UT, USA) software was used for indexing the EBSD patterns. The dimension of the scan areas was 20×20 µm. OIM maps were collected from transverse (shear plane) and longitudinal sections of the samples. They were collected at the sample's center (~0.5 mm from the sample axis), at the middle-radius (~2.7 mm from the sample axis), and at the sample's edge (~4.5 mm from the sample axis). Scan step was 0.1 µm for the initial state and at the central area of the deformed samples, and 0.05 µm at the mid-radius and edge of the HPTE samples.

To minimize misindexing error, nine Kikuchi-bands were used for indexing. To ensure reliability of the EBSD data, all grains comprising of three or fewer pixels were automatically removed from the maps using the grain-dilation and neighbor orientation correlation options of the TSL software (minimal grain tolerance angle was 5°). The camera settings and the Hough parameters allowed us to collect data without significant decrease in pattern quality with the angular resolution ~0.5°. For noise reduction, a lower limit boundary misorientation cut-off value of 2° was applied. A 15° criterion was used to differentiate between low—(LABs) and high—(HABs) angle GBs. The grain size (D) was computed using the equal diameter method. Two misorientation threshold values, of 3° and 15°, were used to define the grain area. The respective mean grain sizes are labeled as D_3 and D_{15}. Specific grain area (S/S_i) distributions by the grain equivalent diameter were plotted for more than 500 grains in each state, where S is the map area and S_i is the area of the grains within the i-th interval.

Grain shape was characterized by the N_x/N_y, the aspect ratio parameter corresponding to the ratio between the number of HAB intersections with vertical (N_x) and horizontal (N_y) secant lines in the transverse and longitude sample sections, respectively. The dislocation density was measured by means of two methods, i.e., EBSD and X-ray diffraction (XRD). The EBSD technique allows to define only geometrically necessary dislocation (GND) density [24], whereas XRD measured total dislocation density, including GNDs and statistically stored dislocations (SSDs). As a first-order estimate, the kernel average misorientation (KAM), which is retrieved directly from EBSD data, was chosen as a measure for the local misorientations. KAM gives the average misorientation around a measurement point with respect to a defined set of nearest and second-nearest neighbor points. The GND density was calculated from the local misorientations, using the option of the OIM software for the angles smaller than 2°. Standard preset for fcc crystal-type slip systems was used in the software set-up. For the calculation of KAM, we considered the slip in (111) crystallographic planes and four possible slip systems along four close-packed <110> directions. We computed the GND density, using the experimental data on the misorientation angle. Additional details regarding GND density calculation can be found elsewhere [25]. The calculation was done for each map pixel, taking into account the data for five nearest neighbors. Orientation gradient was determined as the misorientation between two points over the corresponding length [26].

XRD spectra were collected with a Philips X'Pert powder diffractometer (Malvern Panalytical Ltd., Almelo, The Netherlands), operating in the Bragg–Brentano geometry with the Cu-K$_\alpha$ emission line. The background pattern was calculated by the X'pert HighScore software (Malvern Panalytical Ltd., Malvern, UK). The irradiation area had the dimensions of 5 mm in width and 1 mm in length. This made it possible to obtain XRD spectra from three different locations of the sample along the shear direction (SD) in the longitudinal rod section.

Parameters of the XRD peak profiles, i.e., peak intensity and full width at half maximum, were fitted by the Pseudo-Voigt function. The mean diameter of the crystallites (size of coherently scattering domains) and micro-strain have been estimated from the diffraction peak broadening, including reflections up to (420), in a modified Williamson–Hall method [27]. It is known that in the case of the random distribution of dislocations, the total dislocation density ρ can be defined as [28,29]:

$$\rho = \frac{2\sqrt{3}\left\langle \varepsilon_{hkl}^2 \right\rangle^{1/2}}{D_{hkl}b}, \qquad (2)$$

where D_{hkl} is crystallite size, $<\varepsilon_{hkl}^2>^{1/2}$ is the mean squared micro-strain value, and b is the absolute value of the Burgers vector.

SEM Zeiss LEO1530 at the acceleration voltage of 20 kV coupled with energy-dispersion spectrometry (EDS) was used for the elemental analysis of the samples.

To enhance the spatial and angular resolution of the OIM analysis we carried out the investigations using the FEI Tecnai F20 transmission electron microscope (200 kV) with field emission gun, equipped with the system for automated crystal orientation mapping in the TEM (ACOM-TEM) [30]. For the mapping we used the μ-probe set-up with a beam diameter of ~1–1.5 nm and step size of 5 nm.

A Buehler Micromet-5104 tester was used for the Vickers hardness measurements. A load of 0.2 kg with a dwell time of 15 s was applied to all samples during the hardness measurements. Five indentations spaced 1 mm apart were made along two mutually perpendicular diameters of the specimens. The average value of each measurement was computed from 10 indentations.

Tensile tests were carried out at ambient temperature at a strain rate of 1×10^{-3} s^{-1}, using a Zwick Z100 screw-driven testing machine on cylinder-shaped samples with twist holders (Figure 2b). The tensile samples were machined from the three HPTE-processed billets along the extrusion direction in accordance with ASTM E 8/E 8M-08 requirements.

3. Results

3.1. Microstructure in the Central Area

The OIM analysis was performed for the central, mid-radius, and edge areas of the samples, previously processed by HPTE in v1w1 regime and the corresponding OIM maps are shown in Figure 3. Grains displaying different crystallographic orientations are indicated by different colors, according to the standard stereographic triangle in the pole figure. Corresponding histograms of the grain size distribution and a GB misorientation distribution from the OIM maps are shown in Figure 3. Microstructure parameters, i.e., grain size at respective boundary misorientation threshold (D_3 and D_{15}), average misorientation angle (Θ_{av}), the HABs volume fraction (V_{HAB}), twin boundaries volume fraction (V_{twin}), second-phase particles' size (d_p) and their volume fraction (f), are given in Table 1.

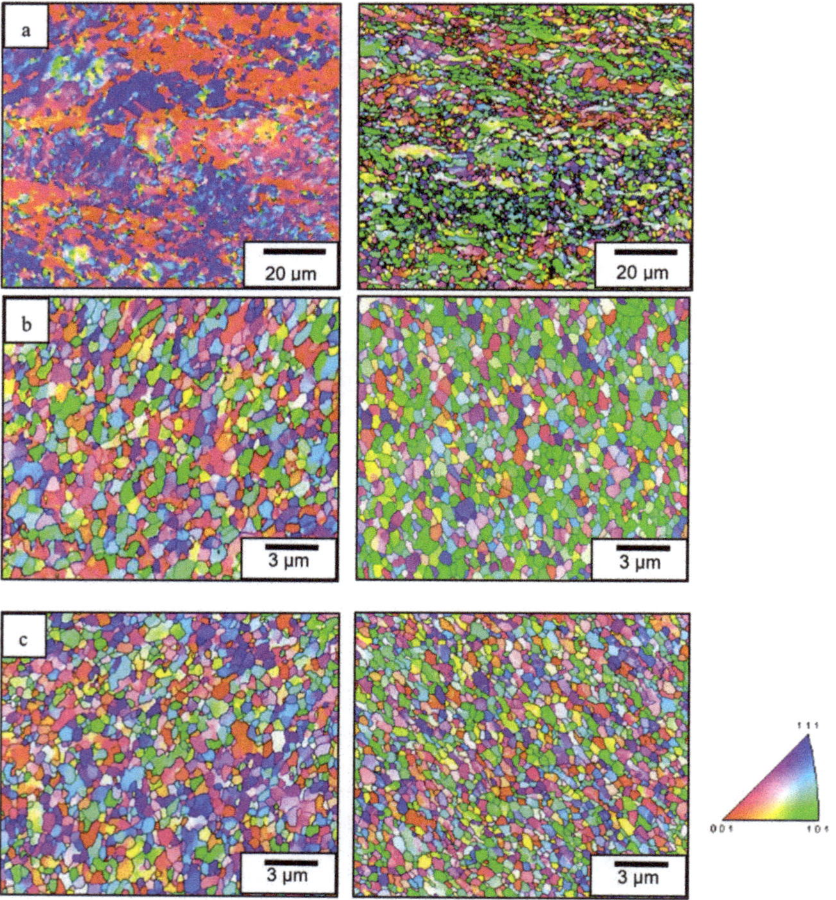

Figure 3. OIM maps from (**left**) transverse and (**right**) longitude sections of the sample areas at (**a**) the center; (**b**) the mid-radius; and (**c**) the edge, after HPTE.

The D_{15} grain size distribution in the transverse section of central area was bimodal with two maxima, at 16.7 µm for coarse grains and at 1.2 µm for fine grains (Figure 4a). These two fractions had a specific area ratio of 1:3. Grain size distribution in the longitude section was of a lognormal type. For this section the average grain size of 1.4 µm matched the histogram maximum (Figure 4b).

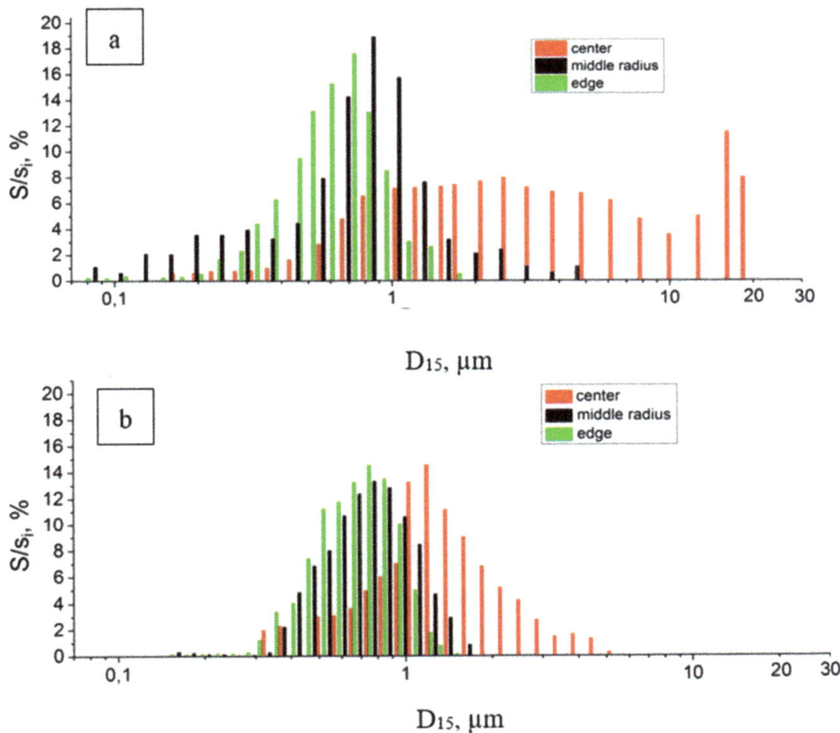

Figure 4. D_{15} grain size distribution by the grains specific area (S/S_i) after HPTE in the (**a**) transverse and (**b**) longitude sample sections.

According to the EBSD data, the GB misorientation histograms were bimodal in both sample sections, with one maximum being close to 2° in the low angle range, and a second one at 47°–48° in the high angle range (Figure 5).

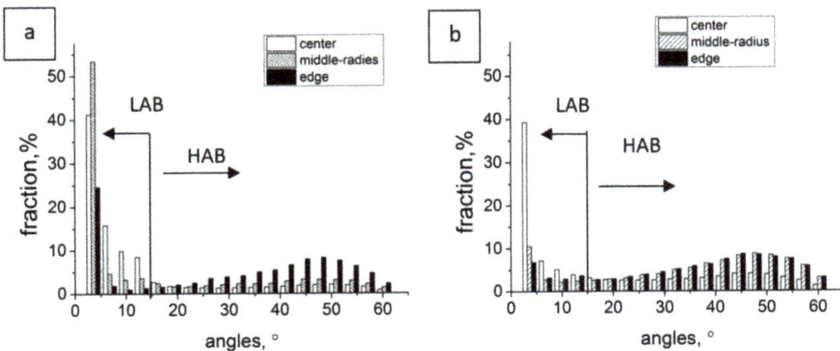

Figure 5. GB misorientation distributions after HPTE in the (**a**) transverse and (**b**) longitudinal sample sections.

In the central part, the HABs volume fraction in the longitudinal cross-section was larger than that of the transverse cross-section (41% vs. 25%, see Table 1 and Figure 5a,b). This difference in the volume fraction can be explained by the grain morphology, i.e., the coarse grains were elongated in the transverse section, containing numerous LABs (Figure 3a). The grain size that was determined at the 3° (D_3) boundary misorientation threshold was 0.8 µm in both cross-sections, whereas D_{15} in the longitudinal section was four times smaller than that in the transverse section (refer to Table 1). Due to this difference in the D15 value, the volume fraction of HABs in the longitudinal section was larger than that in the transverse section.

3.2. Microstructure in the Mid-Radius Area

In contrast to the inhomogeneous microstructure observed in the central part of the processed sample, the grain structure in the mid-radius area was rather uniform. The average grain size of both D_{15} and D_3 were significantly smaller than those in the central area (Table 1) and grain morphology was equiaxed (Figure 3b).

The SEM-BSE images revealed that the volume fraction of the second-phase particles was slightly smaller than that of the central area. Furthermore, no precipitates in the grains after HPTE were seen in TEM bright field (BF) images (Figure 6a).

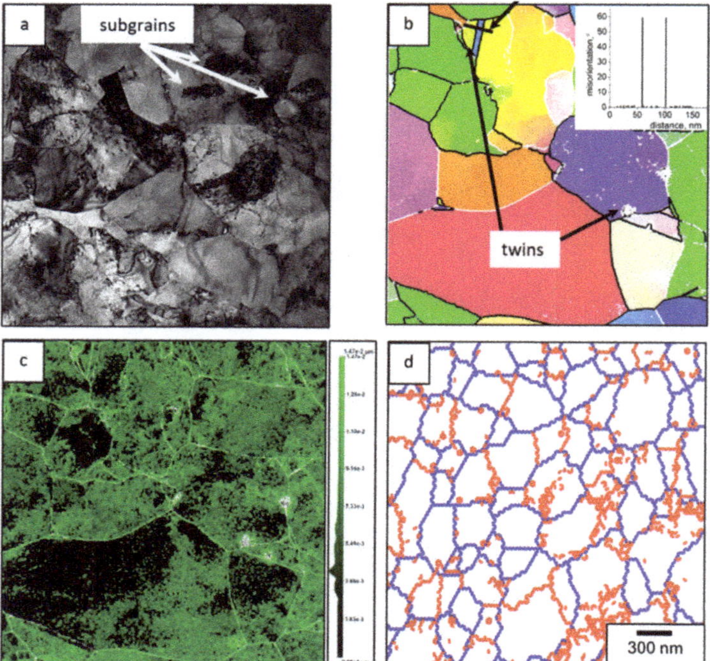

Figure 6. Microstructure of the HPTE Cu in the mid-radius area of the longitudinal section; all the images are at the same scale. (**a**) TEM BF image; (**b**) ACOM TEM OIM map in which LABs ($3° < \Theta < 15°$ are highlighted with white lines) and HABs ($\Theta > 15°$) are highlighted with black lines; (**c**) GND distribution map from ACOM TEM (all GBs are marked with white lines; and (**d**) EBSD OIM GBs network (LABs marked with red lines, HABs marked with blue lines).

The BF TEM image shown in Figure 6a also illustrates the grain substructure of copper after HPTE in the mid-radius area. The majority of triple junctions had contact angles close to 120° (Figure 6a,b). Some grains contained a higher dislocation density compared to that of the adjacent grains, manifested by the diffraction contrast in the image. In general, the TEM observations were consistent with the EBSD maps with respect to the grain size values (Table 2).

Table 2. Vickers hardness (HV), yield strength (YS), ultimate strength (UTS) and elongation to failure (δ) of CP Cu after HPTE and after some other SPD processing methods.

Process Parameters	HV	YS, MPa	UTS, MPa	δ, %	Source
annealing	75	130	230	31.8	this article
HPTE v1w1	125	370	420	32.0	this article
soft rod ASTM		100	220	>5.0	ASTM B49-17
Free-end torsion 1 revolution		312	333	~9	[31]
11 revolutions		387	458	~9	
16 revolutions		412	514	~10	
ECAP e = 3	41	382	404	2.1	[8]
ECAP e = 18	120	358	415	3.6	
HPT $\frac{1}{2}$ turn	133	474	512	2.0	
HPT 10 turns	127	444	487	4.0	

Grain size histograms for transverse and longitudinal sections demonstrated similar lognormal distributions. The maxima in both distributions corresponded to 0.8–0.9 µm in the transverse and to 0.7–0.8 µm in the longitudinal sections (Figure 4). However, in the transverse section the number of very small grains, of sizes less than 0.3 µm, was larger than that present in the longitudinal section (Figure 4). The grain size distribution was broader in the transverse section, with grains in range 3 µm $< D_{15} < 6$ µm, whereas no grains larger than 3 µm were observed in the longitudinal section (Figure 4). From this, we concluded that the grain structure in copper after HPTE in the mid-radius area was quite similar both in transverse and longitudinal sections, which was not the case for the sample's central area.

From GB misorientation distributions, it can be observed that the volume fraction of HABs in longitudinal section was higher than that in the transverse one (49% and 33%, respectively, Table 1). This difference was similar to the difference found in the central area and was most likely due to the same grain refinement mechanism.

The differences in grain size and grain size distribution provide evidence that in the mid-radius area the HPTE creates conditions for efficient grain refinement, resulting in smaller grain size than that in the central area of the sample (Table 1).

Examination of the CP copper microstructure after HPTE in the mid-radius area by means of ACOM TEM provided additional microstructural details. Grain substructure was examined in detail using the BF images like that shown in Figure 6a. A typical orientation map is shown in Figure 6b, whereas a GND spatial distribution is shown in Figure 6c. It should be noted here that twins were observed inside the ultrafine grains, which is the reason for the characteristic peak at 60° present in the boundary misorientation histogram (Figure 6b). The volume fraction of the twin-type boundaries was 14% and this value was greater by 5.5% compared to that obtained by the EBSD method (Table 1). Most probably, this difference could be the result of the different step sizes used in the two methods as well as by higher spatial resolution in ACOM TEM.

The average thickness of the twins was about 40 nm (Figure 6b), and the average twin size, estimated using equal circles method, was about 110 nm, i.e., smaller than the measured grain size, $D_3 = 250$ nm. The nano-size twins and grains less than 50 nm could not be resolved in SEM, using a 50 nm step size. Figure 6a,b shows that a high number of very small subgrains, bounded by LABs, was present in the HPTE-processed sample. In general, the majority of grains, bounded by HABs (and marked with blue lines in Figure 6d) contained one or more LABs.

We can see in the ACOM TEM maps (Figure 6b,c) that many twin-type GBs were formed during the HPTE deformation within the fine grains, in this way dividing them into smaller grains. The GND density differed significantly in neighboring (sub)grains (Figure 6c), which means that both LABs and twin-types GBs provided effective barriers for the dislocation slip and they should be taken into account in the analysis of possible strengthening factors, in particular Hall–Petch strengthening. Since the volume fraction of twin boundaries increases from the sample´s center to the edge (Table 1), we can conclude that twinning was more active at higher strains. From the nano-twins morphology, it is hypothesized that twinning occurred in the newly formed small grains.

3.3. Microstructure at the Edge of the Sample

Structure at the sample´s edge consisted of ultrafine equiaxed grains (Figure 3c). The average grain size (D_{15}) in this location was still higher than D_3 and it became a bit smaller than that in the mid-radius area (Table 1).

Grain size distribution histograms for the transverse and longitudinal sections looked similar to those of the mid-radius area. The distributions were of log-normal type, with a maximum at 0.7 µm in both sections (Figure 4a,b). GB misorientation distributions show that the fractions of HABs in transverse and longitude sections were 61 and 79%, respectively. These values were much higher than the fractions of HABs in both the central and mid-radius areas (Table 1).

3.4. Dislocation Density

From the center to the edge of the HPTE-processed rod, the GND density in copper after HPTE was 10-times higher when compared to that of the annealed material and it gradually increased from 100×10^{12} m^{-2} to 130×10^{12} m^{-2}. (Figure 7). Total dislocation density derived from XRD measurements was higher (150×10^{12} m^{-2} in the central areas and 300×10^{12} m^{-2} at the edge, Figure 7), being that X-ray diffraction data accounted for both GND and SSD.

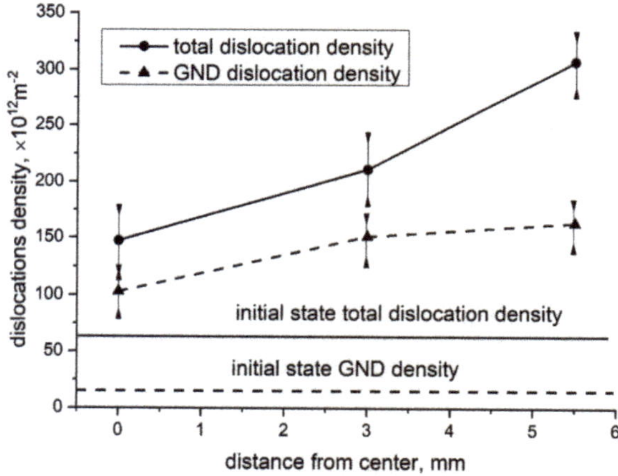

Figure 7. Dislocation density distribution in the CP Cu rod before and after HPTE.

3.5. Microhardness

In agreement with microstructural observations and dislocation density measurements, the microhardness of the HPTE-processed copper increased from initial value of 75 HV to 110 HV in the central area, and to 125 HV at the mid-radius and at the edge of the sample (Figure 8a). It can be seen that at a distance of ~0.4 R from the center, microhardness reached its saturation value of ~125 HV and it remained roughly the same up to the specimen edge (Figure 8a).

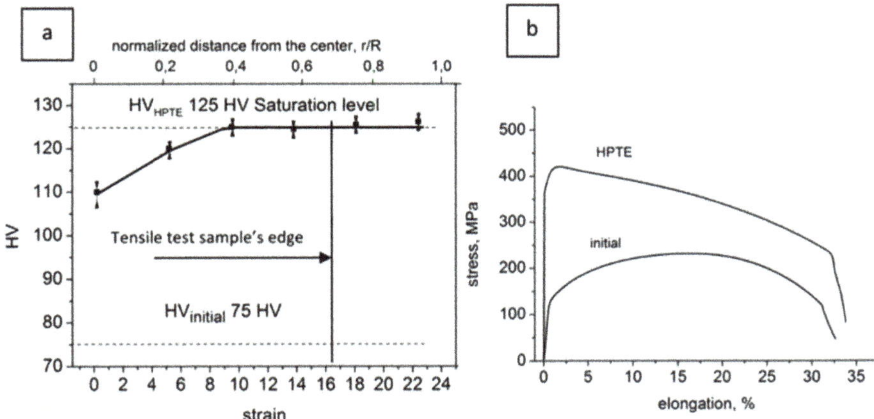

Figure 8. (**a**) Microhardness distribution along the radius and (**b**) tensile curves of Cu before and after HPTE.

Therefore, according to the HV data, the Cu rod samples produced by HPTE consisted of two distinct regions, i.e., an inner and an outer region. There are: (1) a central region that extends up to ~0.4 R = 2.0 mm (corresponding equivalent strain ~8, Figure 8a) with a gradually increasing microhardness; and (2) an outer region (>0.4 R) with saturated microhardness value of about 125 HV. The tensile test samples had the gauge diameter of 6.85 mm, sampling both the central area and half of outer area, as indicated by a vertical line in Figure 8a.

3.6. Tensile Tests

Tensile tests showed that the HPTE method resulted in an increase in the yield strength of Cu by more than twice that of the initial annealed state (370 MPa vs. 130 MPa) and a significant ultimate tensile strength enhancement of 430 MPa, with no changes in total elongation (Figure 8b, Table 2).

A rapid strengthening and a short uniform elongation range characterized the stress–elongation curve of copper after HPTE. A decrease of uniform elongation from ~16% in the annealed state to ~2% after the HPTE deformation was a typical behavior for the materials produced by SPD methods. However, after the deformation was localized, the elongation to fracture reached 30%, which was two times higher than post-necking elongation of the annealed copper (Figure 8b).

4. Discussion

The microstructure study demonstrated that the v1w1 regime of the HPTE deformation, carried out at 100 °C, resulted in the formation of a gradient microstructure in CP copper. This is not surprising since HPTE combines both extrusion and shear deformation modes. The shearing mode gradually increases from the sample's center to the edge (Equation (1)). Consequently, the central area of the HPTE-processed rod is characterized by the presence of a fine-grained structure, whereas in the mid-radius and edge areas are characterized by an ultrafine-grained structure. Apparently, large grains were gradually refined with an increasing strain, resulting in an increased fraction of the fine grains, as shown in Figure 4a. Even in the central area of the HPTE rod the majority of grains (about 80–85%) are the newly refined grains with the sizes in the range of ~2–5 μm (Figure 3a). In this area, new dislocation boundaries were actively formed and the difference between D_{15} and D_3 is significant (Table 1).

At a larger distance from the center, the applied strain increased, and the formation rate of new GBs became slower since both grain sizes, D_3 and D_{15}, were practically the same at the mid-radius and at the edge, indicative of no further grain refinement. This indicates that the saturation of the grain size refinement was reached, which is typical for other SPD methods [32]. Following grain refinement saturation, the misorientations at the already-formed interfaces were growing resulting in an increase of the HABs fraction (Figure 5). Note that this increase occurred at different rates at different grain facets and the difference between D_{15} and D_3 still persisted up to the highest strain reached in our experiment. GB distribution maps like one shown in Figure 6d clearly demonstrate that in all sample areas, the microstructure is complex, i.e., the LABs and HABs represent an interpenetrated network. In other words, a crystallite (grain or subgrain) can be simultaneously bounded by high angle and low angle boundaries. Interestingly, the density of SSD showed an enhanced increase at the edge of the HPTE-processed sample in comparison with that of GND (Figure 6c). This fact suggests that the multiplication rate of SSD was higher than their recovery/annihilation rate at the edge (also in the mid-radius) area, compared to the kinetics of the same processes in the central area. The difference in density for both kinds of dislocations (GND and SSD) is consistent with the published data [33], showing that: (i) in well-annealed metallic materials the GND density is 20%–30% lower than the total dislocation density; and (ii) that both values are strongly correlated.

The microstructure resulting from HPTE processing has similar features with the gradient microstructure of CP Cu after free-end torsion [31,34]. Nevertheless, several distinctions should be mentioned here. HPTE provided a more efficient grain refinement compared to that achievable with free-end torsion. Furthermore, instead of a coarse-grained structure in the center and a fine-grained structure at the periphery, we observed an FG structure at the center and a UFG structure at the periphery after HPTE. After HPTE, we observed a more homogeneous grain structure in the entire sample with no grain elongation, as it was observed after the free-end torsion.

CP Cu samples deformed via the HPTE under the v1w1 regime demonstrated a yield strength and an ultimate strength close to the corresponding values obtained after the free-end torsion with strain 1.3 [31]. The elongation to failure of Cu was reduced for more than 1.5 times compared to the elongation after free end torsion. By contrast, the ductility after HPTE remained at the level of the annealed state.

The microhardness saturation value of 125 HV is very close to the microhardness of oxygen-free (OF) copper deformed by HPTE under the v1w1 regime at room temperature [12], and also close to OF copper microhardness after HPT [8,15] (Table 2). The strength of the HPTE-processed copper is comparable with that of copper after ECAP, and it is not as high as that after HPT [8].

The change of the microhardness along the radius of the Cu rod cross-section allows to analyze the contribution of various hardening mechanisms after HPTE in different sample areas (Figure 8a). In fact, after HPTE treatment the variation of hardness is determined by the distribution of strain in the sample affecting the dislocation density values and the degree of grain refinement which gives rise to Hall–Petch hardening [35]. The inhomogeneity of the strain distribution in the HPTE-processed rod (Equation (1)) leads to the formation of the gradient structure. The tensile sample geometry (Figure 2b) made it possible to conduct mechanical testing on the sample with gradient structure.

We assume that experimentally obtained strength in CP copper samples after HPTE results from the following hardening mechanisms: accumulation of dislocations (σ_ρ), Hall–Petch strengthening and subgrains ($\sigma_{H\text{-}P}$), solid solution strengthening (σ_{SSS}), and second-phase particle strengthening (σ_{SPPS}).

The contribution of stored dislocations to the yield strength can be estimated using the Taylor Equation:

$$\sigma_P = \alpha MGb\sqrt{\rho}, \tag{3}$$

where M is Taylor factor, α is a constant, G is the shear modulus, b is the absolute value of the average Burgers vector, and ρ is the total dislocation density. We can estimate strengthening due to the grain refinement using the Hall–Petch relation:

$$\sigma_{H\text{-}P} = \sigma_0 + kd^{-1/2}, \tag{4}$$

where σ_0 is the friction stress, k is a constant, and d is the mean grain size in the corresponding sample area. The sample contained material from central and mid-radius areas. The microstructure in both areas is complex because the grains are bounded by both HABs and LABs. Therefore, it is not clear which GB misorientation threshold and grain size should be used in Equation (3). Prior in-situ TEM study [36] had shown that in FCC metals with low and middle stacking fault energy, the LABs with $\Theta > 3°$ and special GBs provide effective barriers for the dislocation slip. Therefore, we can assume that the effective grain size, d, in the Hall–Petch equation is equal to D_3. In Hall–Petch strengthening it is also necessary to account for the input from nanotwins. Twin boundaries are the obstacles for dislocations slip. The volume fraction of twin boundaries was 15.4% and the effective grain size d for the Hall–Petch strengthening calculation is $D_{3\,\text{with twins}} = 0.25$ μm, estimated from several ACOM TEM maps. This value is smaller than D_3 obtained from EBSD maps (0.4 μm, Table 1) due to higher angular resolution of ACOM TEM technique as previously discussed.

Solid solution strengthening is complex since the solute atoms can interact with dislocations by the elastic, electrical, short-range and long-range interactions [37]. For the Cu-Al substitutional solid solution, the elastic interaction is most important. This elastic interaction can be associated with local short-range mean normal stress. In the linear approximation, the elastic interaction is assessed as Paierls-Nabarro forces taking into account the atomic radius misfits and alloying element concentrations [37]. As a result, solid solution strengthening leads to an increase in the friction stress σ_0. Solid solution strengthening for the Cu–Al single crystal was measured experimentally and, depending on the concentration of Al in solid solution, a linear dependence of the σ_{sss} was obtained [38]:

$$\sigma_{SSS} = M4c^{\frac{2}{3}},\qquad(5)$$

where c is the concentration of Al in the Cu–Al solid solution.

Second-phase particle strengthening was provided by the large (about 1 micrometer in diameter) particles and it can be calculated using the Orovan's formula for the non-coherent particles:

$$\sigma_{SPPS} = \frac{2\alpha MGb}{d_p\,\sqrt{\pi/6f}},\qquad(6)$$

The parameters (σ_0, k, α, G, b) used for computing the various strengthening mechanisms described above were taken from [22,39,40], whereas the structure parameters D_3, ρ, d_p, f are given in Tables 1 and 3.

Table 3. Measured parameters used for the strengthening calculations.

Parameter	Initial Value	HPTE Center	HPTE Mid-Radius
ρ, $\times10^{12}$ m^{-2}	111	147	211
d, μm	30.0	0.8	0.25
M	3.16	3.15	3.08
d_p, μm	6.1	1.2	1.2
f, $\%$	0.17	0.16	0.16
C, $at\%$	0.5	0.6	0.6

As can be seen in Table 4, the calculated contributions from the solid solution and the precipitation strengthening are small and they provide less than 10% to the YS_{calc} value, while the dislocation strengthening is 3–4 times higher, which provides ~50% to the YS_{calc} in the initial state and only 20% of the YS_{calc} in HPTE-processed copper. The low contribution of dislocation strengthening to YS_{calc} in the HPTE-processed copper is likely related to the elevated temperature of the HPTE processing of 100 °C, which promoted recovery of dislocation density. Therefore, the main contribution (~60%) to the strength of HPTE-processed Cu is due to the GB strengthening via the Hall–Petch mechanism.

Table 4. Contributions (all are given in MPa) of various strengthening mechanisms in CP Cu before and after HPTE. The testing sample contained central and mid-radius areas of the rod.

Area		σ_0	σ_{SSS}	σ_{SPPS}	σ_ρ	$\sigma_{H\text{-}P}$	YS_{calc} *	YS_M **	YS_{exp} ***	HV
initial		30	7.9	3.4	60.7	25.6	127.6		130	750
HPTE	center	30	8.9	17.3	69.8	156.5	282.6	374.6	370	1100
	mid-rad	30	8.8	16.9	83.7	224.5	419.3			1250

* YS_{calc} is an arithmetic sum of the contributions from different strengthening mechanisms. ** YS_M is the yield strength calculated by the mixture rule.*** YS_{exp} is the experimental strength after HPTE.

In order to compare the calculated (YS_{calc}) and the experimentally measured (YS_M) yield strengths of the entire HPTE-processed sample we can use the rule of mixture:

$$YS_M = YS_C V_C + YS_{MR} V_{MR} \tag{7}$$

where YS_C and YS_{MR} are the calculated values of the yield strength of the material from the center and mid-radius area, respectively (Table 4), and V_C and V_{MR} are the volume fractions in these same areas. According to the microhardness distribution along the radius of the HPTE-processed sample (Figure 8a), the hardness is low in a narrow area with the radius of 1.2 mm around the rod axis. Therefore, the volume fraction of this part of the tensile specimen is only 0.12. Consequently, the majority of the volume fraction is contributed by mid-radius area. The calculated value of $YS_M = 374.6$ MPa is in a good agreement with experimentally measured yield strength of 370 MPa. A part of the HPTE-processed rod with high hardness and strength values was removed upon machining of the tensile specimen (Figure 8b). A rapid strengthening and a short uniform elongation range characterized the stress–elongation curve of copper after HPTE. A decrease of uniform elongation from ~16% in the annealed state to ~2% after the HPTE deformation is a typical behavior for the materials produced by SPD methods. However, as the deformation is localized, the elongation to fracture reaches 30%, which is two times higher than post-necking elongation of the annealed copper (Figure 8b). Such a behavior is typical for materials with ultrafine-grained microstructure [41–43] and most likely related to the increased strain rate sensitivity of such materials. Otherwise, the strength of the entire specimen would be even higher. Therefore, HPTE processing of CP copper provides large elongation to failure and high strengthening mostly due to the Hall–Petch mechanism. The uniform elongation (~2%) had not been improved despite the presence of gradient microstructure as proposed in [7]. Most likely, this approach could not be realized in the HPTE-processed copper because both central and outer layers were sufficiently strengthened, and the material demonstrated tensile behavior similar to that of the homogeneous UFG materials.

5. Conclusions

1. It was shown that HPTE equipment can be used to produce CP Cu rods combining both high strength and large elongation to failure.
2. HPTE under the v1w1 regime provides the formation of a gradient structure with fine-grained microstructure close to the rod axis and ultrafine-grained microstructure in the rest of the rod.
3. Tensile behavior of the CP copper after HPTE with gradient structure is similar to that of the CP copper after SPD techniques providing a homogeneous structure.
4. The strength of the gradient HPTE Cu sample follows the rule of mixture.
5. The main contribution to the strength of copper processed by HPTE at 100 °C is GB strengthening via Hall–Petch mechanism. Low angle boundaries and nanotwins serve as effective obstacles for dislocation slip and they should be taken into account regarding the grain size in Hall–Petch equation.
6. The potential of the HPTE method for obtaining a high strength in bulk structural materials was demonstrated. Similar strength properties were found earlier after conventional HPT treatment; however, HPT cannot produce bulk materials required for industrial use.

Author Contributions: Conceptualization, H.H. and Y.I.; methodology, D.N., A.M. and Y.I.; investigation, D.N. and A.M.; writing—original draft preparation, D.N.; writing—review and editing, Y.I. and A.M.; supervision, Y.I. and H.H.; project administration, H.H.; funding acquisition, Y.I. and H.H.

Funding: This work was financially supported by German Research Foundation (DFG) under Grants Nos. IV98/8-1, HA1344/39-1.

Acknowledgments: Authors are grateful to R. Kulagin and Y. Beygelzimer for fruitful discussion of the material's mechanical properties and to V. Provenzano for the careful editing of the manuscript. This work was partly performed with the support of the Karlsruhe Nano Micro Facility (KNMF), a Helmholtz Research Infrastructure at the Karlsruhe Institute of Technology (KIT).

Conflicts of Interest: The authors declare no conflict of interest.

References

1. Valiev, R.Z.; Estrin, Y.; Toth, L.S.; Lowe, T.C. Special Issue: Bulk Nanostructured Materials. *Adv. Eng. Mater.* **2015**, *17*, 1708–1709. [CrossRef]
2. Hoppel, H.W.; Kautz, M.; Xu, C.; Murashkin, A.; Langdon, T.G.; Valiev, R.Z.; Mughrabi, H. An overview: Fatigue behaviour of ultrafine-grained metals and alloys. *Int. J. Fatigue* **2006**, *28*, 1001–1010. [CrossRef]
3. Segal, V.M. Materials processing by simple shear. *Mater. Sci. Eng. A* **1995**, *197*, 157–164. [CrossRef]
4. Furukawa, M.; Horita, Z.; Nemoto, M.; Langdon, T.G. Review: Processing of metals by equal-channel angular pressing. *J. Mater. Sci.* **2001**, *36*, 2835–2843. [CrossRef]
5. Valiev, R.Z.; Langdon, T.G. Principles of equal-channel angular pressing as a processing tool for grain refinement. *Prog. Mater. Sci.* **2006**, *51*, 881–981. [CrossRef]
6. Estrin, Y.; Vinogradov, A. Extreme grain refinement by severe plastic deformation: A wealth of challenging science. *Acta Mater.* **2013**, *61*, 782–817. [CrossRef]
7. Fang, T.H.; Li, W.L.; Tao, N.R.; Lu, K. Revealing Extraordinary Intrinsic Tensile Plasticity in Gradient Nano-Grained Copper. *Science* **2011**, *331*, 1587–1590. [CrossRef]
8. Alawadhi, M.Y.; Sabbaghianrad, S.; Huang, Y.; Langdon, T.G. Direct influence of recovery behaviour on mechanical properties in oxygen-free copper processed using different SPD techniques: HPT and ECAP. *J. Mater. Res. Technol.* **2017**, *6*, 369–377. [CrossRef]
9. Beygelzimer, Y.; Prilepo, D.; Kulagin, R.; Grishaev, V.; Abramova, O.; Varyukhin, V.; Kulakov, M. Planar Twist Extrusion versus Twist Extrusion. *J. Mater. Process. Technol.* **2011**, *211*, 522–529. [CrossRef]
10. Richert, J. A New Method for Unlimited Deformation of Metals and Alloys. *Aliminum* **1986**, *62*, 604–607.
11. Pardis, N.; Talebanpour, B.; Ebrahimi, R.; Zomorodian, S. Cyclic expansion-extrusion (CEE): A modified counterpart of cyclic extrusion-compression (CEC). *Mater. Sci. Eng. A-Struct. Mater. Prop. Microstruct. Process.* **2011**, *528*, 7537–7540. [CrossRef]
12. Ivanisenko, Y.; Kulagin, R.; Fedorov, V.; Mazilkin, A.; Scherer, T.; Baretzky, B.; Hahn, H. High Pressure Torsion Extrusion as a new severe plastic deformation process. *Mater. Sci. Eng. A* **2016**, *664*, 247–256. [CrossRef]
13. Mizunuma, S. Large straining behavior and microstructure refinement of several metals by torsion extrusion process. In *Mater Sci Forum*; Horita, Z., Ed.; Trans Tech Publications Ltd.: Stafa-Zurich, Switzerland, 2006; Volume 503–504, pp. 185–190.
14. Fedorov, V.; Ivanisenko, J.; Baretzky, B.; Hahn, H. Vorrichtung und Verfahren zur Umformung von Bauteilen aus Metallwerkstoffen. German Patent DE 10 2013 213 072.4, European Patent EP 2 821 156 B1, 2015.
15. Edalati, K.; Imamura, K.; Kiss, T.; Horita, Z. Equal-Channel Angular Pressing and High-Pressure Torsion of Pure Copper: Evolution of Electrical Conductivity and Hardness with Strain. *Mater. Trans.* **2012**, *53*, 123–127. [CrossRef]
16. Khereddine, A.Y.; Larbi, F.H.; Kawasaki, M.; Baudin, T.; Bradai, D.; Langdon, T.G. An examination of microstructural evolution in a Cu-Ni-Si alloy processed by HPT and ECAP. *Mater. Sci. Eng. A* **2013**, *576*, 149–155. [CrossRef]
17. Li, J.H.; Li, F.G.; Zhao, C.; Chen, H.; Ma, X.K.; Li, J. Experimental study on pure copper subjected to different severe plastic deformation modes. *Mater. Sci. Eng. A* **2016**, *656*, 142–150. [CrossRef]
18. Balasundar, I.; Raghu, T. On the die design for Repetitive Upsetting—Extrusion (RUE) process. *Int. J. Mater. Form.* **2013**, *6*, 289–301. [CrossRef]

19. Wu, X.L.; Yang, M.X.; Yuan, F.P.; Wu, G.L.; Wei, Y.J.; Huang, X.X.; Zhu, Y.T. Heterogeneous lamella structure unites ultrafine-grain strength with coarse-grain ductility. *Proc. Natl. Acad. Sci. USA* **2015**, *112*, 14501–14505. [CrossRef]

20. Wu, X.L.; Zhu, Y.T. Heterogeneous materials: A new class of materials with unprecedented mechanical properties. *Mater. Res. Lett.* **2017**, *5*, 527–532. [CrossRef]

21. Rabkin, D.M.; Ryabov, V.R.; Lozovskaya, A.V. Preparation and properties of copper-aluminum intermetallic compounds. *Powder Met. Met. Ceram.* **1970**, *9*, 695. [CrossRef]

22. Davis, J.R.; Committee, A.S.M.I.H. *Copper and Copper Alloys*; ASM International: Novelty, OH, USA, 2001.

23. Omranpour, B.; Kulagin, R.; Ivanisenko, Y.; Garcia Sanchez, E. Experimental and numerical analysis of HPTE on mechanical properties of materials and strain distribution. *IOP Conf. Series: Mater. Sci. Eng.* **2017**, *194*, 012047. [CrossRef]

24. El-Dasher, B.S.; Adams, B.L.; Rollett, A.D. Viewpoint: Experimental recovery of geometrically necessary dislocation density in polycrystals. *Scri. Mater.* **2003**, *48*, 141–145. [CrossRef]

25. Calcagnotto, M.; Ponge, D.; Demir, E.; Raabe, D. Orientation gradients and geometrically necessary dislocations in ultrafine grained dual-phase steels studied by 2D and 3D EBSD. *Mater. Sci. Eng. A* **2010**, *527*, 2738–2746. [CrossRef]

26. Field, D.P.; Trivedi, P.B.; Wright, S.I.; Kumar, M. Analysis of local orientation gradients in deformed single crystals. *Ultramicroscopy* **2005**, *103*, 33–39. [CrossRef] [PubMed]

27. Markmann, J.; Yamakov, V.; Weissmuller, J. Validating grain size analysis from X-ray line broadening: A virtual experiment. *Scr. Mater.* **2008**, *59*, 15–18. [CrossRef]

28. Williamson, G.K.; Smallman, R.E., III. Dislocation densities in some annealed and cold-worked metals from measurements on the X-ray debye-scherrer spectrum. *Philos. Mag. A J. Theor. Exp. Appl. Phys.* **1956**, *1*, 34–46. [CrossRef]

29. Smallman, R.E.; Westmacott, K.H. Stacking faults in face-centred cubic metals and alloys. *Philos. Mag. A J. Theor. Exp. Appl. Phys.* **1957**, *2*, 669–683. [CrossRef]

30. Kobler, A.; Kübel, C. Challenges in quantitative crystallographic characterization of 3D thin films by ACOM-TEM. *Ultramicroscopy* **2017**, *173*, 84–94. [CrossRef]

31. Guo, N.; Song, B.; Yu, H.B.; Xin, R.L.; Wang, B.S.; Liu, T.T. Enhancing tensile strength of Cu by introducing gradient microstructures via a simple torsion deformation. *Mater. Des.* **2016**, *90*, 545–550. [CrossRef]

32. Pippan, R.; Scheriau, S.; Taylor, A.; Hafok, M.; Hohenwarter, A.; Bachmaier, A. Saturation of Fragmentation During Severe Plastic Deformation. *Annu. Rev. Mater. Res.* **2010**, *40*, 319–343. [CrossRef]

33. Kundu, A.; Field, D.P. Geometrically Necessary Dislocation Density Evolution in Interstitial Free Steel at Small Plastic Strains. *Metall. Mater. Trans. A* **2018**, *49*, 3274–3282. [CrossRef]

34. Wang, C.P.; Li, F.G.; Li, J.H.; Dong, J.Z.; Xue, F.M. Microstructure evolution, hardening and thermal behavior of commercially pure copper subjected to torsion deformation. *Mater. Sci. Eng. A* **2014**, *598*, 7–14. [CrossRef]

35. Hall, E.O. The Deformation and Ageing of Mild Steel: III Discussion of Results. *Proc. Phys. Soc. Sect. B* **1951**, *64*, 747–753. [CrossRef]

36. Fujita, H.; Toyoda, K.; Mori, T.; Tabata, T.; Ono, T.; Takeda, T. Dislocation Behavior in the Vicinity of Grain Boundaries in FCC Metals and Alloys. *Trans. Jpn. Inst. Met.* **1983**, *24*, 195–204. [CrossRef]

37. Stremel, M.A. *Prochnost Splavov. Chast 2*; MISIS: Moscow, Russia, 1997; Volume 2, p. 528.

38. Kratochvíl, P.; Neradová, E. Solid solution hardening in some copper base alloys. *Czechoslov. J. Phys. B* **1971**, *21*, 1273–1278. [CrossRef]

39. Hansen, N. Hall–Petch relation and boundary strengthening. *Scr. Mater.* **2004**, *51*, 801–806. [CrossRef]

40. Huang, C.X.; Wang, K.; Wu, S.D.; Zhang, Z.F.; Li, G.Y.; Li, S. Deformation twinning in polycrystalline copper at room temperature and low strain rate. *Acta Mater.* **2006**, *54*, 655–665. [CrossRef]

41. Valiev, R. Nanostructuring of metals by severe plastic deformation for advanced properties. *Nat. Mater.* **2004**, *3*, 511–516. [CrossRef]

42. Meyers, M.A.; Mishra, A.; Benson, D.J. Mechanical properties of nanocrystalline materials. *Prog. Mater. Sci.* **2006**, *51*, 427–556. [CrossRef]

43. Wang, Y.M.; Ma, E. Strain hardening, strain rate sensitivity, and ductility of nanostructured metals. *Mater. Sci. Eng. A* **2004**, *375–377*, 46–52. [CrossRef]

Article

Texture Hardening Observed in Mg–Zn–Nd Alloy Processed by Equal-Channel Angular Pressing (ECAP)

Jitka Stráská [1],*, Peter Minárik [1], Stanislav Šašek [1], Jozef Veselý [1], Jan Bohlen [2], Robert Král [1] and Jiří Kubásek [3]

[1]   Department of Physics of Materials, Charles University, 121 16 Prague, Czech Republic;
     peter.minarik@mff.cuni.cz (P.M.); sasekstanislav@seznam.cz (S.Š.); vesely@gjh.sk (J.V.);
     rkral@met.mff.cuni.cz (R.K.)
[2]   Helmholtz-Zentrum Geesthacht, Magnesium Innovation Centre, 215 02 Geesthacht, Germany;
     jan.bohlen@hzg.de
[3]   Department of Metals and Corrosion Engineering, University of Chemistry and Technology,
     166 28 Prague, Czech Republic; Kubasek.jiri@gmail.com
*   Correspondence: straska.jitka@gmail.com; Tel.: +420-95155-1613

Received: 27 November 2019; Accepted: 17 December 2019; Published: 24 December 2019

Abstract: The addition of Nd significantly improves the mechanical properties of magnesium alloys. However, only limited amounts of Nd or other rare earth (RE) elements should be used due to their high price. In this study, a low-alloyed Mg–1% Zn–1% Nd (ZN11) alloy was designed and processed by hot extrusion and subsequent equal-channel angular pressing (ECAP) in order to achieve a very fine-grained condition with enhanced strength. The microstructure, texture, and mechanical properties were thoroughly studied. The microstructure after 8 passes through ECAP was homogeneous and characterized by an average grain size of 1.5 μm. A large number of tiny secondary phase precipitates were identified as ordered Guinier–Preston (GP) zones. Detailed analysis of the Schmid factors revealed the effect of the texture on deformation mechanisms. ECAP processing more than doubled the achieved yield compression strength (YCS) of the ZN11 alloy. Significant strengthening by ECAP is caused by grain refinement and the formation of ordered Guinier–Preston zones and particles of a secondary γ-phase.

Keywords: magnesium; severe plastic deformation; equal-channel angular pressing; microstructure; deformation tests; texture; schmid factor

---

1. Introduction

Magnesium materials usually have very low density and high specific strength. The intended applications of lightweight magnesium products are primarily in the automotive or aerospace industries [1]. The addition of alloying elements determines the fundamental properties of the alloys. In our study, we focused on Mg–RE–Zn alloys, namely a Mg–Nd–Zn alloy.

Zn is often utilized as an alloying element in magnesium alloys, because it can improve mechanical properties of Mg alloys by solid solution strengthening [2]. However, for practical purposes in structural parts, the strength and fracture toughness of binary Mg–Zn alloys should be further improved, including at elevated temperatures [3]. It has been reported that rare earth (RE) elements can improve the tensile properties of Mg alloys both at room temperature and at elevated temperatures [4]. Even a low additional content of Nd can result in significant increases in hardness and yield strength of ternary Mg–Zn–Nd alloys after adequate thermal treatment [3]. The increase in hardness and strength is caused by the formation of plate-shaped Guinier–Preston (GP) zones and Mg_3Nd phases [5]. The addition of

Zn to Mg–RE alloys decreases the required concentrations of RE additions for achieving mechanical properties similar to binary Mg–RE alloys. This is beneficial for the adoption of RE-containing Mg alloys in industrial applications (such as the automotive industry) since RE elements are relatively expensive [6]. Yang et al. studied the mechanical properties and microstructure of Mg–4.5 Zn–x Nd (x = 0, 1, and 2 wt.%) alloys, and the best mechanical properties were obtained in the alloy with smaller amounts of Nd (1 wt.% of Nd) [3]. We also considered Nie's thorough review in which he showed that increased content of Zn (>1.3 wt.%) in Mg–Nd–Zn alloys can cause detrimental effects on the possible precipitation hardening [7]. Therefore, the presented study focuses on low-alloyed magnesium alloy ZN11 (Mg–1 wt.% Zn–1 wt.% Nd).

Solid solution strengthening and precipitation hardening are very effective and important in all magnesium alloys. Additionally, grain boundary strengthening is capable of further improving mechanical properties. Severe plastic deformation techniques are very useful methods for preparing ultrafine-grained materials.

Equal-channel angular pressing (ECAP) is a severe plastic deformation (SPD) technique based on repetitive simple shear. The facility for ECAP is comparatively simple and it is easily scaled-up to produce materials with reasonably large dimensions [8]. Various magnesium alloys were successfully prepared by ECAP, and the obtained microstructures were ultrafine-grained or even nanostructured (depending on the ECAP process parameters, especially ECAP temperature, and on the processed material—the amount and type of alloying elements and previous thermo-mechanical treatments) [9–11]. Zhao et al. investigated Mg–3.0Nd–0.4Zn–0.5Zr alloy, prepared by integrated extrusion and ECAP, and an ultrafine-grained microstructure with an average grain size of ~0.5 μm was obtained [12]. Zhang et al. prepared a Mg–2.0Nd–0.2 wt.% Zn alloy by friction stir processing, which resulted in grain refinement to ~2 μm [13]. Analogous grain refinement to the grain size of ~2 μm was achieved by Dvorský et al. in a similar Mg–3.0Nd–0.5Zn alloy, processed by hot extrusion only [14].

Properties of low-alloyed Mg–Nd–Zn alloys processed by the SPD method have not yet been properly studied.

In this contribution, the low-alloyed magnesium alloy ZN11—prepared by extrusion and ECAP—was investigated with a focus on microstructure, texture, secondary phases, and mechanical properties.

2. Materials and Methods

The experimental magnesium alloy ZN11 (Mg–1 wt.% Zn–1 wt.% Nd) was cast in Helmholtz-Zentrum Geesthacht, Germany, and extruded at 400 °C, with an extrusion ratio of 30 and a crosshead speed of 1 mm/s. Bars with the cross-section 10×10 mm^2 were machined from the extruded rod. Samples with a length of 10 cm were processed by ECAP, with the die having the inner angle of 90°, following route B$_C$ [15]. The processing temperatures and pressing speeds for ZN11 after extrusion are shown in Table 1. Each pass of the ECAP process was performed at the lowest possible temperature, because deformation temperature has a very strong effect on the resulting microstructure and mechanical properties, as was shown in the AX41 magnesium alloy by Krajňák et al. in [16] and for other materials [17–22]. Note that further decrease of the processing temperature would lead to occurrence of cracking and billet segmentation. Four conditions were investigated in this study: as-extruded (EX) condition, and conditions after one pass through ECAP (1P), four passes (4P), and eight passes (8P).

Table 1. ECAP (Equal-Channel Angular Pressing) procedure parameters of the investigated alloy.

ECAP Pass Number	1	2	3	4	5	6	7	8
Processing Temperature (°C)	340	330	320	310	300	290	280	280
Pressing Speed (mm/min)	5	7	10	10	10	10	10	10

The microstructure of all samples was investigated by scanning and transmission electron microscopy (SEM, TEM). An SEM Zeiss Auriga Compact (Zeiss, Oberkochen, Germany), equipped with an EDAX electron backscattered diffraction (EBSD) camera (EDAX Inc., Mahwah, NJ, USA), was used for the investigation of the grain structure. EBSD analysis was performed with a scan size of 400×400 μm^2 and a step size of 0.4 μm in the case of the extruded sample, and a scan size of 100×100 μm and step size of 0.1 μm in the case of the samples processed by ECAP. The raw data were partially cleaned in OIM TSL software (EDAX Inc., Mahwah, NJ, USA) using one step of confidence index (CI) standardization and one step of grain dilatation. Note that only points with CI > 0.1 were used for calculations. Analysis regarding occurrence of twinning in the pre-deformed sample was performed by in-built feature in OIM TSL software. Only one type of twin, tension twin $\{10\bar{1}2\} \langle 10\bar{1}1 \rangle$ (the misorientation angle between the matrix and the twin–86.4° around $\langle a \rangle$-axis), was observed. TEM JEOL 2200FS (JEOL USA Inc., Peabody, MA, USA), operating at 200 kV, was used for secondary phase analysis. Samples for EBSD measurements and thin TEM discs were mechanically prepared and subsequently ion-polished using a Leica EM RES102 (Leica Microsystems GmbH, Wien, Austria). The texture of the initial as-extruded condition was investigated by an X-ray PANalytical XPert MRD diffractometer (XRD) (Malvern Panalytical Ltd, Grovewood Road Malvern, UK). Cu$_K$ radiation and polycapillaries in the primary beam were employed during the measurements. The full pole figures were calculated using MTEX software (TU Chemnitz, Chemnitz, Germany) [23].

Mechanical properties were studied using the Vickers microhardness measurement and compression and tensile deformation tests. Microhardness measurements were performed by a Qness Q10 device (Qness GmbH, Colling, Austria) with a load of 0.5 kg (HV 0.5) for 10 s. Microhardness was measured on a cross-section plane. More than 100 experimental points (indents) were evaluated for each specimen. Compression and tensile deformation tests were performed using a universal testing device (Instron 5882, Instron, Coronation Road, Buckinghamshire, UK). Compression tests were performed on cuboid specimens with the aspect ratio of 2:3 ($4 \times 4 \times 6$ mm^3). The samples were machined with the deformation axis parallel to the extrusion direction (ED), normal direction (ND) and transverse direction (TD), see Figure 1. Note that compression tests along ND and TD were performed only for the 8P sample. Tensile deformation tests were performed only for the 8P sample along ED. Bone-shaped flat samples with a rectangular cross-section of 4×1.4 mm^2 and a gauge length of 10 mm were machined. At least three samples for each combination of deformation axis, deformation direction and materials condition were tested. Finally, the true stress–strain curves were calculated from the engineering stress–strain curves (see [24], for example).

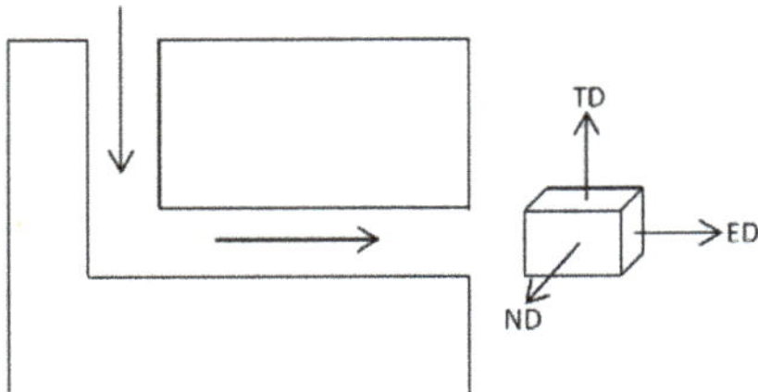

Figure 1. Mutual orientation of the equal-channel angular pressing (ECAP) die and the billet.

3. Results

3.1. As-Extruded Condition

The initial, as-extruded, microstructure of the investigated alloy was studied by EBSD, TEM, and XRD. Figure 2a shows the EBSD orientation map revealing the grain structure of the material. It is clear that extrusion resulted in a significant grain refinement, and the formation of a homogeneous, fully recrystallized microstructure. The grain size—determined as the average value weighted by the area fraction—was ~11 μm.

The presence of small secondary phase particles is clearly seen in the TEM micrograph in Figure 2b. Many grain boundaries are decorated by small particles in a necklace-like formation. Local chemical analysis and selected area electron diffraction confirmed that the particles are stable γ-phase Mg$_3$ (Nd, Zn), face-centered cubic (FCC) structure, $Fm\bar{3}m$, with $a = 0.70$ nm [7]. The size of the γ-phase particles varied from ~100 nm to 1 µm. In addition, Figure 2b discloses that the grain interiors are almost strain-free.

XRD measurement was performed to acquire information about the texture formation during the extrusion. Extrusion of magnesium alloys usually leads to the formation of the typical $\left(10\bar{1}0\right)$ fiber texture, which is often strong (explained theoretically by Mayama et al. [25]). However, Figure 3 shows that the texture observed in the investigated ZN11 alloy was completely different. This kind of texture has already been observed in Mg–RE alloys after extrusion, and is referred to as "rare earth texture" [26]. In the investigated sample, this texture component is weak, and is represented by grains which have their c-axis tilted from ED by approximately 45°.

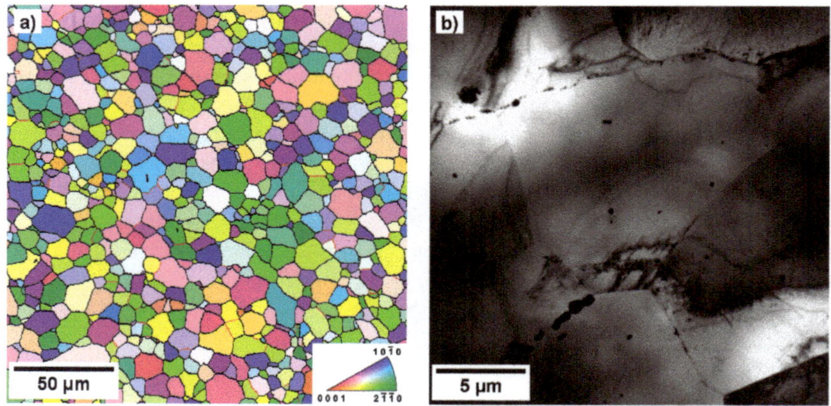

Figure 2. (**a**) Electron backscattered diffraction (EBSD) orientation map and (**b**) transmission electron microscopy (TEM) micrograph of the as-extruded (EX) sample (cross-section).

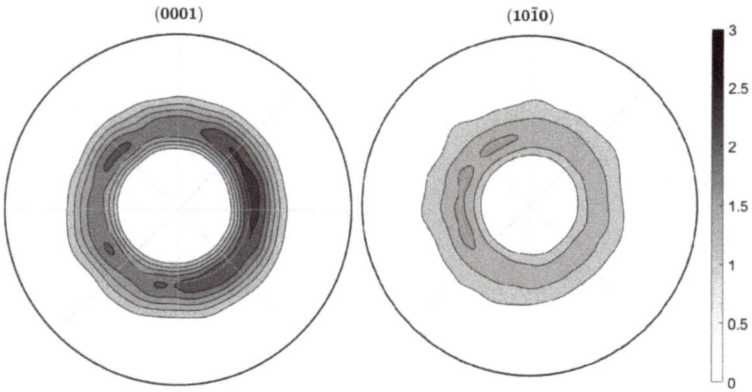

Figure 3. X-ray diffractometer (XRD) pole figures of the EX sample (cross-section).

3.2. Microstructure after ECAP

Significant grain refinement was observed in specimens processed by an increasing number of ECAP passes. The corresponding EBSD orientation maps of 1P, 4P, and 8P samples are shown in Figure 4a–c, respectively. As is typical for many other Mg alloys, the first pass resulted in the formation of a non-homogeneous microstructure. Grain refinement occurred particularly along the former grain boundaries, and large grains were heavily deformed, as manifested by the variation of colors in individual grains. The large grains of up to 20 µm in diameter were surrounded by small grains with a size of 3 µm. Increasing strain imposed on the material by successive ECAP passes resulted in the higher degree of refinement, and ultimately in the homogenization of the microstructure. The average grain size determined in the 4P and 8P samples was ~2.6 and ~1.5 µm, respectively.

Figure 5 shows the pole figure measured in the 8P sample. The texture exhibits the typical basal slip component, which corresponds to grains with c-axes tilted by 45° from the ED. However, a significant volume fraction of grains was oriented with the c-axis perpendicular to the processing direction. As a result, the basal slip texture component reached a texture strength of only ~9. Note that the texture strength of this component was found to be more than 15 in low-alloyed Mg alloys without rare earth elements [27,28].

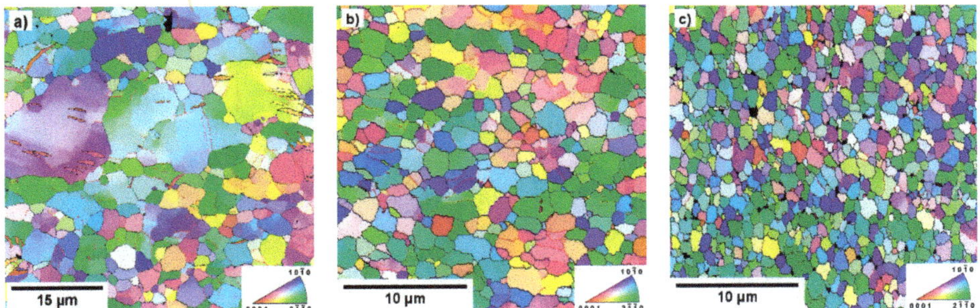

Figure 4. EBSD orientation maps of (**a**) 1P sample, (**b**) 4P sample, and (**c**) 8P sample (cross-section).

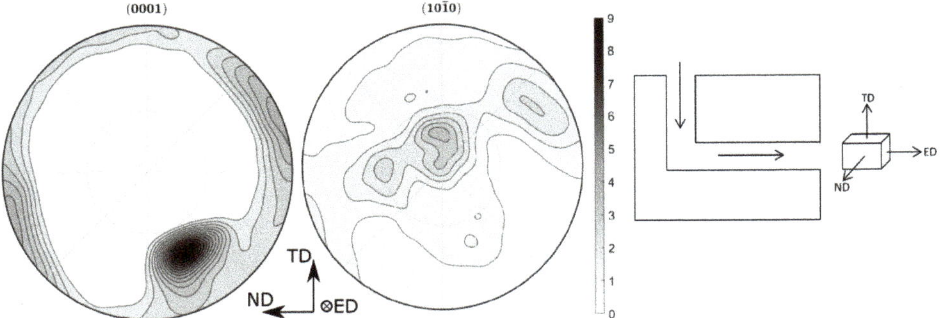

Figure 5. EBSD pole figure of the 8P sample (cross-section).

TEM investigation of the microstructure of specimens after different numbers of ECAP passes is consistent with the grain refinement observed by EBSD (compare Figures 6 and 7 with Figure 4a–c). In contrast to the as-extruded condition, the distribution of secondary phase particles in the ECAPed specimens was rather homogeneous, that is, they are distributed both along grain boundaries as well as in grain interiors. The size of the particles did not significantly change during the ECAP straining. Interestingly, significant residual strain was observed in the 8P sample (see Figure 7a). Moreover, higher magnification shows that the magnesium matrix contains also a large number of tiny secondary phase precipitates (see Figure 7b). They were identified as ordered Guinier–Preston (GP) zones (hcp, $a = 0.556$ nm, monolayer $(0001)_\alpha$ disc) and their presence has already been reported elsewhere [7]. It should be noted, however, that these GP zones were not observed in all grains of the 8P sample.

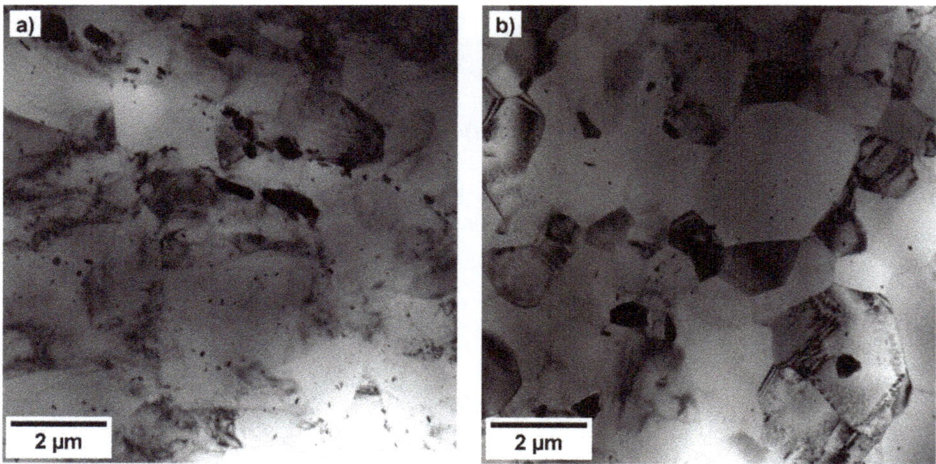

Figure 6. TEM images of (**a**) 1P sample and (**b**) 4P sample.

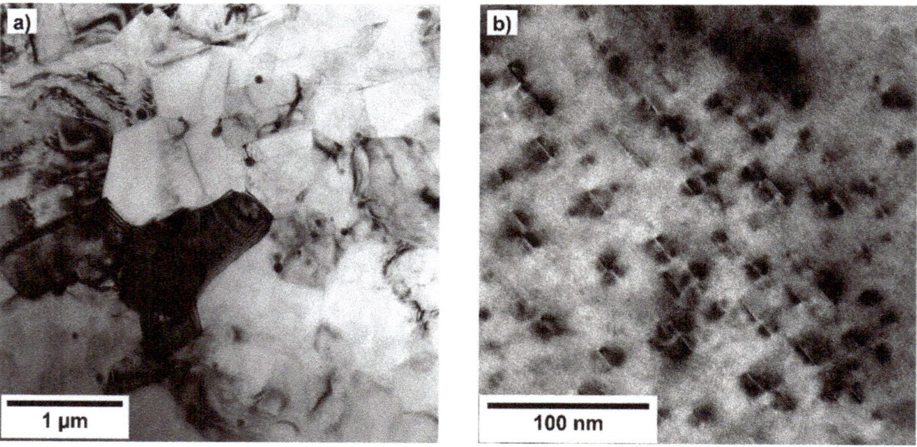

Figure 7. TEM images of (**a**) 8P sample and (**b**) detail of ordered Guinier–Preston zones.

3.3. Mechanical Properties

A significant change in the microstructure after ECAP alters the mechanical strength of the investigated material. In this report, we investigated microhardness as a multiaxial loading as well as compression/tensile tests as a uniaxial loading.

The resulting microhardness measurements shown in Table 2 reveal that the mechanical strength increased with the increasing number of ECAP passes, and in the final condition (8P), the microhardness was 1.5 times higher than in the as-extruded condition. Additionally, the yield compression strength (YCS) determined from flow curves (shown in Figure 8a) increased more rapidly (see Table 2). The YCS of the 8P sample was roughly 2.4 times higher than in the case of the as-extruded condition. The differences in the microhardness evolution and the evolution in ultimate tensile strength in uniaxial loading are usually caused by the changes in texture. The flow curves of all samples deformed in compression exhibited an S-shape character, which indicates the activation of twinning [29]. Moreover, the sharp yield point was observed in the 8P sample, which is quite unusual in magnesium-based materials. In order to reveal the possible effect of texture, compression tests in two mutually perpendicular directions (ED, transverse direction (TD)) were performed, and a tension test was performed along the processing direction (ED). The resulting flow curves are shown in Figure 8b. The tensile curve did not exhibit the S-shape character, suggesting that twinning does not occur in tension in the processing direction (ED). However, the strain anisotropy was very low despite the different characters of the flow curves.

Note that unlike ED flow curves, the flow curves along TD and normal direction (ND) did not exhibit the sharp yield point. In order to reveal the microstructural changes that led to the appearance of the sharp yield point, EBSD was performed on a sample deformed to 4% of plastic deformation along its ED. The resulting EBSD parent matrix/twin map is shown in Figure 9. The microstructure is heavily twinned, and the twinned grains are aligned in stripes declined by ~45° from the loading ED.

Table 2. Results of mechanical tests.

Sample	Microhardness (HV)	Yield (Compression) Strength (YCS) (MPa)	σ_{max} (MPa)
ZN11 EX	48 ± 1	96 ± 5	268 ± 9
ZN11 1P	52 ± 1	117 ± 6	322 ± 16
ZN11 4P	63 ± 1	201 ± 8	348 ± 28
ZN11 8P	75 ± 2	230 ± 15 (compression) 218 ± 14 (tension)	377 ± 22 (compression) 335 ± 20 (tension)

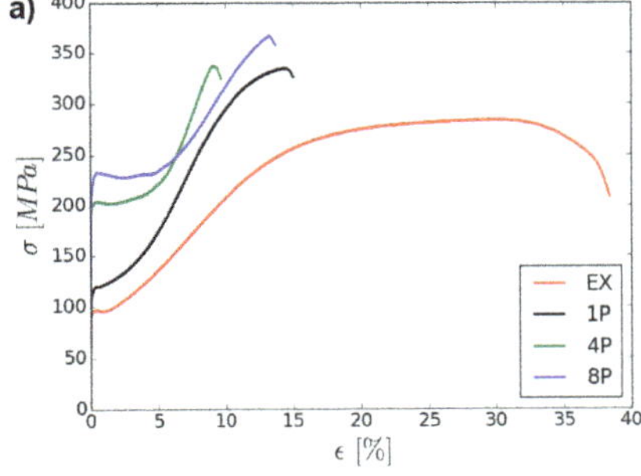
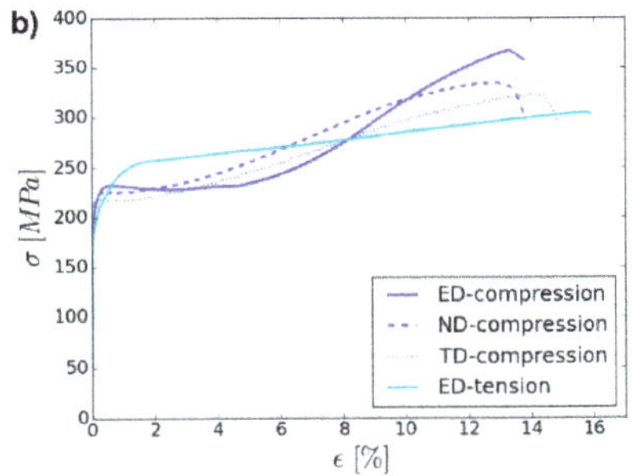

Figure 8. True stress–true strain compression curves of (**a**) all investigated conditions measured along the processing direction (extrusion direction (ED)) and (**b**) of the 8P sample measured in ED, normal direction (ND), and transverse direction (TD), see Figure 1.

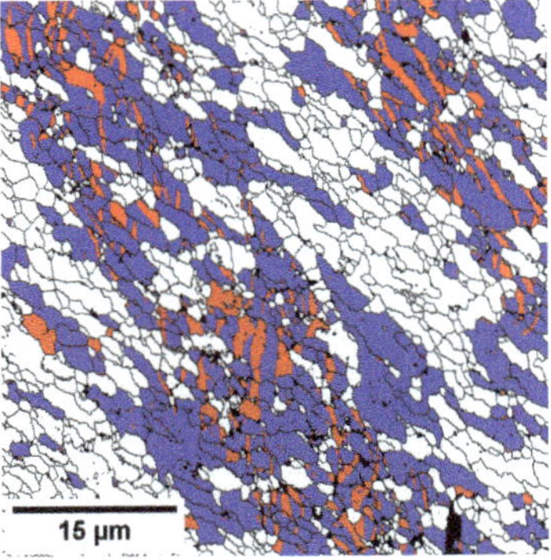

Figure 9. Parent/twin map calculated from EBSD, measured on a longitudinal section of the 8P sample, deformed in compression up to 4% of plastic deformation. Loading direction is horizontal. (Blue color corresponds to parent grains, red color corresponds to twins.)

4. Discussion

The as-extruded condition exhibited a fully recrystallized and homogeneous microstructure with relatively small grains and weak texture. Comparable texture was previously observed in the same alloy [30] and also in the binary Mg–Nd alloy [31], and is directly connected with the presence of the rare earth elements in the material. TEM investigation revealed that the secondary phase particles are mainly formed along the grain boundaries and belong to the stable γ-phase. Their inhomogeneous distribution probably resulted from the slow cooling rate after the extrusion.

Processing by ECAP resulted in additional grain refinement. Homogeneous microstructures with an average grain size of 1.5 μm were achieved after eight passes of the ECAP, which is common for similarly processed magnesium alloys [27,28,32–35]. However, the studied ZN11 alloy showed enhanced strength despite the comparable grain refinement. As mentioned above, the measured increase in microhardness was by a factor of 1.5, while YCS increased by a factor of 2.5. A larger relative effect on yield stress than on microhardness is rather common; however, note that the initial material has a different grain structure in different studies [27,28,32,36]. Therefore, the effect of the processing on the enhanced microhardness can be analyzed with the use of the Hall–Petch (HP) plot shown in Figure 10.

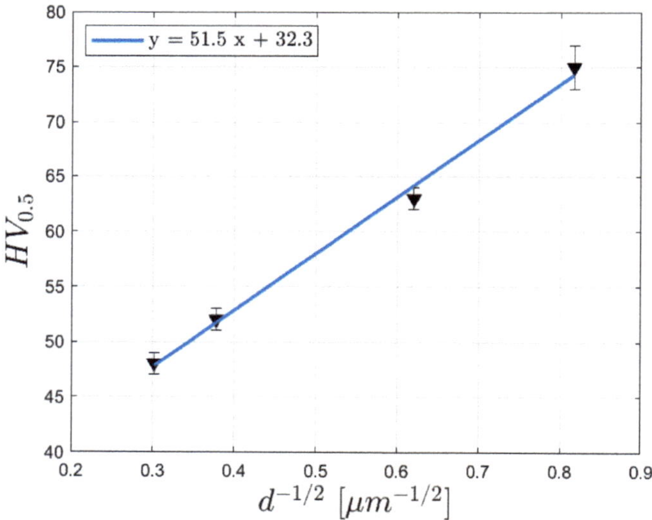

Figure 10. Hall–Petch plot showing the dependence of the microhardness on (grain size)$^{-1/2}$. (Grain size was calculated from the average density of grain boundaries in the case of bimodal grain size distribution.)

The HP plot has a linear character, which indicates the grain boundary validity of the Hall–Petch relation for the studied processed samples. However, the HP slope coefficient calculated from Figure 10 was ~50 HV.μm$^{1/2}$, which is a higher value than in the Mg–Al–Zn- and Mg–Al–RE-based alloys processed similarly [28,37]. Due to lower alloying, the studied alloy has lower microhardness in its coarse-grained condition, and the effect of grain refinement is enhanced. Additionally, other strengthening mechanisms may act concurrently with the grain refinement. For instance, dislocation density observed in the 8P sample was significant, contrary to the previously studied alloys in [28,37]. Furthermore, TEM analysis showed that ECAP processing of the studied alloy caused the formation of GP zones. Consequently, the higher HP coefficient can be explained by the synergic effect of these three phenomena acting together. However, the contribution of each strengthening factor cannot be differentiated from the investigated samples.

On the other hand, the increase in YCS was significantly higher than the increase in microhardness. This discrepancy can be explained by the texture evolution after ECAP. Note that the formation of the "rare earth texture" after the extrusion causes texture softening during uniaxial deformation along the extrusion direction [30]. A high volume fraction of grains had their c-axis rotated ~45° from the processing direction. Therefore, the activation of a basal slip system during uniaxial deformation along the extrusion direction is significantly facilitated. Figure 11a,b shows calculated distributions of the Schmid factor (SF) for all major deformation mechanisms for both extrusion and 8P conditions, when loaded in compression along the processing direction. The calculation was performed in MTEX software from raw EBSD data [23]. The SF for each point and each deformation mode was calculated and the resulting histograms were plotted.

A relatively sharp distribution of the SF for basal slip, with a maximum at SF = 0.5, turned to broad distribution, with maximum at SF = 0 after ECAP. It should be stressed that this texture development in the investigated alloy is quite uncommon compared to other Mg alloys. Usually, the initial as-extruded condition has a sharp $(10\bar{1}0)$ fiber texture, causing low values of SF for basal slip but high values of SF for twinning. ECAP processing results in the formation of a sharp and strong basal slip texture component, representing grains having their c-axis rotated by ~45° from all major axes. Consequently, significant texture softening occurs, which usually overwhelms grain boundary strengthening [27,28]. Therefore, the reason for the significantly higher increase of YCS as compared to *HV* values in the investigated alloy is the different texture evolution during ECAP. In magnesium alloys, the texture is typically formed during ECAP by the pronounced activation of a basal slip, or by combined activation of basal and non-basal slip systems [28,32]. The texture development in the investigated alloy can be explained according to the recently reported model on a ZN12 alloy. It was shown that a prismatic slip system has a relatively high activity during the hot rolling of the ZN12 alloy, and its effect on the texture is maintained by retarded recrystallization kinetics [38]. High activity of prismatic slip systems during ECAP causes a rotation of grains with their c-axis perpendicular to the processing direction (for details, see [28], for example). Consequently, significant grain boundary strengthening is even more promoted by the texture strengthening resulting from the processing.

However, a high volume fraction of grains having their c-axis perpendicular to the processing direction is favorable for the activation of tensile twinning during compression along the processing direction. This is also clearly visible in Figure 11b, where the distribution of the SF for tensile twinning is shown as well. High activity of twinning during the compression test also explains the formation of the sharp yield point in the ED-curve. It was shown previously that the formation of twin bands in the microstructure (Figure 9) can cause a drop in stress [39]. In order to compare changes in the deformation direction with the calculated SF for twinning, see Figures 11 and 12, which prove unambiguously that the probability of activation twinning in tension is significantly inferior to that in compression.

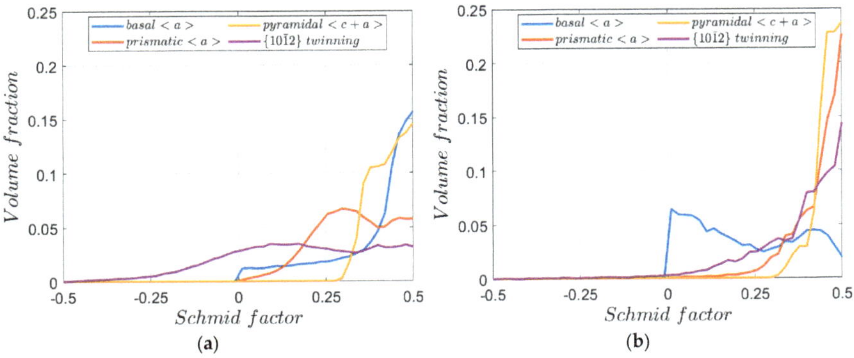

Figure 11. Schmid factor (SF) distribution calculated for major deformation mechanisms during the compression loading along the ED of (**a**) as-extruded sample (EX) and (**b**) 8P sample.

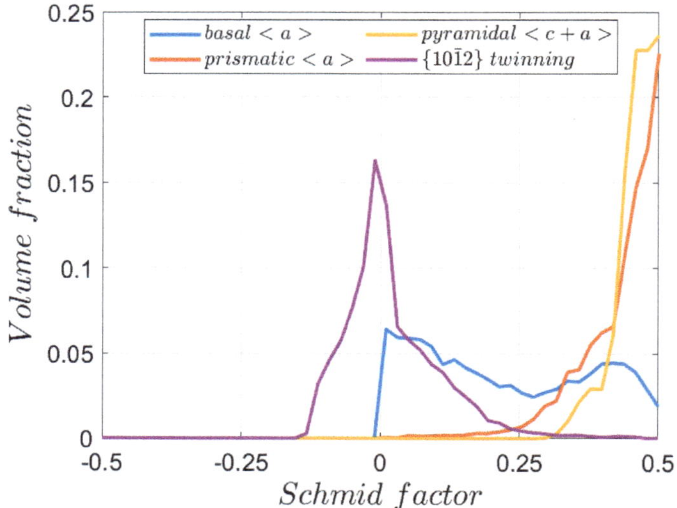

Figure 12. Schmid factor distribution calculated for major deformation mechanisms during the tensile loading along the ED of the 8P sample.

5. Conclusions

Low-alloyed magnesium alloy ZN11 was processed by hot extrusion and ECAP. The microstructure, texture, and mechanical properties were studied in detail and the main results can be summarized as follows:

The microstructure of the sample after 8 ECAP passes is fine-grained (with an average grain size of 1.5 µm) and contains significant residual strain and a large number of tiny secondary phase precipitates, which were identified as ordered Guinier–Preston zones.

The mechanical strength increased with an increasing number of ECAP passes. In the specimen after 8 ECAP passes, the microhardness was 1.5 times higher than that of the as-extruded condition. However, the yield compression strength (YCS) increased more rapidly. The YCS of the 8P sample was roughly 2.4 times higher than that of the as-extruded condition.

The effect of texture on deformation mechanisms was studied and Schmid factors were calculated. Measurements and calculations exposed the significant activity of tensile twinning $\{10\bar{1}2\}$ during compression loading in the ED, whereas work-hardening predominantly mediated by slip activity was observed during tensile loading in the extrusion direction.

Author Contributions: Conceptualization and methodology, J.S. and P.M.; investigation, P.M., S.Š. and J.V.; resources, J.B. and R.K.; writing—original draft preparation, J.S. and P.M.; visualization, P.M. and S.Š.; project administration, J.S. and J.K. All authors have read and agreed to the published version of the manuscript.

Funding: This research was funded by GACR project number 19-08937S. P.M. acknowledges partial financial support by ERDF under project No. CZ.02.1.01/0.0/0.0/15 003/0000485.

References

1. Dziubińska, A.; Gontarz, A.; Dziubiński, M.; Barszcz, M. The forming of magnesium alloy forgings for aircraft and automotive applications. *Adv. Sci. Technol. Res. J.* **2016**, *10*, 158–168. [CrossRef]

2. Cáceres, C.H.; Blake, A. The Strength of Concentrated Mg–Zn Solid Solutions. *Phys. Status Solidi* **2002**, *194*, 147–158. [CrossRef]

3. Yang, J.; Wang, J.; Wang, L.; Wu, Y.; Wang, L.; Zhang, H. Microstructure and mechanical properties of Mg–4.5Zn–xNd (x = 0, 1 and 2, wt%) alloys. *Mater. Sci. Eng. A* **2008**, *479*, 339–344. [CrossRef]

4. Lü, Y.; Wang, Q.; Zeng, X.; Ding, W.; Zhai, C.; Zhu, Y. Effects of rare earths on the microstructure, properties and fracture behavior of Mg–Al alloys. *Mater. Sci. Eng. A* **2000**, *278*, 66–76. [CrossRef]

5. Gärtnerová, V.; Trojanová, Z.; Jäger, A.; Palček, P. Deformation behaviour of Mg–0.7 wt.% Nd alloy. *J. Alloys Compd.* **2004**, *378*, 180–183. [CrossRef]

6. Javaid, A.; Hadadzadeh, A.; Czerwinski, F. Solidification behavior of dilute Mg-Zn-Nd alloys. *J. Alloys Compd.* **2019**, *782*, 132–148. [CrossRef]

7. Nie, J.-F. Precipitation and Hardening in Magnesium Alloys. *Metall. Mater. Trans. A* **2012**, *43*, 3891–3939. [CrossRef]

8. Valiev, R.Z.; Langdon, T.G. Principles of equal-channel angular pressing as a processing tool for grain refinement. *Prog. Mater. Sci.* **2006**, *51*, 881–981. [CrossRef]

9. Mostaed, E.; Hashempour, M.; Fabrizi, A.; Dellasega, D.; Bestetti, M.; Bonollo, F.; Vedani, M. Microstructure, texture evolution, mechanical properties and corrosion behavior of ECAP processed ZK60 magnesium alloy for biodegradable applications. *J. Mech. Behav. Biomed. Mater.* **2014**, *37*, 307–322. [CrossRef]

10. Minárik, P.; Veselý, J.; Král, R.; Bohlen, J.; Kubásek, J.; Janeček, M.; Stráská, J. Exceptional mechanical properties of ultra-fine grain Mg-4Y-3RE alloy processed by ECAP. *Mater. Sci. Eng. A* **2017**, *708*, 193–198. [CrossRef]

11. Cabibbo, M.; Paoletti, C.; Minárik, P.; Král, R.; Zemková, M. Secondary phase precipitation and thermally stable microstructure refinement induced by ECAP on Mg-Y-Nd (WN43) alloy. *Mater. Lett.* **2019**, *237*, 5–8. [CrossRef]

12. Zhao, S.; Guo, E.; Cao, G.; Wang, L.; Lun, Y.; Feng, Y. Microstructure and mechanical properties of Mg-Nd-Zn-Zr alloy processed by integrated extrusion and equal channel angular pressing. *J. Alloys Compd.* **2017**, *705*, 118–125. [CrossRef]

13. Zhang, W.; Tan, L.; Ni, D.; Chen, J.; Zhao, Y.-C.; Liu, L.; Shuai, C.; Yang, K.; Atrens, A.; Zhao, M.-C. Effect of grain refinement and crystallographic texture produced by friction stir processing on the biodegradation behavior of a Mg-Nd-Zn alloy. *J. Mater. Sci. Technol.* **2019**, *35*, 777–783. [CrossRef]

14. Dvorský, D.; Kubásek, J.; Vojtěch, D.; Voňavková, I.; Veselý, M.; Čavojský, M. Structure and mechanical characterization of Mg-Nd-Zn alloys prepared by different processes. *IOP Conf. Ser. Mater. Sci. Eng.* **2017**, *179*, 012018. [CrossRef]

15. Nakashima, K.; Horita, Z.; Nemoto, M.; Langdon, T.G. Development of a multi-pass facility for equal-channel angular pressing to high total strains. *Mater. Sci. Eng. A* **2000**, *281*, 82–87. [CrossRef]

16. Krajňák, T.; Minárik, P.; Stráská, J.; Gubicza, J.; Máthis, K.; Janeček, M. Influence of equal channel angular pressing temperature on texture, microstructure and mechanical properties of extruded AX41 magnesium. *J. Alloys Compd.* **2017**, *705*, 273–282. [CrossRef]

17. Shin, D.H.; Pak, J.-J.; Kim, Y.K.; Park, K.-T.; Kim, Y.-S. Effect of pressing temperature on microstructure and tensile behavior of low carbon steels processed by equal channel angular pressing. *Mater. Sci. Eng. A* **2002**, *323*, 409–415. [CrossRef]

18. Chen, Y.C.; Huang, Y.Y.; Chang, C.P.; Kao, P.W. The effect of extrusion temperature on the development of deformation microstructures in 5052 aluminium alloy processed by equal channel angular extrusion. *Acta Mater.* **2003**, *51*, 2005–2015. [CrossRef]

19. Huang, W.H.; Yu, C.Y.; Kao, P.W.; Chang, C.P. The effect of strain path and temperature on the microstructure developed in copper processed by ECAE. *Mater. Sci. Eng. A* **2004**, *366*, 221–228. [CrossRef]

20. Málek, P.; Cieslar, M.; Islamgaliev, R.K. The influence of ECAP temperature on the stability of Al–Zn–Mg–Cu alloy. *J. Alloys Compd.* **2004**, *378*, 237–241. [CrossRef]

21. Goloborodko, A.; Sitdikov, O.; Kaibyshev, R.; Miura, H.; Sakai, T. Effect of pressing temperature on fine-grained structure formation in 7475 aluminum alloy during ECAP. *Mater. Sci. Eng. A* **2004**, *381*, 121–128. [CrossRef]

22. Wang, Y.Y.; Sun, P.L.; Kao, P.W.; Chang, C.P. Effect of deformation temperature on the microstructure developed in commercial purity aluminum processed by equal channel angular extrusion. *Scr. Mater.* **2004**, *50*, 613–617. [CrossRef]

23. Bachmann, F.; Hielscher, R.; Schaeben, H. Texture Analysis with MTEX—Free and Open Source Software Toolbox. *Solid State Phenom.* **2010**, *160*, 63–68. [CrossRef]

24. Gottstein, G. *Physical Foundations of Materials Science*; Springer: Berlin/Heidelberg, Germany, 2004; ISBN 978-3-540-40139-1.
25. Mayama, T.; Noda, M.; Chiba, R.; Kuroda, M. Crystal plasticity analysis of texture development in magnesium alloy during extrusion. *Int. J. Plast.* **2011**, *27*, 1916–1935. [CrossRef]
26. Stanford, N.; Barnett, M.R. The origin of "rare earth" texture development in extruded Mg-based alloys and its effect on tensile ductility. *Mater. Sci. Eng. A* **2008**, *496*, 399–408. [CrossRef]
27. Janeček, M.; Yi, S.; Král, R.; Vrátná, J.; Kainer, K.U. Texture and microstructure evolution in ultrafine-grained AZ31 processed by EX-ECAP. *J. Mater. Sci.* **2010**, *45*, 4665–4671. [CrossRef]
28. Minárik, P.; Král, R.; Čížek, J.; Chmelík, F. Effect of different c/a ratio on the microstructure and mechanical properties in magnesium alloys processed by ECAP. *Acta Mater.* **2016**, *107*, 83–95. [CrossRef]
29. Dobroň, P.; Chmelík, F.; Yi, S.; Parfenenko, K.; Letzig, D.; Bohlen, J. Grain size effects on deformation twinning in an extruded magnesium alloy tested in compression. *Scr. Mater.* **2011**, *65*, 424–427. [CrossRef]
30. Nascimento, L.; Yi, S.; Bohlen, J.; Fuskova, L.; Letzig, D.; Kainer, K.U. High cycle fatigue behaviour of magnesium alloys. *Procedia Eng.* **2010**, *2*, 743–750. [CrossRef]
31. Minárik, P.; Drozdenko, D.; Zemková, M.; Veselý, J.; Čapek, J.; Bohlen, J.; Dobroň, P. Advanced analysis of the deformation mechanisms in extruded magnesium alloys containing neodymium or yttrium. *Mater. Sci. Eng. A* **2019**, *759*, 455–464. [CrossRef]
32. Krajňák, T.; Minárik, P.; Gubicza, J.; Máthis, K.; Kužel, R.; Janeček, M. Influence of equal channel angular pressing routes on texture, microstructure and mechanical properties of extruded AX41 magnesium alloy. *Mater. Charact.* **2017**, *123*, 282–293. [CrossRef]
33. Figueiredo, R.B.; Langdon, T.G. Factors influencing superplastic behavior in a magnesium ZK60 alloy processed by equal-channel angular pressing. *Mater. Sci. Eng. A* **2009**, *503*, 141–144. [CrossRef]
34. Kang, Z.; Zhou, L.; Zhang, J. Achieving high strain rate superplasticity in Mg–Y–Nd–Zr alloy processed by homogenization treatment and equal channel angular pressing. *Mater. Sci. Eng. A* **2015**, *633*, 59–62. [CrossRef]
35. Figueiredo, R.B.; Langdon, T.G. Grain refinement and mechanical behavior of a magnesium alloy processed by ECAP. *J. Mater. Sci.* **2010**, *45*, 4827–4836. [CrossRef]
36. Krajňák, T.; Minárik, P.; Stráská, J.; Gubicza, J.; Máthis, K.; Janeček, M. Influence of the initial state on the microstructure and mechanical properties of AX41 alloy processed by ECAP. *J. Mater. Sci.* **2019**, *54*, 3469–3484. [CrossRef]
37. Stráská, J.; Janeček, M.; Čížek, J.; Stráský, J.; Hadzima, B. Microstructure stability of ultra-fine grained magnesium alloy AZ31 processed by extrusion and equal-channel angular pressing (EX–ECAP). *Mater. Charact.* **2014**, *94*, 69–79. [CrossRef]
38. Zeng, X.; Minárik, P.; Dobroň, P.; Letzig, D.; Kainer, K.U.; Yi, S. Role of deformation mechanisms and grain growth in microstructure evolution during recrystallization of Mg-Nd based alloys. *Scr. Mater.* **2019**, *166*, 53–57. [CrossRef]
39. Hazeli, K.; Cuadra, J.; Vanniamparambil, P.A.; Kontsos, A. In situ identification of twin-related bands near yielding in a magnesium alloy. *Scr. Mater.* **2013**, *68*, 83–86. [CrossRef]

 *metals*

Article

On the Influence of Ultrasonic Surface Mechanical Attrition Treatment (SMAT) on the Fatigue Behavior of the 304L Austenitic Stainless Steel

Clément Dureau [1,2,3,*], Marc Novelli [1,2,4], Mandana Arzaghi [3], Roxane Massion [1,2], Philippe Bocher [4], Yves Nadot [3] and Thierry Grosdidier [1,2]

[1]   Laboratoire d'Etude des Microstructures et de Mécanique des Matériaux, Université de Lorraine, CNRS UMR 7239, 7 rue Félix Savart, 57073 Metz, France; marc.novelli@univ-lorraine.fr (M.N.); roxane.massion@univ-lorraine.fr (R.M.); thierry.grosdidier@univ-lorraine.fr (T.G.)
[2]   LABoratoire d'EXcellence Design des Alliages Métalliques pour Allègement de Structures, Université de Lorraine, 7 rue Félix Savart, 57073 Metz, France
[3]   Institut Pprime, CNRS-ENSMA, Université de Poitiers, UPR CNRS 3346, Physics and Mechanics of Materials Department, ENSMA, Téléport 2, 1 Avenue Clément Ader, BP 40109, 86961 Futuroscope Chasseneuil CEDEX, France; mandana.arzaghi@isae-ensma.fr (M.A.); yves.nadot@ensma.fr (Y.N.)
[4]   Ecole de Technologie Supérieure de Montréal, 1100 Rue Notre-Dame Ouest, Montréal, QC H3C 1K3, Canada; philippe.bocher@etsmtl.ca
*   Correspondence: clement.dureau@univ-lorraine.fr

Received: 25 November 2019; Accepted: 1 January 2020; Published: 8 January 2020

Abstract: The potential of ultrasonic surface mechanical attrition treatment (SMAT) at different temperatures (including cryogenic) for improving the fatigue performance of 304L austenitic stainless steel is evaluated along with the effect of the fatigue loading conditions. Processing parameters such as the vibration amplitude, the size, and the material of the shot medias were fixed. Treatments of 20 min at room temperature and cryogenic temperature were compared to the untreated material by performing rotating–bending fatigue tests at 10 Hz. The fatigue limit was increased by approximately 30% for both peening temperatures. Meanwhile, samples treated for 60 min at room temperature were compared to the initial state in uniaxial fatigue tests performed at R = −1 (fully reversed tension–compression) at 10 Hz, and the fatigue limit enhancement was approximately 20%. In addition, the temperature measurements done during the tests revealed a negligible self-heating ($\Delta t < 50$ °C) of the run-out specimens, whereas, at high stress amplitudes, temperature changes as high as 300 °C were measured. SMAT was able to increase the stress range for which no significant local self-heating was reported on the surface.

Keywords: surface mechanical attrition treatment (SMAT); ultrasonic shot peening (USP); cryogenic temperature; 304L austenitic stainless steel; rotating–bending fatigue; tension–compression fatigue

---

1. Introduction

In the high cycle fatigue regime, cracks mostly nucleate at the surface of the loaded workpiece. Therefore, the surface or near-surface features such as surface roughness and residual stresses affect the fatigue life of a component. The surface roughness has an influence essentially at the crack initiation stage [1]. Indeed, the presence of surface irregularities leads to high local stresses, which may create a high amount of plastic deformation locally, eventually contributing to crack-type defect creation depending on the intrinsic behavior of the studied material. Furthermore, the presence of residual stresses may affect the crack propagation behavior [2]. While a tensile stress loading promotes crack opening and faster propagation, the introduction of compressive residual stresses can delay or stop crack propagation with the consequence of increasing the total fatigue life of mechanical components.

In this context, plastic deformation treatments such as burnishing [3], laser shock-peening [4,5], or hammering [6,7] were used on stainless steels to delay early fatigue crack initiation and growth and increase part performance.

In addition, surface mechanical attrition treatment (SMAT) and ultrasonic shot peening (USP) were used to impart severe plastic deformation (SPD) at the surface of components. The effects of SMAT and similar kind of surface severe plastic deformation techniques on the surface and near-surface microstructure modifications and associated mechanical properties were reviewed in several papers [8–12]. It is noticeable that positive effects on the fatigue behavior were observed on different materials such as Fe- [13–15], Ti- [16], or Mg-based [17] alloys. In some cases, the use of these treatments pushed the initiation site underneath the peened surface, thus enhancing the fatigue limit significantly [18,19]. At the same time, the peening energy also strongly influences the treated surface roughness [20] and integrity, which can decrease the fatigue resistance of the components in spite of the nanostructured surface layer and higher compressive residual stresses [16,21,22].

The fatigue resistance of 316L austenitic stainless steels was enhanced by nearly 30% (from 300 to 380 MPa) due to the presence of a superficial refined microstructure delaying crack initiations and a high compressive residual stress impeding crack propagation [13].

The effect of SMAT treatments on the fatigue behavior of the 304L was investigated at room temperature (RT), as well as elevated temperature, by Kakiuchi et al. [23]; the shot-peened specimens exhibited higher fatigue strengths than the untreated ones at both RT and 300 °C. By decreasing or removing the surface topology induced by severe shot peening through a repeening or grinding process, the fatigue life of a shot peened low-alloyed steel can be further improved as shown in the work of Bagherifard et al. [20].

The idea of practicing cryogenic SMAT on 304L comes from the fact that some Fe–Ni–Cr austenitic stainless steels, such as 304L, are metastable and undergo stress- or strain-induced martensitic transformations through severe plastic deformation. Accordingly, the volume fraction of martensite in the SMAT samples increases at lower temperatures [24–26]. Although no definitive explanation was given for the influence of martensite (α') prior to the cyclic loadings [27], its presence is usually reported to be beneficial regarding tensile strength [28]. Also, for a given SMAT processing condition, the use of cryogenic temperature was shown to decrease the surface roughness [25,26], a modification that may result in improved fatigue performance.

In this context, the present paper has two major goals: (i) to investigate the effect of lowering the SMAT temperature on the fatigue behavior of a 304L stainless steel, and (ii) to analyze differences that may exist between rotating–bending and fully reversed uniaxial tension–compression fatigue tests on a 304L stainless steel.

2. Materials and Methods

2.1. Sample Preparation and Surface Treatments

Prior to machining, the 304L cylindrical specimens were fully annealed at 900 °C for 40 min resulting in an average grain size of approximately 40 μm. The fatigue behavior was studied under rotating–bending (RB) and tension–compression (TC) tests. For RB, 6-mm-diameter samples with a gauge length of 25 mm were machined and ground in order to remove the grooves generated by the machining procedure. For TC, 9-mm-diameter samples with a gauge length of 12.5 mm were machined and then polished to a mirror finish. The SMAT was carried out using a machine developed by SONATS company (Carquefou, France) [29]. The samples were treated using a vibrating amplitude of 60 μm with ø2 mm 100Cr6 steel shots. For rotating–bending, the SMAT was carried out for 20 min at two temperatures: (i) at room temperature (RT), or (ii) under cryogenic conditions (CT) at about −100 °C. For tension–compression samples, SMAT was done for 60 min at room temperature.

2.2. Fatigue Tests

Samples were tested in two loading conditions, rotating–bending (RB) and tension-compression (TC). For RB, the fatigue behavior was studied with a loading frequency of 10 Hz and a load ratio of R = −1, in air, at room temperature using an R. R. Moore rotating–bending fatigue test bench (Instron, Norwood, MA, USA). The fatigue limit was estimated from the highest stress amplitude at which no failure occurred up to 10^6 cycles.

For TC, fully reversed tension–compression (R = −1) fatigue tests were carried out with a loading frequency of 10 Hz, in air, at room temperature on an 810 MTS (MTS systems, Eden Prairie, MN, USA) servo-hydraulic machine equipped with a dynamic loading cell with a maximum capacity of ±100 kN.

2.3. Sample Characterizations

The specimen surface aspect was observed using an LEXT OLS4000 confocal microscope (Olympus, Tokyo, Japan), and the R_q surface roughness parameter was measured using a Mitutoyo SJ-400 probe (Mitutoyo, Kawasaki, Japan). The surface and subsurface hardening states were analyzed on a cross-section using a microhardness test with a measurement step of 50 µm. Phase identification (γ austenite and α' martensite) was done by EBSD (Electron Back-Scattered Diffraction) technique on specimen cross-sections in order to link the hardening state with the observed microstructural features.

On the RB samples, the surface residual stresses were evaluated using a Proto iXRD X-ray diffraction apparatus (PROTO Manufacturing Inc., Taylor, MI, USA), operating under a 20 kV tension and a 3 mA intensity, before cyclic loading, as well as after fatigue tests on the run-out samples. Two X-ray wavelengths were used to measure the residual stresses in each phase: (i) a chromium tube ($Cr_{K\alpha}$ = 2.291 Å) for the {211} α' martensite, and (ii) a manganese tube ($Mn_{K\alpha}$ = 2.103 Å) for the {311} austenite. Residual stresses were calculated by the classical $\sin^2 \psi$ method [30].

On the TC samples, the surface residual stresses were evaluated with a Pulstec µ-X360 X-ray diffraction equipment (Pulstec, Hamamatsu, Japan) operating under a 30 kV tension and a 1.5 mA current. A chromium tube ($Cr_{K\alpha}$ = 2.291 Å and $Cr_{K\beta}$ = 2.085 Å) was selected for the measurements, and the $\cos \alpha$ method [30] was used for the calculation of residual stresses in both γ and α' phases before and after fatigue tests.

In order to quantify the self-heating of the specimens during TC tests, a few samples were painted with black heat-resistant paint to allow a good emissivity, and their surface temperature was measured using an Infratech infrared camera (Infratech, Kennesaw, GA, USA).

Testing conditions and surface treatments used are summarized in Table 1.

Table 1. Summary of all testing and surface treatment conditions with their associated designations. SMAT—surface mechanical attrition treatment.

Testing Condition	Rotating–Bending, Air, Room Temperature, 10 Hz, R = −1 (RB)			Tension–Compression, Air, Room Temperature, 10 Hz, R = −1 (TC)	
Surface Treatment Condition	Ground (G)	SMAT for 20 min at room temperature (20 min RT)	SMAT for 20 min at cryogenic temperature (20 min CT)	Polished (mirror finish) (P)	SMAT for 60 min at room temperature (60 min RT)
Fatigue Sample's Name	RB_G	RB_20 min RT	RB_20 min CT	TC_P	TC_60 min RT

3. Results

3.1. Surface Modifications

Several representative examples of the surface topology, as well as their respective measured roughness values, are displayed in Figure 1. Despite a low roughness resulting from the surface grinding (R_q = 0.32 µm), some machining grooves still remained visible on the surface of the RB_G samples (Figure 1a). These grooves were completely removed after 20 min of SMAT at RT (Figure 1b) but not completely for the cryogenic treatment (Figure 1c). Both types of peening conditions induced a considerable increase of the surface roughness (from 0.32 to 3.42 and 1.97 µm). However, SMAT for 20 min at CT significantly reduced the surface roughness compare to the SMAT for the same duration at RT; a reduction of 42% can be noticed (Figure 1c). This is consistent with previous studies [26], which showed that, as the material becomes harder at low temperature, the steel beads impacting the specimen surface generate less pronounced craters. The effect of SMAT is also visible for the TC specimens (Figure 1d,e). Due to mirror-finish polishing, the TC_P samples exhibited an extremely low roughness (R_q = 0.022 µm). The surface aspect of the SMAT for 60 min at RT sample (Figure 1e) was rather similar to the one obtained after 20 min (Figure 1b), except that the longer peening duration generated a reduction in roughness of about 50% (R_q = 1.755 µm).

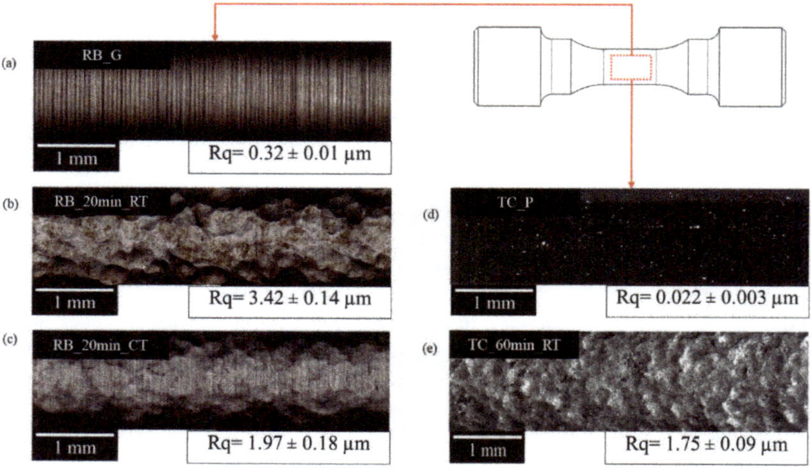

Figure 1. Observation of the lateral surface of cylindrical samples with their corresponding R_q roughness values: (**a**) ground condition, (**b**) after surface mechanical attrition treatment (SMAT) for 20 min at room temperature, (**c**) after SMAT for 20 min in cryogenic condition and (**d**) polished condition, and (**e**) after SMAT for 60 min at room temperature.

The hardness values are plotted as a function of the depth for different samples in Figure 2. The horizontal black dotted line defines the initial hardness (210 HV). The SMAT treatment increased the surface and subsurface hardness for all treatment duration and temperature conditions; maximum hardness value was located near the treated surface followed by a gradual decrease toward the specimen core until it reached the initial material hardness. The profile of the curve for the cryogenic treatment had some specific characteristics. Compared to the 20 min room temperature treatment, the use of cryogenic temperature substantially increased the hardness along the first 200 µm, while the hardness was comparatively lower at higher depths. In addition, if the hardened depth is defined as the depth at which the initial hardness of the material is reached, one can see that this depth was 30% lower for the sample treated at CT than for the two other conditions.

The EBSD maps (Figure 2b) show that a large amount of martensite was formed in the CT specimen, whereas this phase is hardly detectable in the sample treated for 20 min at RT. The crossover between the hardness evolution curves (seen in Figure 2a) corresponds to the depth at which the martensitic phase transformation was triggered in the CT processed sample (i.e., about 200 µm below the surface).

It is also interesting to highlight the fact that increasing the RT treatment time from 20 to 60 min did not drastically modify the hardened depth (~500 μm) but increased the maximum subsurface hardness by approximately 20%. Ultimately, the same top surface hardness value as for the CT treatment (~550 HV) was achieved.

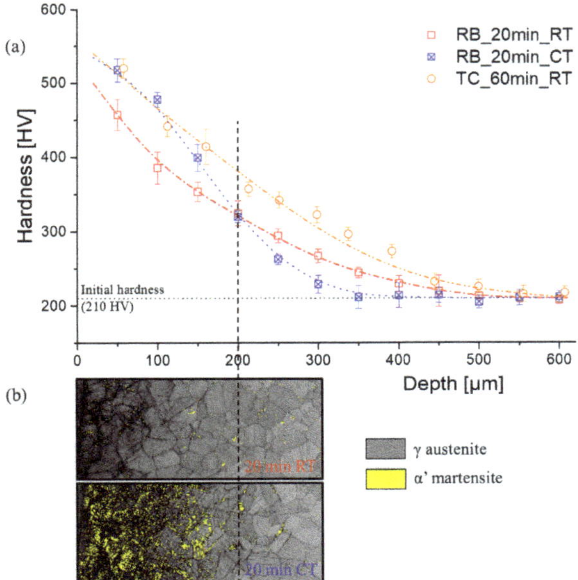

Figure 2. (**a**) Hardness measurements done on a cross-section of the (i) rotating–bending (RB) specimens treated at room temperature (RT; red) and under cryogenic conditions (CT; blue), and (ii) tension–compression (TC) specimens tested at RT (orange) (0 μm corresponds to the outer surface of the sample). (**b**) Corresponding EBSD band contrast maps with α' martensite distributions.

3.2. Fatigue Properties

The Wöhler curves associated with the three conditions and both types of cyclic loadings are shown in Figure 3. After 10^6 cycles, if failure did not happen, the tests were considered to be run-out tests and a horizontal arrow is attached to them on the graph. Squares are used for rotating–bending, and circles denote tension–compression. Full symbols illustrate the untreated samples, whereas hollow symbols are used for room temperature SMAT and crossed hollow symbols are used for cryogenic treatments. Except for RB_20 min CT (due to the lack of data), the Stromeyer expression (1) [31] was chosen to fit the experimental results and the constants obtained are given in Table 2.

$$\sigma_a = \sigma_D + \left(\frac{c}{N}\right)^{\left(\frac{1}{m}\right)},$$ (1)

where σ_D is a value determined from the experimental data, c and m are fitting parameters, σ_a is the stress amplitude, and N is the number of cycles. Nonlinear least squares curve-fitting following a Levenberg–Marquardt algorithm was used to determine c and m.

As can be seen in Figure 3, a similar trend was found between RB_G and TC_P, and, as reported in Table 2, the fatigue limits of the initial material were 250 MPa in rotating–bending and 205 MPa in tension–compression, representing a −18% difference. The effect of SMAT on the fatigue life of 304L was not the same depending on the considered fatigue life range; it is more pronounced in the high cycle fatigue regime than in the low cycle one. After SMAT, a rather high increase of the fatigue limits was obtained, the improvements being +28% in RB and +17% in TC. Even though the phases distributions, microstructures, hardness gradients, and residual stresses were different in RB_20 min RT and RB_20 min CT samples, their fatigue behavior was almost identical. Also, the obtained *m* parameter was significantly higher for the SMAT conditions compared to the initial states due to the smooth form of their S–N curves.

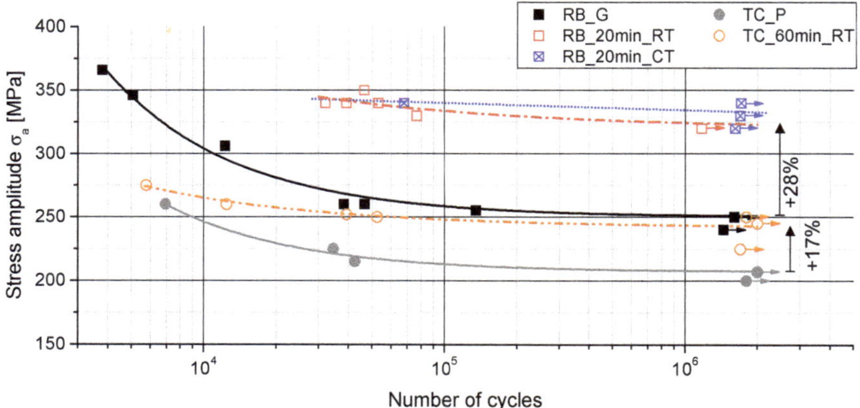

Figure 3. S–N curves obtained at 10 Hz for (i) RB_G (filled black squares), RB_20 min RT (hollow red squares), and RB_20 min CT (crossed hollow blue squares) treated specimens, and for (ii) TC_P (filled gray circles) and TC_60 min RT (hollow orange circles) samples.

Table 2. Estimated fatigue limits and fitted constants c and m of Stromeyer expression (1).

Testing Condition	σ_D (MPa)	c	m
RB_G	250	1.360×10^6	1.231
RB_20 min RT	320	2.225×10^7	2.060
RB_20 min CT	320	-	-
TC_P	205	8.756×10^5	1.218
TC_60 min RT	240	2.208×10^6	1.715

3.3. Surface Residual Stresses

The values of the axial surface residual stresses measured before and after fatigue tests (on non-broken specimens) are compared in Figure 4. The residual stress on the surface of the RB_G sample was approximately −400 MPa (compression) showing that, before SMAT, rather high compressive residual stresses were created by machining followed by grinding, as in the work of Velásquez et al. [32]. After SMAT, the magnitude of the compressive residual stresses was significantly enhanced. However, the temperature at which the SMAT was done only slightly increased the residual stresses, +10% for the γ austenite and +3% for α′ martensite, compared to the SMAT done at RT. It is interesting to note that higher residual stresses were reached in α′ martensite than in the γ austenite (approximately 200 MPa more in martensitic phase than in the austenitic one) which may be attributed to the difference in Young's modulus of these two phases, as explained in the work of Spencer et al. [28].

For polished TC specimens, due to the fact that the first stages of polishing were similar to grinding in terms of removing the top surface layer, the measured compressive residual stress (−340 MPa) was almost as high as for the ground RB specimens, and no martensite was detected. For the TC sample treated by SMAT for 60 min, no austenite was found, and only the characteristic {211} plan of α′ martensite diffracted, which resulted in a measured residual stresses value of −920 MPa. The magnitude of the residual stresses was significantly higher after 60 min of RT SMAT (TC sample) than after 20 min (RB samples).

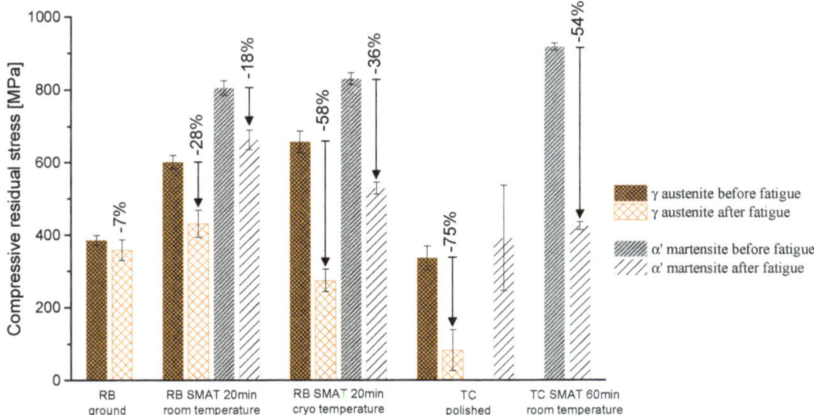

Figure 4. Measured residual stresses in the longitudinal direction on the sample surfaces done by X-ray diffraction. Different colors and textures were used to illustrate austenite or martensite and pre- or post-cycled measurements.

Concerning the residual stresses after rotating–bending fatigue, the comparison of the fine and spare textured bars in Figure 4 clearly indicates that the residual stress relaxation after fatigue loading was rather pronounced in the austenitic phase of the SMAT samples (30% and 60% for the RB_20 min RT and RB_20 min CT samples, respectively) but quite limited for the RB_G sample (only a few tens of MPa). The extent of the relaxation for the samples treated by SMAT at cryogenic temperature was much more pronounced than in the RT processed ones, and this applied to both phases. Indeed, for the RB_20 min CT sample, the relaxations were roughly twice those recorded for the RB_20 min RT sample. Decreases of −58% and −36% were measured, respectively, in the austenitic and martensitic phases for the RB_20 min CT, while they were only −28% and −18% for the RB_20 min RT condition.

For the tension–compression fatigue tests, a high relaxation of the residual stresses occurred in the austenitic phase of the polished sample (about −75%), and a diffraction peak of martensite was indexed after 2×10^6 cycles at a stress amplitude of 207 MPa. For the sample treated at RT for 60 min, approximately half of the residual stress was relaxed in run-out tests, leading to a very similar measured value compared to the initial state run-out specimens.

Finally, a rather large error bar can be seen on the graph for the α' martensite, formed during fatigue loading of TC_20 min RT. The reason is that the amount of martensite was low, and the number of diffracting planes was not high enough to ensure a good confidence on the residual stress measurement. All other measurements showed an acceptable error level.

3.4. Fracture Surfaces

Fracture surfaces of different samples that underwent a similar number of cycles to failure ($\sim 5 \times 10^4$ cycles) were investigated, and different initiation and propagation mechanisms were observed depending on the type of loading and surface treatment.

For instance, under rotating–bending, as shown in Figure 5a–c, all samples exhibited multiple crack initiation spots. These spots corresponded to surface initiation sites for all the tested conditions, and one example of the RB_20 min TC sample is displayed in Figure 6a. After initiation, these cracks could grow simultaneously, culminating in ductile fracture of the sample. This crack propagation behavior was the same for samples both with and without SMAT. Comparatively, under tension–compression loading, for the polished condition, not so many initiation spots were visible, and the main section reduction resulted from the propagation of a single crack (Figure 5d). After SMAT, the TC_60 min RT sample exhibited multiple initiation sites (Figure 5e), but only one main crack (initiation zone highlighted in red) propagated substantially into the core of the sample. It seems that this crack initiated at the surface (as shown in Figure 6b) and that, after a certain number of cycles, the crack propagation path changed between the SMAT-affected sub-surface and the core of the specimen, leading to a water drop-shaped zone on the fracture surface (Figure 5e). Higher-resolution fracture surface analyses are needed to gain a better insight into the crack propagation behavior in both the RB and TC samples, because it is more reasonable to address fatigue damage assessment by estimating fatigue crack growth rates.

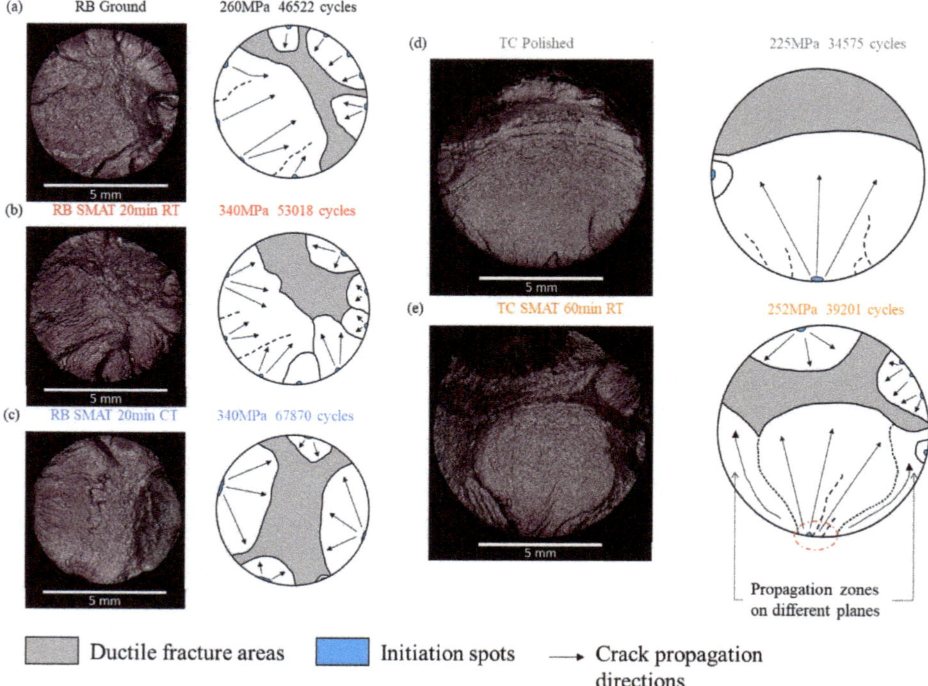

Figure 5. Typical fracture surfaces of (**a**) ground sample tested in rotating–bending, (**b**) sample tested in rotating–bending after 20 min of SMAT at room temperature, (**c**) sample tested in rotating–bending after 20 min of cryogenic SMAT, (**d**) polished samples tested in tension–compression, and (**e**) sample tested in tension–compression after 60 min of SMAT at room temperature.

3.5. Self-Heating

During cyclic loading, self-heating of the specimens may occur, depending on the stress amplitudes and strain rates imposed during each cycle, which may introduce a high amount of plastic deformation. The temperature measurements showed that the self-heating was homogeneous in the middle of the gage section of the run-out samples. For the broken samples; however, a localized and sharp increase of temperature was observed before the final breakdown. Typical chronograms from these measurements are plotted in Figure 7a, and a temperature variation of almost 50 °C can be observed before rupture (see red curve). The maximum homogeneous temperature values as a function of the applied stress were measured for samples tested in TC (see Figure 7b). The maximum surface temperature evolution with the applied stress amplitude followed an almost linear trend for both polished and 60 min of SMAT at RT conditions. In the fatigue limit stress range, the polished samples (~205 MPa) exhibited a maximum measured temperature of approximately 50 °C, whereas, for 60 min RT SMAT (~245 MPa), this value was higher than 100 °C. Nevertheless, the measured temperature values for the highest stress amplitudes were comparable: 320 °C for the initial state loaded at 260 MPa and 280 °C for the treated samples loaded at 275 MPa. After SMAT, in order to obtain the same level of self-heating, 15% higher stress amplitudes were required.

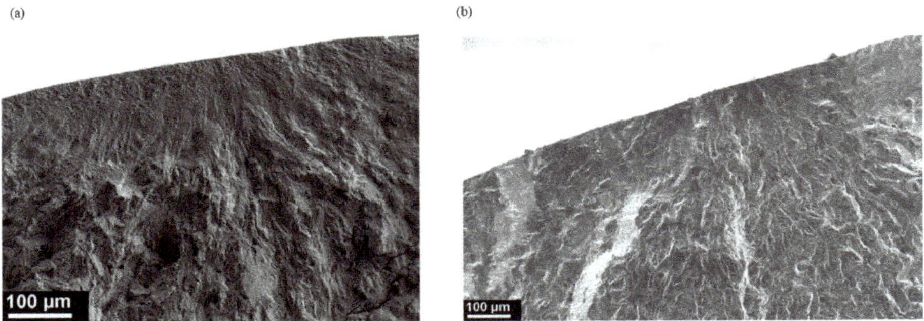

Figure 6. Secondary electron SEM pictures of surface initiation sites of (**a**) sample tested in rotating–bending after 20 min of cryogenic SMAT (corresponding to Figure 5c), and (**b**) sample tested in tension–compression after 60 min of SMAT at room temperature (corresponding to Figure 5e).

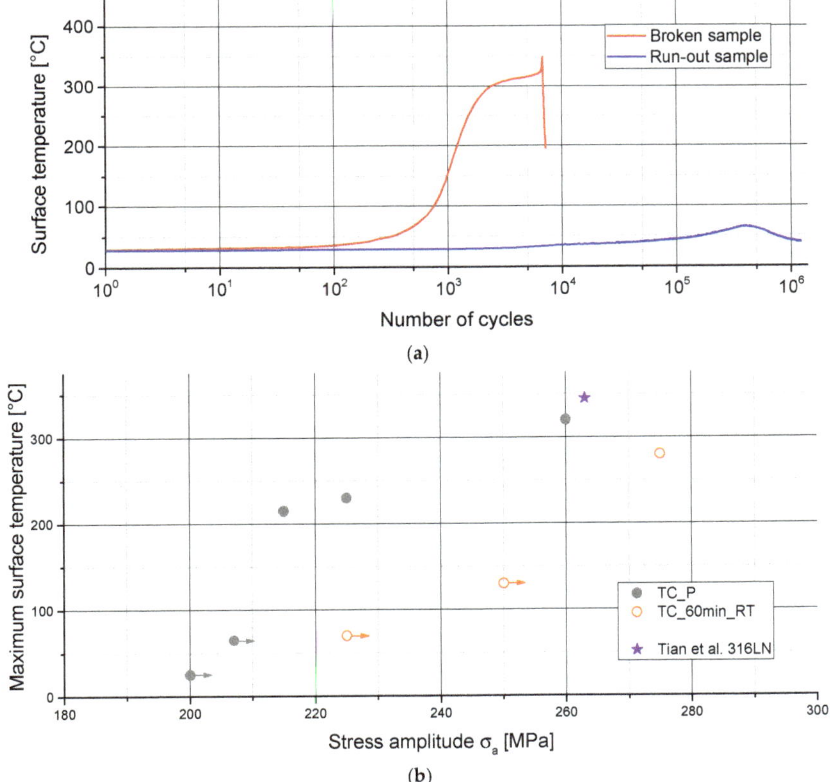

Figure 7. (**a**) Typical self-heating chronograms measured during TC loading, for breaking samples (in red) and run-out samples (in blue). (**b**) Maximum stabilized temperature as a function of σ_a.

4. Discussion

Consistently with the results from the literature [13,15,23], the use of SMAT was found to be an efficient way to increase the fatigue resistance of the 304L stainless steel. Enhancements of 28% and 17% were observed in rotating–bending and tension–compression, respectively. Nevertheless, self-heating of the specimens during testing was not anticipated. The temperature evolutions with applied load presented in Figure 7b show almost linear trends with similar positive slopes for both polished and SMAT samples. In the case of the polished samples, a good accordance with the fatigue results obtained on 316 LN stainless steel by Tian et al. [33] can be noted, especially around 260 MPa where the number of cycles to failure and self-heating values were very similar. Also, for the low stress amplitudes related to the endurance limit, such as 200 MPa and 207 MPa, the maximum surface temperature measured during the fatigue tests was less than 70 °C, representing a self-heating that did not exceed 50 °C. After SMAT, the same level of self-heating required a 15% higher applied stress amplitude, confirming the difference of cyclic plastic strain accumulation between the SMAT and the untreated samples for the same applied stress amplitude. Indeed, the surface strain-hardening generated by SMAT pushed the onset of the plasticity at the surface to a higher stress range and delayed the generation of significant self-heating. Regarding the thermal exposure during the tests, the work of Kakuichi et al. [23] showed that the exposure to 300 °C during rotating–bending fatigue at high and low stress amplitudes does not relax the surface residual stresses of a 304L sample ultrasonic shot-peened at room temperature. This would mean that the observed self-heating mainly behaves as a mechanical property reducer and that the key factor governing the residual stress relaxation is the introduction of supplementary plastic strains.

Meanwhile, the presence and magnitude of residual stresses also play a role in RB and TC fatigue behavior and, as shown in Figure 4, different residual stresses values were measured before and after fatigue tests at the surface of specimens. Before fatigue, the different residual stress states are directly linked to the different studied conditions, whereas, after fatigue, relaxations depend on several testing parameters such as the type of loading or the applied stress amplitude. For example, it was shown in Figure 4 that the residual stress relaxation in austenite for the ground specimen after rotating–bending fatigue was only approximately −10% compared to −75% for the tension-compression testing condition. For the same level of stress amplitudes, due to the stress gradients during RB tests, plastic strains can mainly be introduced at the surface region of the specimen, whereas the whole volume of the TC sample can accommodate plastic strains during TC tests, resulting in more residual stress relaxation. The initial residual stresses can also favor the onset of plasticity, especially when significant compressive residual stresses are present in a part that is mechanically loaded at $R = -1$. The stress applied during the compressive stage of a fatigue test is added to the residual stress, reaching the material elastic limit at or below the surface, resulting in the relaxation of the residual stresses. Both elements can justify the specific behavior in the tension–compression testing condition. At this stage, the sequence of these different mechanisms is not known, as only surface residual stress relaxations were evaluated before and after the tests. Further investigations of the residual stress gradient evolution during fatigue loading are needed to better explain the behavior of the processed part during fatigue tests. Such an investigation will contribute to understanding when, how, and in which way the residual stress gradients relax during fatigue testing, finally providing a better idea of the role of the residual stresses in the fatigue behavior of 304L stainless steel treated by SMAT.

Concerning the comparison of untreated RB with untreated fully reversed ($R = -1$) TC fatigue tests, the gradient in the applied load within the specimens in RB is commonly used to explain a lower fatigue limit (approximately −10%) in TC [34]; however, in this study, a decrease of −18% of the fatigue limit was observed. The volume effect of the tested samples can be considered to explain this difference. Indeed, the samples tested in RB had a gauge of 6 mm in diameter and 25 mm in length, whereas the TC ones had a diameter of 9 mm with a length of 12.5 mm. Pogorestskii et al. [35] showed that, for 40Kh steel loaded in four-point rotating–bending, larger gauge length and diameter had a detrimental effect on the fatigue limit, and they defined two different coefficients: −0.18 MPa/mm for the length and −4.4 MPa/mm for the radius. If similar behavior can be considered for the 304L stainless steel, an increase of 3 mm in diameter and a decrease of 12.5 mm in length would result in an overall decrease of approximately 5 MPa for the RB fatigue limit. The difference in fatigue limit between RB and TC would consequently be reduced to −15%, which is very similar to the −14.9% or the −11.9% proposed by Palin-Luc et al. [36] on 30NCD16 and XC18 steels, respectively.

For the SMAT conditions, a decrease of approximately −25% of the fatigue limit in TC compared to the RB one was obtained in Table 2 (from 320 to 240 MPa). This difference was significantly larger than for the initial state and could be due to the difference in treatment duration and resulting microstructure and mechanical property gradients. Similarly to the work of Sun et al. [37], one can consider that, after SMAT, the obtained functionally grade material behaves like a composite material composed of a SMAT-affected layer and a bulk core. Considering that the mechanical properties are proportional to the hardness, and by using a simple mixing law, the fatigue limit enhancement by SMAT can be determined as follows:

$$Enhancement\ [\%] = \left(\frac{f_{SMAT} \times HV_{SMAT} + f_{bulk} \times HV_{bulk}}{HV_{bulk}} - 1 \right) \times 100, \qquad (2)$$

where f_{SMAT} is the volume fraction of the SMAT-affected layer and HV_{SMAT} is its mean hardness, where $f_{bulk} = (1 - f_{SMAT})$ and $HV_{bulk} = 210$ (as shown in Figure 2a). The data used for the calculation and the obtained estimations are summarized in Table 3. The differences in diameter and hardened depth led to a higher fraction of SMAT-affected layer in the case of RB_20 min RT (16%) than the two other conditions (11.3% for RB_20 min CT and 11.8% for TC_60 min RT). Nevertheless, the difference of mean hardness in the SMAT layer resulted in similar fatigue limit estimations (~267 MPa) for both RB conditions and a lower value for TC (~220 MPa). The estimated fatigue limits were compared to the experimental ones, and the respective errors were calculated.

Table 3. Different SMAT conditions with their respective affected volume fraction, mean hardness, and corresponding measured and estimated fatigue limits.

Testing Condition	Sample's Diameter (mm)	SMAT Affected Depth (mm)	f_{SMAT} (%)	HV_{SMAT} (HV)	Estimated σ_D (MPa)	Measured σ_D (MPa)	Error (%)
RB_20 min RT	6	0.5	16	298.7	266.9	320	−16.6
RB_20 min CT	6	0.35	11.3	345.3	268.2	320	−16.2
TC_60 min RT	9	0.55	11.8	330.8	219	240	−8.8

It is worth remembering that the effect of residual stresses and their variation during fatigue tests were not taken into account here. This approximation underestimates the fatigue limit enhancement by SMAT for all tested conditions. As shown in Table 3, the error of approximation was two-fold higher in RB than in TC. Even by considering the differences in sample geometry and hardening state resulting from the treatment conditions, a difference remains which can only be explained by the loading condition difference. An explanation can be that the most solicited area in RB is the SMAT-affected layer that has enhanced mechanical properties and is under significant compressive residual stress state. These facts suggest that the effect of SMAT would be more significant in RB.

In the case of rotating–bending, the use of cryogenic SMAT provided the same effect in terms of fatigue resistance as the SMAT at room temperature. Thus, despite potential beneficial modifications such as a lower roughness, a higher martensite fraction, and a slightly higher compressive residual stresses, the use of SMAT at cryogenic temperature did not bring the desired additional improvement. It is likely that the beneficial modifications induced by CT SMAT were counter-balanced by other factors linked to the sub-surface or to the surface modifications such as the lower hardness below 200 μm (see Figure 2a) compared to the room temperature treatment. Indeed, as shown in Table 3 for CT SMAT, the high surface hardness (~520 HV) together with the limited affected depth (~350 μm) led to a very similar estimation of the fatigue limit compared to the RT SMAT that was characterized by a lower surface hardness (~450 HV) and a significantly higher affected depth (~500 μm).

5. Conclusions

Different surface mechanical attrition treatments (SMAT) were used in order to investigate their effects on the fatigue behavior of a 304L austenitic stainless steel. The metastable nature of this stainless steel makes it a good candidate for cryogenic treatment. The focus was put on (i) the potential of cryogenic SMAT treatment, and (ii) the nature of the cyclic loading condition during fatigue tests.

1. The use of SMAT for 20 min at cryogenic temperature (CT) led to a lower roughness, a higher martensitic fraction, a higher surface hardness, and higher residual stresses in both phases (γ austenite and α' martensite).

2. As tested in rotating–bending (RB), the use of SMAT at CT did not bring the expected potential additional improvement in terms of fatigue. Indeed, the fatigue life enhancement was approximately 30% for both cryogenic and room temperature (RT) SMAT compared to the initial ground state.

3. In tension–compression (TC), the use of SMAT at RT for 60 min on polished samples also led to an increase of approximately 20% of the fatigue limit compared to the initial state. However, the fatigue limit was found to be significantly smaller than the 20-min SMAT samples tested in RB (25% less), even if higher hardness and residual stress were reported after treatment. This could be due to more significant relaxations of the surface residual stresses coupled with sample size and SMAT-affected volume ratio effects.

4. The primary and secondary cracks were initiated at the surface of the samples, but the observed crack propagation profiles were different depending on the loading conditions.

5. A negligible homogeneous self-heating ($\Delta t < 50\ °C$) of the specimens tested in tension-compression at low stress amplitudes was measured, whereas, at high stress amplitudes, much higher local temperature changes ($\Delta t > 300\ °C$) were measured on the surface of the samples. The SMAT process was able to delay the onset of these local temperature increases to higher stress amplitude values (+10%).

Author Contributions: Conceptualization, M.A. and T.G.; methodology, M.A., P.B., Y.N. and T.G.; validation, C.D. and M.N.; formal analysis, M.A., P.B., Y.N. and T.G.; investigation, C.D. and M.N.; resources, M.A., P.B., Y.N. and T.G.; data curation, C.D. and M.N.; writing—original draft preparation, C.D. and M.N.; writing—review and editing, M.A., R.M., P.B., Y.N. and T.G.; supervision, M.A. and T.G.; funding acquisition, M.A., P.B., Y.N. and T.G. All authors have read and agreed to the published version of the manuscript.

Funding: This study was supported by the French State through the program "Investment in the future" operated by the National Research Agency (ANR), referenced by ANR-11-LABX-0008-01 (Labex DAMAS). M.N. was supported by a Canadian–French Frontenac scholarship during his stay in ETS-Montréal.

Conflicts of Interest: The authors declare no conflict of interest.

References

1. Maiya, P.S.; Busch, D.E. Effect of surface roughness on low-cycle fatigue behavior of type 304 stainless steel. *Metall. Trans. A* **1975**, *6*, 1761. [CrossRef]

2. Webster, G.A.; Ezeilo, A.N. Residual stress distributions and their influence on fatigue lifetimes. *Int. J. Fatigue* **2001**, *23*, 375–383. [CrossRef]

3. Huang, H.W.; Wang, Z.B.; Lu, J.; Lu, K. Fatigue behaviors of AISI 316L stainless steel with a gradient nanostructured surface layer. *Acta Mater.* **2015**, *87*, 150–160. [CrossRef]

4. Chen, G.; Gao, J.; Cui, Y.; Gao, H.; Guo, X.; Wu, S. Effects of strain rate on the low cycle fatigue behavior of AZ31B magnesium alloy processed by SMAT. *J. Alloys Compd.* **2018**, *735*, 536–546. [CrossRef]

5. Nikitin, I.; Altenberger, I. Comparison of the fatigue behavior and residual stress stability of laser-shock peened and deep rolled austenitic stainless steel AISI 304 in the temperature range 25–600 °C. *Mater. Sci. Eng. A* **2007**, *465*, 176–182. [CrossRef]

6. Yasuoka, M.; Wang, P.; Zhang, K.; Qiu, Z.; Kusaka, K.; Pyoun, Y.S.; Murakami, R.I. Improvement of the fatigue strength of SUS304 austenite stainless steel using ultrasonic nanocrystal surface modification. *Surf. Coat. Technol.* **2013**, *218*, 93–98. [CrossRef]

7. Cherif, A.; Pyoun, Y.; Scholtes, B. Effects of ultrasonic nanocrystal surface modification (UNSM) on residual stress state and fatigue strength of AISI 304. *J. Mater. Eng. Perform.* **2010**, *19*, 282–286. [CrossRef]

8. Bagheri, S.; Guagliano, M. Review of shot peening processes to obtain nanocrystalline surfaces in metal alloys. *Surf. Eng.* **2009**, *25*, 3–14. [CrossRef]

9. Schulze, V. Characteristics of Surface Layers Produced by Shot Peening. In Proceedings of the Eighth International Conference on Shot Peening ICSP-8 in Garmisch-Partenkirchen DGM, Garmisch-Partenkirchen, Germany, 16–20 September 2002; pp. 145–160.

10. Azadmanjiri, J.; Berndt, C.C.; Kapoor, A.; Wen, C. Development of surface nano-crystallization in alloys by surface mechanical attrition treatment (SMAT). *Crit. Rev. Solid State Mater. Sci.* **2015**, *40*, 164–181. [CrossRef]

11. Grosdidier, T.; Novelli, M. Recent developments in the application of surface mechanical attrition treatments for improved gradient structures: Processing parameters and surface reactivity. *Mater. Trans.* **2019**, *60*, 1344–1355. [CrossRef]

12. Bagherifard, S. Enhancing the structural performance of lightweight metals by shot peening. *Adv. Eng. Mater.* **2019**, *21*, 1801140. [CrossRef]

13. Roland, T.; Retraint, D.; Lu, K.; Lu, J. Fatigue life improvement through surface nanostructuring of stainless steel by means of surface mechanical attrition treatment. *Scr. Mater.* **2006**, *54*, 1949–1954. [CrossRef]

14. Uusitalo, J.; Karjalainen, L.P.; Retraint, D.; Palosaari, M. Fatigue properties of steels with ultrasonic attrition treated surface layers. In *Materials Science Forum*; Trans Tech Publications: Baech, Switzerland, 2009; Volume 604, pp. 239–248.

15. Zhou, J.; Sun, Z.; Kanouté, P.; Retraint, D. Effect of surface mechanical attrition treatment on low cycle fatigue properties of an austenitic stainless steel. *Int. J. Fatigue* **2017**, *103*, 309–317. [CrossRef]

16. Kumar, S.A.; Raman, S.G.S.; Narayanan, T.S. Influence of surface mechanical attrition treatment duration on fatigue lives of Ti–6Al–4V. *Trans. Indian Inst. Met.* **2014**, *67*, 137–141. [CrossRef]

17. Nikitin, I.; Scholtes, B.; Maier, H.J.; Altenberger, I. High temperature fatigue behavior and residual stress stability of laser-shock peened and deep rolled austenitic steel AISI 304. *Scr. Mater.* **2004**, *50*, 1345–1350. [CrossRef]

18. Torres, M.A.S.; Voorwald, H.J.C. An evaluation of shot peening, residual stress and stress relaxation on the fatigue life of AISI 4340 steel. *Int. J. Fatigue* **2002**, *24*, 877–886. [CrossRef]

19. Masaki, K.; Ochi, Y.; Matsumura, T. Initiation and propagation behaviour of fatigue cracks in hard-shot peened Type 316L steel in high cycle fatigue. *Fatigue Fract. Eng. Mater. Struct.* **2004**, *27*, 1137–1145. [CrossRef]

20. Bagherifard, S.; Guagliano, M. Fatigue behavior of a low-alloy steel with nanostructured surface obtained by severe shot peening. *Eng. Fract. Mech.* **2012**, *81*, 56–68. [CrossRef]

21. Tian, J.W.; Villegas, J.C.; Yuan, W.; Fielden, D.; Shaw, L.; Liaw, P.K.; Klarstrom, D.L. A study of the effect of nanostructured surface layers on the fatigue behaviors of a C-2000 superalloy. *Mater. Sci. Eng. A* **2007**, *468*, 164–170. [CrossRef]

22. Pandey, V.; Chattopadhyay, K.; Srinivas, N.S.; Singh, V. Role of ultrasonic shot peening on low cycle fatigue behavior of 7075 aluminium alloy. *Int. J. Fatigue* **2017**, *103*, 426–435. [CrossRef]

23. Kakiuchi, T.; Uematsu, Y.; Hasegawa, N.; Kondoh, E. Effect of ultrasonic shot peening on high cycle fatigue behavior in type 304 stainless steel at elevated temperature. *J. Soc. Mater. Sci.* **2016**, *65*, 325–330. [CrossRef]

24. Olson, G.B.; Cohen, M. A mechanism for the strain-induced nucleation of martensitic transformations. *J. Less-Common Met.* **1972**, *28*, 107–118. [CrossRef]

25. Novelli, M.; Fundenberger, J.J.; Bocher, P.; Grosdidier, T. On the effectiveness of surface severe plastic deformation by shot peening at cryogenic temperature. *Appl. Surf. Sci.* **2016**, *389*, 1169–1174. [CrossRef]

26. Novelli, M.; Bocher, P.; Grosdidier, T. Effect of cryogenic temperatures and processing parameters on gradient-structure of a stainless steel treated by ultrasonic surface mechanical attrition treatment. *Mater. Charact.* **2018**, *139*, 197–207. [CrossRef]

27. Topic, M.; Tait, R.B.; Allen, C. The fatigue behaviour of metastable (AISI-304) austenitic stainless steel wires. *Int. J. Fatigue* **2007**, *29*, 656–665. [CrossRef]

28. Spencer, K.; Embury, J.D.; Conlon, K.T.; Véron, M.; Bréchet, Y. Strengthening via the formation of strain-induced martensite in stainless steels. *Mater. Sci. Eng. A* **2004**, *387*, 873–881. [CrossRef]

29. Sonats' Company Website. Available online: https://sonats-et.com/ (accessed on 25 November 2019).
30. Delbergue, D.; Texier, D.; Levesque, M.; Bocher, P. Comparison of two X-ray residual stress measurement methods: Sin2 ψ and cos α, through the determination of a martensitic steel X-ray elastic constant. In Proceedings of the 10th International Conference on Residual Stresses (ICRS10), Sydney, Australia, 3–7 July 2016.
31. Stromeyer, C.E. The determination of fatigue limits under alternating stress conditions. *Proc. R. Soc. Lond.* **1914**, *90*, 411–425. [CrossRef]
32. Velásquez, J.P.; Tidu, A.; Bolle, B.; Chevrier, P.; Fundenberger, J.J. Sub-surface and surface analysis of high speed machined Ti–6Al–4V alloy. *Mater. Sci. Eng. A* **2010**, *527*, 2572–2578. [CrossRef]
33. Tian, H.; Liaw, P.K.; Fielden, D.E.; Brooks, C.R.; Brotherton, M.D.; Jiang, L.; Mansur, L.K. Effects of frequency on fatigue behavior of type 316 low-carbon, nitrogen-added stainless steel in air and mercury for the spallation neutron source. *Metall. Mater. Trans. A* **2006**, *37*, 163–173. [CrossRef]
34. Papadopoulos, I.V.; Panoskaltsis, V.P. Invariant formulation of a gradient dependent multiaxial high-cycle fatigue criterion. *Eng. Fract. Mech.* **1996**, *55*, 513–528. [CrossRef]
35. Pogoretskii, R.G. Effect of test piece length on the fatigue strength of steel in air. *Mater. Sci.* **1966**, *1*, 63–66. [CrossRef]
36. Palin-Luc, T.; Lasserre, S. An energy based criterion for high cycle multiaxial fatigue. *Eur. J. Mech. A Solids* **1998**, *17*, 237–251. [CrossRef]
37. Sun, Z.; Chemkhi, M.; Kanoute, P.; Retraint, D. Fatigue properties of a biomedical 316L steel processed by surface mechanical attrition. *IOP Conf. Ser. Mater. Sci. Eng.* **2004**, *63*, 012021. [CrossRef]

Article

Improving Strength and Ductility of a Mg-3.7Al-1.8Ca-0.4Mn Alloy with Refined and Dispersed Al₂Ca Particles by Industrial-Scale ECAP Processing

Ce Wang [1], Aibin Ma [1,2,*], Jiapeng Sun [1], Xiaoru Zhuo [1], He Huang [1], Huan Liu [1,3,*], Zhenquan Yang [1] and Jinghua Jiang [1]

[1] College of Mechanics and Materials, Hohai University, Nanjing 211100, China
[2] Suqian Research Institute, Hohai University, Suqian 223800, China
[3] Ocean and Coastal Engineering Research Institute, Hohai University, Nantong 226300, China
* Correspondence: aibin-ma@hhu.edu.cn (A.M.); liuhuanseu@hhu.edu.cn (H.L.);
   Tel.: +86-025-8378-7239 (A.M.)

Received: 24 June 2019; Accepted: 8 July 2019; Published: 9 July 2019

Abstract: Tailoring the morphology and distribution of the Al₂Ca second phase is important for improving mechanical properties of Al₂Ca-containing Mg-Al-Ca based alloys. This work employed the industrial-scale multi-pass rotary-die equal channel angular pressing (RD-ECAP) on an as-cast Mg-3.7Al-1.8Ca-0.4Mn (wt %) alloy and investigated its microstructure evolution and mechanical properties under three different processing parameters. The obtained results showed that RD-ECAP was effective for refining the microstructure and breaking the network-shaped Al₂Ca phase. With the increase of the ECAP number and decrease of the processing temperature, the average sizes of Al₂Ca particles decreased obviously, and the dispersion of the Al₂Ca phase became more uniform. In addition, more ECAP passes and lower processing temperature resulted in finer α-Mg grains. Tensile test results indicated that the 573 K-12p alloy with the finest and most dispersed Al₂Ca particles exhibited superior mechanical properties with tensile yield strength of 304 MPa, ultimate tensile strength of 354 MPa and elongation of 10.3%. The improved comprehensive mechanical performance could be attributed to refined DRX grains, nano-sized Mg₁₇Al₁₂ precipitates and dispersed Al₂Ca particles, where the refined and dispersed Al₂Ca particles played a more dominant role in strengthening the alloys.

Keywords: Mg-3.7Al-1.8Ca-0.4Mn alloy; Al₂Ca phase; equal channel angular pressing; refinement; mechanical properties

---

1. Introduction

Developing strong and ductile magnesium alloys has been one of the main research areas in the last twenty years, as they exhibit great potential in aerospace, military, new energy automobile, medical instrument and electronic applications [1–5]. High strength magnesium alloys with ultimate tensile strength higher than 400 MPa have already been successfully prepared with the addition of abundant rare-earth (RE) elements, owing to the intense second phase (precipitate) strengthening effect [6–8]. However, the RE addition impairs the lightweight advantage of magnesium alloys, and increased their cost as well, making them difficult to use for cost-sensitive industry applications [9–11]. Therefore, design and preparation of reinforced RE-free magnesium alloys have recently regained people's attention.

Mg-Al based (AZ series) alloys are one of the most commonly used commercial magnesium alloy series, whose strengthening precipitate is mainly Mg₁₇Al₁₂ phase [12–14]. Mg₁₇Al₁₂ possesses a relatively low melting point (730 K) and tends to be softened above 473 K, thereby exhibiting a weak strengthening effect, especially at high temperatures [15,16]. In the last few years, the Mg-Al-Ca (-Mn) alloys, which were employed as typical heat-resistant casting magnesium alloys formerly, have been researched for high strength wrought Mg alloys. Ultrahigh strength characteristics involving tensile yield strength exceeding 400 MPa have already been obtained for high-Ca-content Mg-Al-Ca-Mn alloys [17–21]. However, excessive Ca addition intrinsically deteriorated the ductility of Mg alloys, whose elongations were always lower than 5%, even after remarkable grain refinement [17–20]. To balance the strength and ductility, Ca content should be decreased in these alloys.

With the increase of Ca content in Mg-Al-Ca alloys, $Mg_{17}Al_{12}$ phase turns into three laves phases in sequence, namely, Al_2Ca (C15), $(Mg,Al)_2Ca$ (C36) and Mg_2Ca (C14) phases, respectively [17]. Among these intermetallic compounds, Al_2Ca phase possesses the highest melting point, approximately 1352 K [22]. Studies of the first-principles calculation also proved that Al_2Ca phase exhibits the best thermal stability for the three laves phases [23]. Moreover, Al_2Ca phase could form with low or moderate Ca addition, which might facilitate a maximum combination of strength and ductility. So far, most Al_2Ca-containing magnesium alloys were prepared by hot extrusion, and they usually exhibited remarkably improved strength [24–26]. Although Al_2Ca network phases were crushed after extrusion, large particles remained and the broken particles were not uniformly distributed within a α-Mg matrix. As a consequence, local stress concentration was easy to generate at the junctions of brittle second phase particles and α-Mg matrix during a tensile test, resulting in nucleation and propagation of microcracks at an early time [19]. Therefore, it is essential to develop an effective method to refine Al_2Ca phases and increase their dispersibility, in order to improve the ductility of these high strength alloys.

Our previous studies have already shown that the multi-pass equal channel angler pressing (ECAP) is effective to refine large Al-Si particles, and network-shaped second phases in magnesium alloys [27–33]. As for Al_2Ca containing alloys, no attempt of ECAP processing has been reported. Therefore, in the present work, we prepared an Al_2Ca-dominated Mg-3.7Al-1.8Ca-0.4Mn (wt %) alloy first, and then systematically investigated its microstructural evolution and mechanical properties under an industrial-scale ECAP processing at three different processing parameters. By refining and dispersing Al_2Ca second phase particles, we successfully prepared a high strength and high ductility Al_2Ca-containing magnesium alloy block.

2. Materials and Methods

The raw materials for preparation of studied Mg-3.7Al-1.8Ca-0.4Mn (wt %) alloy were pure Mg, pure Al, Mg-30 wt % Ca and Mg-10 wt % Mn master alloys. The designed alloy was prepared by semi-continuous casting method, with an ingot diameter of 90 mm. Then large cuboid samples with dimension of 50 mm × 50 mm × 100 mm were cut from the center of the ingot for further industrial-scale ECAP. To explore the influence of processing parameters on refinement of Al_2Ca-containing alloys, three ECAP routes were proposed, i.e., 4 passes at 623 K, 12 passes at 623 K, and 12 passes at 573 K. The ECAP die employed was a self-design rotary-die (RD) with die angle of 90 ° and outer arc angle of 0°, which was described detailedly in our previous work [33]. Before ECAP, the sample was inserted in the die and both were heated and kept at a set temperature for 20 min within an induction heating furnace. Then multi-pass RD-ECAP were performed successively by an automatic control system without intermediate heating. This ECAP process is time-saving, and the total processing time for 12 passes was less than 15 min.

Metallographic specimens were then cut from the cast and ECAP specimens, mechanically grinded by #80, #400, #1000 and #2000 SiC abrasive papers, polished and etched with a 4 mL nitric acid and 96 mL ethanol mixed solution. Then, microstructure characterizations of the alloys were carried out by the X-ray diffractometer (XRD, Cu-Kα, Bruker D8, DISCOVER, Bruker Corporation, Karlsruhe, Germany), optical microscope (OM, Olympus BX51M, Olympus, Tokyo, Japan), and a scanning electron microscope (SEM, Sirion 200, FEI Company, Hillsboro, OR, USA) equipped with an X-ray energy dispersed spectrometer (EDS, Genesis 60S, FEI Company, Hillsboro, OR, USA). To further identify various phases, observation of the transmission electron microscope (TEM, Tecnai G2, FEI Company,

Hillsboro, OR, USA) was conducted. The TEM samples were prepared by mechanical grinding and ion thinning. To reveal the grain size distributions of ECAP alloys, the electron back-scatted diffraction (EBSD) analyses were conducted (SEM, Hitachi S-3400N, Hitachi, Tokyo, Japan). Moreover, the average particle sizes were measured by counting at least 100 particles in three SEM images through the Image-J software (Image 1.48, NationalInstitutes of Health, Bethesda, MD, USA). To evaluate the mechanical properties, tensile tests were performed via a CMT5105 electronic universal testing machine (MTS, Shenzhen, China) with a ram speed of 0.5 mm/min at room temperature. The dumbbell shaped specimens with gauge dimension of 7.5 mm × 2 mm × 2 mm were cut from the center section of ECAP samples with the loading direction parallel to the extrusion direction. For each processing situations, three tensile specimens were employed.

3. Results and Discussions

3.1. Microstructure of As-Cast Alloy

Figure 1 shows the XRD patterns of as-cast alloy and three ECAP alloys. Two phases, α-Mg and Al_2Ca phases, are indexed from the diffraction peaks for all alloys, which suggests no obvious phase transformation occurs during hot deformation. Since the Ca/Al ratio of this alloy is lower than 0.8, Al_2Ca phase becomes the main second phase [24]. Figure 2a,b show the optical and SEM micrographs of as-cast Mg-3.7Al-1.8Ca-0.4Mn alloy. Dendritic α-Mg grains and continuous-network interdendritic second phases (marked by A in Figure 2b) could be observed. Seen from the enlargement of the network phase shown in Figure 2c, it exhibits a lamellar shape, consistent with the morphology of a typical eutectic structure. The corresponding EDS result (Figure 2e) reveals high contents of Al and Ca elements, indicating they could have a Al_2Ca + α-Mg eutectic structure. Figure 3 shows the TEM image of the eutectic structure. Black and white stripes are arranged alternately. The index of the corresponding selected area electron diffraction (SAED) pattern of the black stripes (inset of Figure 3) further demonstrates the existence of Al_2Ca phase.

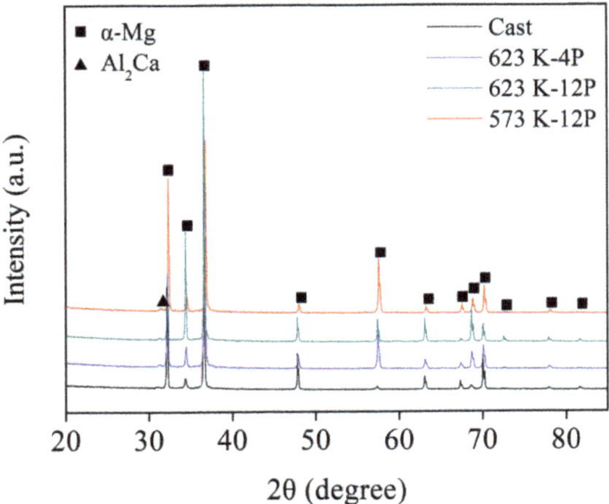

Figure 1. X-ray diffractometer (XRD) patterns of as-cast and ECAP-ed alloys.

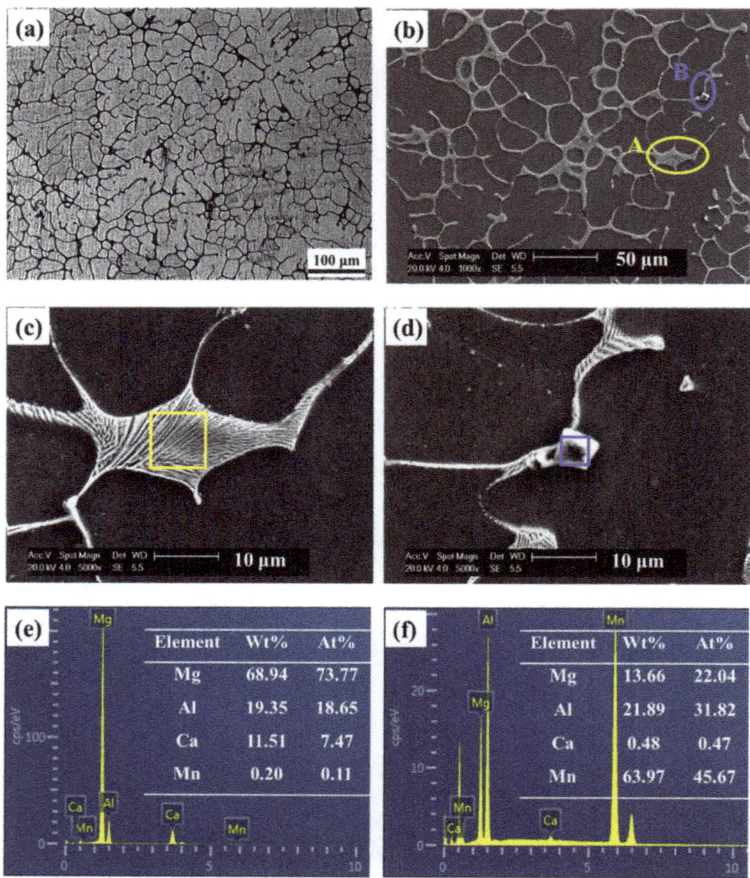

Figure 2. Microstructure of the as-cast alloy. (**a**) Optical image; (**b,c,d**) scanning electron microscope (SEM) images and enlargements of (**c**) the eutectic structure and (**d**) cubic particle; The corresponding EDS results of (**e**) the eutectic and (**f**) the particle.

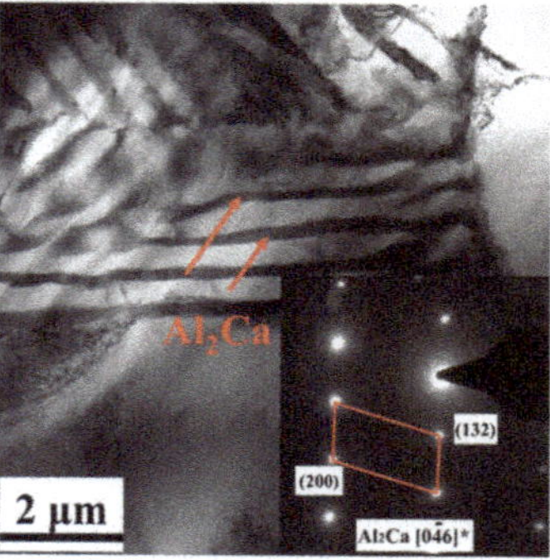

Figure 3. Transmission electron microscope (TEM) micrograph of the eutectic structure and the corresponding selected area electron diffraction (SAED) pattern of the black lamellar phases.

Moreover, some cubic particles (marked by B in Figure 2b) are observed near the eutectic network. EDS analysis on one particle (Figure 2d) was performed and the result (Figure 2f) suggested that it contains much higher Mn and Al elements, which should be Al_8Mn_5 phase [19]. However, Al-Mn particles were not detected from the XRD patterns due to the fact that the content is extremely low.

3.2. Microstructure of ECAP Alloys

Figure 4 shows the SEM images of three Mg-3.7Al-1.8Ca-0.4Mn ECAP alloys processed at different temperatures and passes. Seen from the low-magnification images of Figure 4a,c,e, the continuous network eutectic structure became distorted after ECAP, and more ECAP numbers and a lower processing temperature resulted in a finer microstructure and narrower dendrite spacing. Moreover, as can be seen from the enlarged SEM images of Figure 4b,d,f, the distorted Al_2Ca dendritic phases were broken into ultrafine particles (with particle size lower than 1 μm). However, the average sizes of broken Al_2Ca particles are different in three ECAP alloys. High ECAP numbers refined the Al_2Ca particles from 0.6 μm to 0.5 μm for 623K-4p and 623K-12p alloys, while lower ECAP temperature was more effective to refine these particles, and the average size of Al_2Ca particles was approximately 0.3 μm in 573 K-12p alloy.

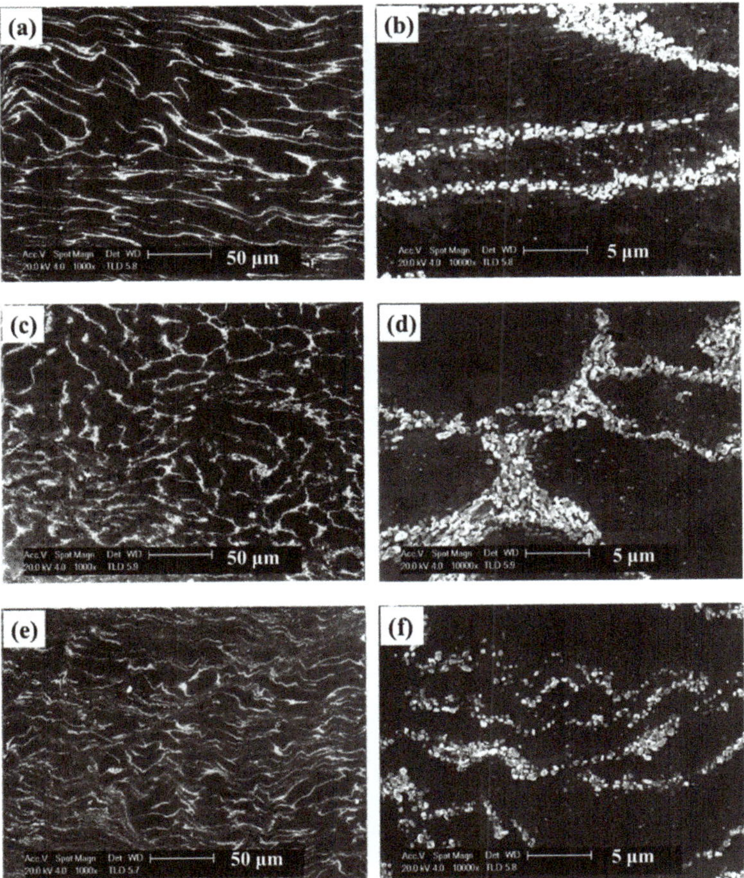

Figure 4. SEM images of ECAP alloys processed at (**a**,**b**) 623 K-4P, (**c**,**d**) 623 K-12P and (**e**,**f**) 573 K-12P with (**a**,**c**,**e**) low and (**b**,**d**,**f**) high magnifications.

To characterize the grain size evolution of studied alloy during ECAP at different processing parameters, Figure 5 shows the inverse pole figure maps and grain size distribution histograms of

ECAP alloys along the extrusion direction. The 623 K-4p alloy showed an uneven grain size distribution. Both coarse and fine grains were identified, suggesting an incomplete dynamic recrystallized (DRX) microstructure. After 12p ECAP at either 623 K or 573 K, homogeneous grain size distribution was observed (Figure 5c,d), indicating a high degree of DRX. Furthermore, Figure 6 shows the area fraction of DRX grains and un-DRX grains in three ECAP alloys. It is apparent that the fraction of DRX grains in two 12p alloys were much higher than the 4p alloy.

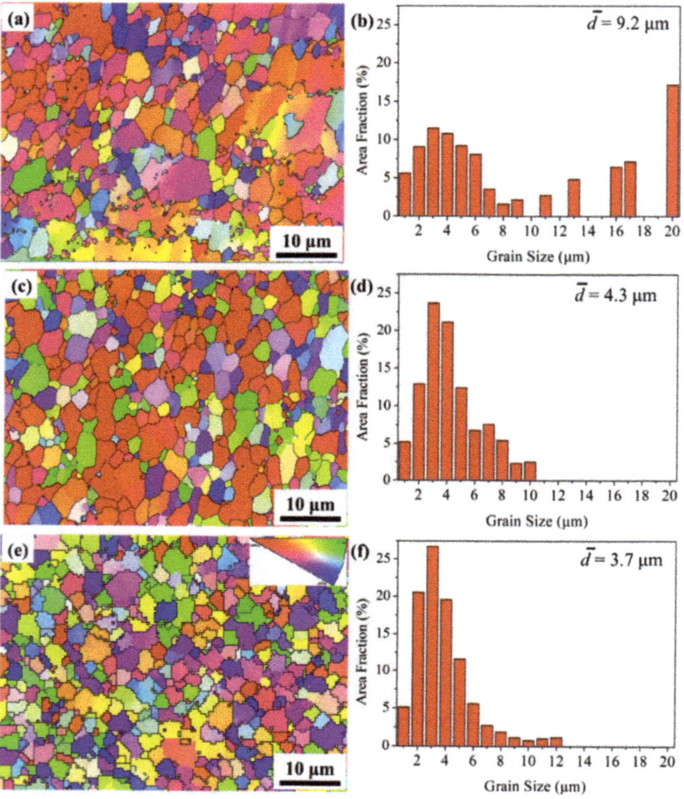

Figure 5. Electron back-scatted diffraction (EBSD) maps and its corresponding grains distributions of ECAP alloys processed from (**a**,**b**) 623 K-4P, (**c**,**d**) 623 K-12P and (**e**,**f**) 573 K-12P.

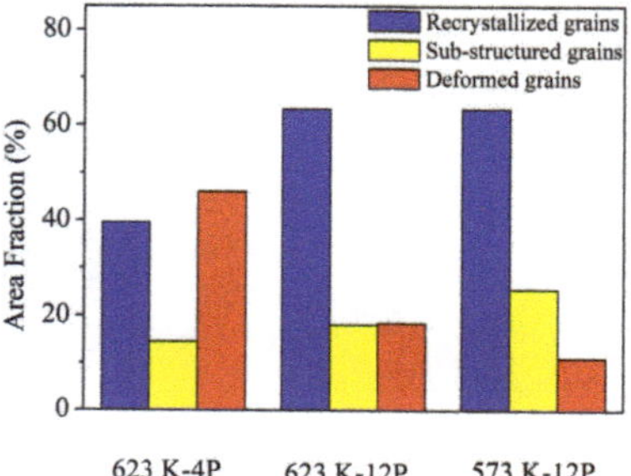

Figure 6. Area fraction of the recrystallized and un-recrystallized grains in three ECAP alloys.

Owing to the high DRX ratios, the average grain sizes of two 12p alloys were smaller than 4p alloy. The average grain sizes were estimated to be 9.2 µm, 4.3 µm and 3.7 µm for 623 K-4p, 623 K-12p and 573 K-12p alloys, respectively. It can be seen that higher ECAP numbers and lower processing temperature contributed to refinement and uniformity of α-Mg grains. Two factors resulted in finer grains for 573 K-12p alloy. For one thing, lower processing temperature could restrain the migrations of grain boundaries. For another, fine and dispersedly distributed Al$_2$Ca particles are more effective to hinder the growth of DRX grains. Figure 7 shows the TEM images of DRX regions in three ECAP alloys. High density of dislocations and sub-structures were observed in 623 K-4p alloy (Figure 7a). When the ECAP number increased to 12, the density of dislocations declined obviously, and the boundaries of DRX grains became distinct, either in 623 K and 573 K processed alloys. Overall, the grain sizes of DRX grains in these alloys are in good agreement with EBSD results.

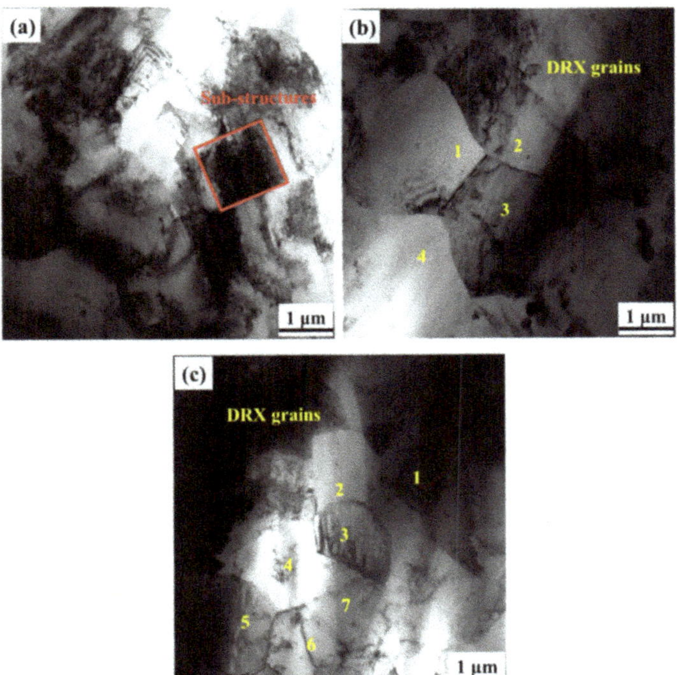

Figure 7. Ddynamic recrystallized (DRX) grains and sub-structured in (**a**) 623 K-4P, (**b**) 623 K-12P, (**c**) 573 K-12P alloys.

As can be seen from the Kernel Average Misorientation (KAM) maps shown in Figure 8, the distribution of dislocation density varied for three ECAP alloys (shown in green and yellow colors). It is shown in Figure 8a that fine DRX grains and coarse deformation grains were mixed in 623 K-4p alloy. Plenty of dislocations existed within coarse deformed grains, while fine DRX grains had little dislocations within, because the operation of DRX consumed dislocations. After increasing ECAP number to 12, more DRX fine grains and less deformation grains suggested a lower density of dislocations (Figure 8b). Moreover, compared with Figure 8b,c, it can be concluded that lower temperature processing led to stronger strain accumulation for ECAP alloys at the same pass (shown in green color).

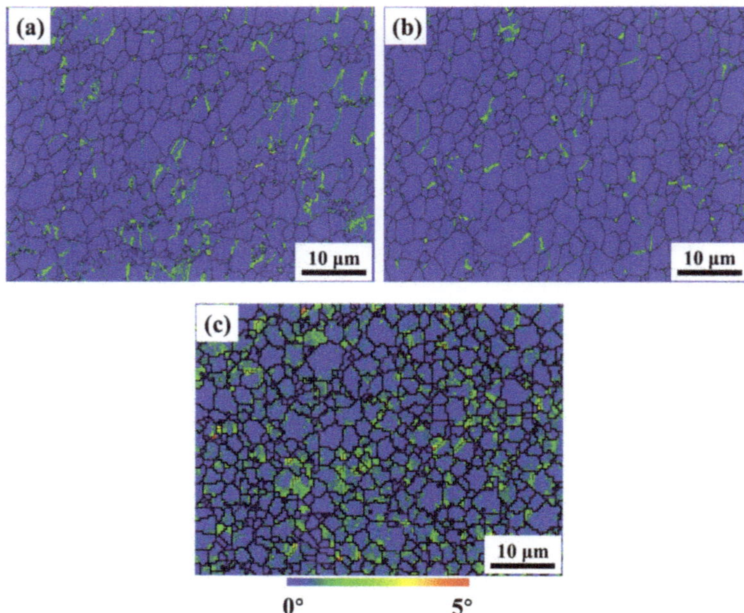

Figure 8. Kernel Average Misorientation (KAM) maps of alloys processed by (**a**) 623 K-4P, (**b**) 623 K-12P and (**c**) 573 K-12P.

Figure 9 shows the inverse pole figures of three ECAP alloys. A typical fiber texture (i.e., c-axes perpendicular to the extrusion direction) was formed for 623 K-4p alloy, and its maximum texture intensity was 2.40 (Figure 9a). Increasing processing number to 623 K, the orientations (c-axes) of grains exhibited a tendency to be parallel to extrusion direction, and the maximum intensity was 4.89 (Figure 9b). With the decrease of processing temperature to 573 K, the maximum texture intensity declined to 2.80. Similar texture evolutions have also been found in other recrystallized Mg-RE and AZ series magnesium alloys [34,35].

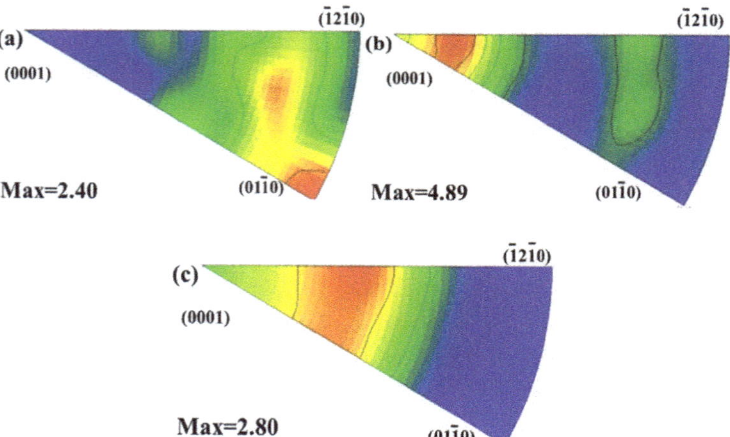

Figure 9. Inverse pole figures of ECAP alloys processed from (**a**) 623 K-4P, (**b**) 623 K-12P and (**c**) 573 K-12P.

Figure 10 exhibits the TEM observation of Al_2Ca second phase in three deformed alloys. In early ECAP passes at 623 K, lamellar Al_2Ca eutectic phase was partially broken into fine particles, as can be seen from the Al_2Ca laths and particles shown in Figure 10a,b, respectively. Since Al_2Ca is a brittle phase, its refining mechanism is more like a mechanical crushing process. Therefore, the crush of Al_2Ca eutectic phase was incomplete and uneven owing to the low deformation strains at low ECAP pass. When the ECAP number was increased to 12, the Al_2Ca phase broke thoroughly into submicron particles (Figure 10b,c), though they were still aggregated.

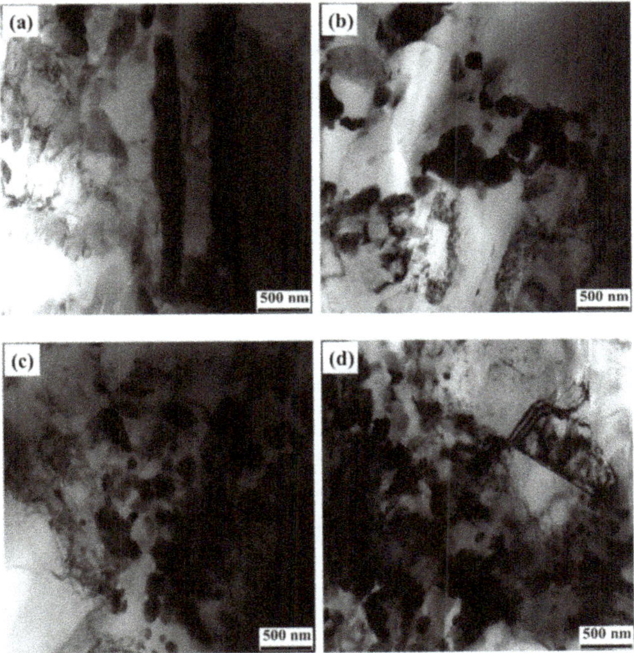

Figure 10. TEM images of Al_2Ca phase particles in (**a**,**b**) 623 K-4P, (**c**) 623 K-12P and (**d**) 573 K-12P alloys.

Furthermore, TEM observations demonstrated that abundant nano-sized precipitates were observed within un-DRX α-Mg grains (including sub-grains) of all ECAP alloys (Figure 11), while the density of precipitates was relatively low in DRX grains. These precipitates were dynamically precipitated during hot deformation, which was commonly observed in Mg-Al-Ca based alloys. Most of the precipitates had a spherical shape (marked by yellow arrows), and a few exhibited rod shapes (red arrows). Seen from Figure 11a, the diameters of these spherical precipitates were around 5–15 nm in 623 K-4p alloy. With increased ECAP numbers, the density of precipitates increased for 623 K-12p alloy, and both rod-like and spherical precipitates were observed. The rod-like precipitates are usually larger than spherical particles, exhibiting a diameter of 20–30 nm and a length of 30–60 nm. After 12 passes of ECAP at 573 K, precipitates with diameter of 10–20 nm were also detected (Figure 11c), and the precipitation density was the highest for three deformed alloys. Furthermore, Figure 11d shows the corresponding SAED patterns of the precipitates. The diffraction patterns exhibit near-ring characteristic, and index of the diffraction rings demonstrates that the precipitates are $Mg_{17}Al_{12}$ phases. The $Mg_{17}Al_{12}$ precipitates are usually reported in AZ91 alloys [36]. As for Mg-Al-Ca alloys, Al_2Ca phase served as the main precipitates during hot deformation or aging in most cases [24], and precipitation of $Mg_{17}Al_{12}$ particles was barely reported. Taking into account the alloy composition and microstructure of this studied alloy, most Ca elements were concentrated within the eutectic phases, and Al elements were more inclined to be enriched in some grains, which caused the precipitation of $Mg_{17}Al_{12}$ phases under the interaction of heat and strain during multi-pass ECAP.

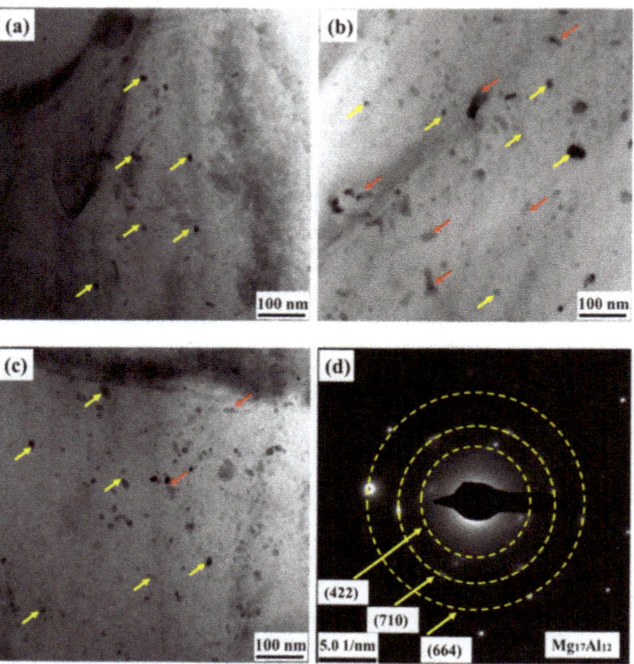

Figure 11. TEM micrographs of the precipitates in (**a**) 623 K-4P, (**b**) 623 K-12P, and (**c**)573 K-12P alloys, as well as (**d**) the corresponding SAED patterns of the precipitates.

3.3. Microstructure of ECAP Alloys

Figure 12 shows the tensile mechanical properties of as-cast and ECAP alloys. The as-cast alloy possessed low mechanical properties with tensile yield strength (TYS) of 70 MPa, ultimate tensile strength (UTS) of 161 MPa and poor ductility (elongation of 5.1%). After the ECAP process, all samples exhibited a remarkable improvement in both strength and ductility. Generally, an increase in the ECAP numbers improved the strength and ductility simultaneously, while decreases in ECAP temperatures further enhanced the alloy, but impaired the ductility slightly. The 573 K-12p alloy showed the highest strength, with TYS of 304 MPa and UTS of 354 MPa, together with a moderate ductility. In addition, by comparing with other commercial Mg alloys and Mg-RE based alloys prepared by ECAP (Figure 12c) [27–32,37–39], this 573 K-12p alloy exhibits a relatively high specific yield strength, which is comparable to those Mg-RE ECAP alloys with high RE contents. Overall, this non-RE low-alloying ECAP alloy with refined and dispersed Al_2Ca particles exhibits great potential as high performance magnesium alloys.

Figure 13 shows the SEM fractograph of as-cast and ECAP alloys after tensile tests. Seen from Figure 13a, many laminated cleavage steps appear on the fracture surface of the as-cast alloy, suggesting a brittle fracture. Owing to the large eutectic structures at grain boundaries, microcracks tend to be formed around the eutectic, which leads to early break and poor ductility of the as-cast alloy. After ECAP, a few dimples and laminated cleavage steps are observed on fracture surface of 623 K-4p alloy (Figure 13b), indicating a mixed fracture mechanism. As for two 12p alloys (Figure 13c,d), although the fracture surfaces become fine and flat, some cleavage steps still exist.

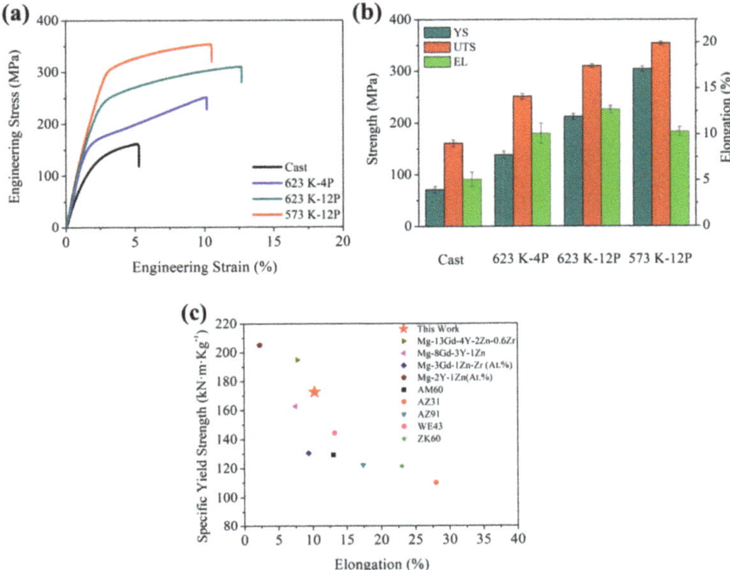

Figure 12. Mechanical properties of Mg-Al-Ca-Mn alloys. (**a**) Typical tensile curves of as-cast and ECAP alloys; (**b**) The summary of yield strength (YS), ultimate tensile strength (UTS) and elongation of various alloys; (**c**) Comparison of specific yield strength and elongation between this work and other high performance ECAP Mg alloys [27–32,37–39].

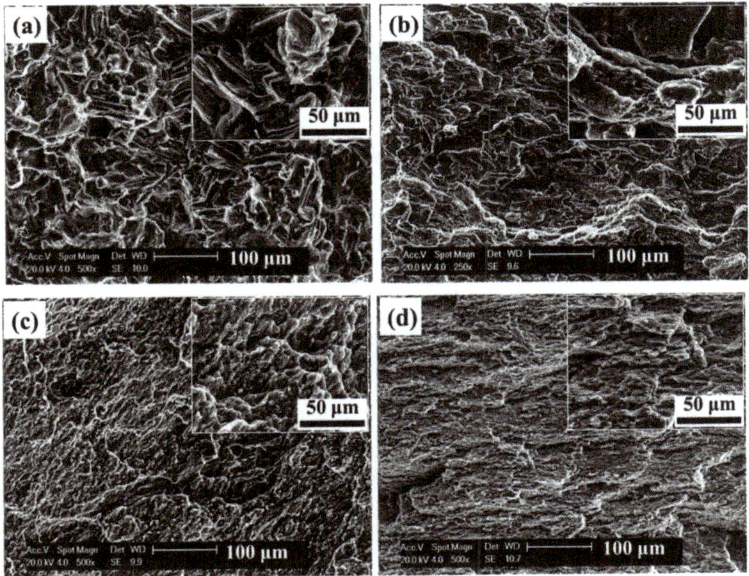

Figure 13. SEM images of fracture surfaces of (**a**) As-cast, (**b**) 623 K-4P, (**c**) 623 K-12P and (**d**) 573 K-12P alloys. Insets of the images are the corresponding enlarged micrographs.

3.4. Influence of Al$_2$Ca Size and Distribution on Mechanical Properties

Based on above tensile results, it can be confirmed that multi-pass ECAP is beneficial for improving the mechanical properties of an Al$_2$Ca-containing alloy. Considering the microstructural evolutions, the improved strength of ECAP alloys could be mainly ascribed to three factors, i.e., refined DRX grains, dynamically precipitated $Mg_{17}Al_{12}$ precipitates, and dispersed Al_2Ca particles. To quantitatively describe the contributions of various strengthening factors, Table 1 lists the characteristic parameters of DRX grains and second phase particles.

Table 1. Microstructural characteristics of ECAP-ed alloys.

Microstructural Characteristics	623 K-4P	623 K-12P	573 K-12P
Grain (μm)	9.2 ± 1.9	4.3 ± 1.5	3.7 ± 1.2
Al_2Ca (μm)	0.6 ± 0.2	0.5 ± 0.2	0.3 ± 0.1
Ribbonlike Al_2Ca arm spacing (μm)	17 ± 4	11 ± 2	8 ± 2
$Mg_{17}Al_{12}$ (nm)	12 ± 4	22 ± 7	14 ± 6
$Mg_{17}Al_{12}$ volume fraction (%)	2.5%	6.7%	6.1%

The contribution of fine grain strengthening could be estimated from the Hall-Petch equation:

$$\sigma_y = \sigma_0 + kd^{-0.5}, \tag{1}$$

where σ_y is yield strength, σ_0 is material constants, k is Hall-Petch slope and d is grain size. The value of k was employed as 170 MPa·μm$^{1/2}$ on the basis of average grains in this work [40]. Accordingly, the contribution of grain refinement strengthening on tensile yield strength could be estimated to be 55 MPa, 97 MPa and 105 MPa for 623 K-4p, 623 K-12p, and 573 K-12p alloys, respectively.

In case of $Mg_{17}Al_{12}$ nano-particles precipitated at grain internal, they could pin dislocation movement and strengthen the alloy by Orowan strengthening mechanism [41]. The increment of YS associated with $Mg_{17}Al_{12}$ particles have already been given in an early report [41], and the equation is shown in Equation (2).

$$\Delta\sigma_{Orowan} = \beta \frac{0.4u_m b}{\pi\left(d\sqrt{\pi/4f_v} - 1\right)} \frac{\ln(d/b)}{\sqrt{1-v_m}}, \tag{2}$$

where β is constant (1.25), μ_m is shear modulus (16.5 GPa), b is Burgers vector (0.32 nm), d is the average size of $Mg_{17}Al_{12}$, f_v is the volume fraction of $Mg_{17}Al_{12}$, and v_m is Poisson ratio (0.35). Moreover, is should be noticed from above TEM observations that the nano-sized $Mg_{17}Al_{12}$ particles were most dynamically precipitated in the un-DRX grains, and they were seldom seen in DRX grains, which has also been reported in other Mg-Al-Ca-Mn alloys [19]. Therefore, we assumed that the Orowan strengthening mechanism caused by $Mg_{17}Al_{12}$ precipitates is only activated in un-DRX grains, and it could be calculated by the following equation:

$$\Delta\sigma_{Mg17Al12} = (1 - f_{DRX})\Delta\sigma_{Orowan}, \tag{3}$$

where f_{DRX} is the volume fraction of DRX grains showed in Figure 6.

As long as the strengthening caused by grain refinement and $Mg_{17}Al_{12}$ precipitates was estimated, the rest of strengthening effect should be caused by the Al_2Ca second phase strengthening, and it is calculated by Equation (4),

$$\Delta\sigma_{Al_2Ca} = \sigma_{YS} - \sigma_{Cast} - \sigma_{Hall-Petch} - \sigma_{Mg17Al12}, \tag{4}$$

where σ_{YS} is the tested tensile yield strength of ECAP alloys, and σ_{Cast} is the tested tensile yield strength of cast alloy.

Figure 14a displays the quantitative contributions of grain refinement, dynamic precipitated $Mg_{17}Al_{12}$ precipitates and dispersed Al_2Ca second phase particles, to TYS values of three ECAP alloys. It is apparent that for 623 K-4P and 12p alloys, fine grain strengthening plays a more important role than other two strengthening factors. With the dispersion of Al_2Ca phase in 573 K-12p alloy, the strengthening contribution of the Al_2Ca phase exceeds the other two factors. To intuitively describe the strengthening effect of a dispersed Al_2Ca second phase, Figure 14b shows the relationship between $\Delta\sigma_{Al2Ca}$ and the distance of ribbonlike Al_2Ca arm spacing [42]. It is obvious that with the decrease of Al_2Ca arm spacing, the contribution of the Al_2Ca phase increases remarkably. Therefore, to further improve the mechanical properties of Al_2Ca containing Mg-Al-Ca-Mn alloys, additional effort should be focused on the refining and dispersion of Al_2Ca second phase particles.

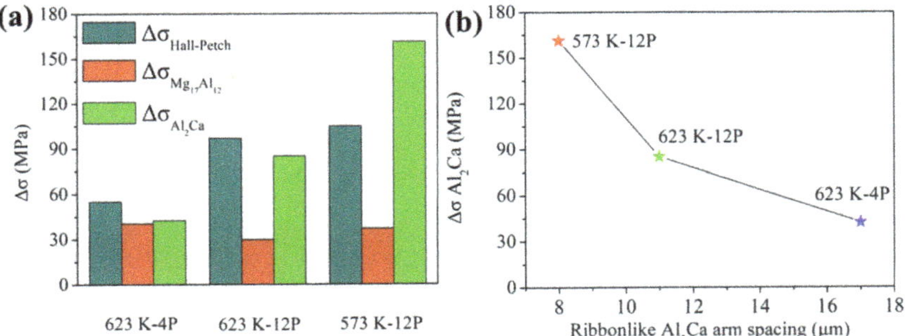

Figure 14. (a) The contributions of various strengthening factors to tensile yield strength of ECAP alloys. (b) The relationship between the ribbonlike Al_2Ca arm spacing and the contribution of Al_2Ca particles to tensile yield strength of ECAP alloys.

4. Conclusions

The microstructure evolutions and mechanical properties of a Mg-3.5Al-1.7Ca-0.4Mn (wt %) alloy during ECAP at different processing numbers and temperatures were systematically investigated. The main conclusions can be drawn as follows:

(1) The microstructure of as-cast Mg-3.5Al-1.7Ca-0.4Mn (wt %) alloy was composed of interdendritic α-Mg grains, network Al_2Ca phase and a few Al_8Mn_5 particles. During multi-pass ECAP, the Al_2Ca network phase was gradually broken into ultrafine particles. More ECAP passes and lower processing temperature resulted in finer sizes and a more dispersed distribution of Al_2Ca particles, as well as finer α-Mg grains.

(2) Multi-pass ECAP simultaneously enhanced the tensile strength and ductility of Mg-3.5Al-1.8Ca-0.4Mn alloy. The 573 K-12p processed alloy exhibited superior mechanical properties with a tensile yield strength of 304 MPa, ultimate tensile strength of 354 MPa and elongation of 10.3%. The improved comprehensive mechanical performance could be attributed to refined DRX grains, nano-sized $Mg_{17}Al_{12}$ precipitates and dispersed Al_2Ca particles. Moreover, quantitative contribution analysis suggested that the refined and dispersedly distributed Al_2Ca particles played a dominant role in strengthening the Al_2Ca containing alloys.

Author Contributions: A.M., H.L. and J.J. designed the project and guided the research; C.W., J.S. and H.L. prepared the manuscript; C.W., H.H. and Z.Y. performed the experiment and analyzed the data; X.Z., H.H. and C.W. prepared the figures; A.M., H.L. and J.J. reviewed the manuscript.

Funding: The authors are grateful to the financial aid from the Natural Science Foundation of Jiangsu Province of China (BK20160869), the National Natural Science Foundation of China (51774109 and 51501039), the Fundamental Research Funds for the Central Universities (2018B48414), the Key Research and Development Project of Jiangsu Province of China (BE2017148), and the Nantong Science and Technology Project (JC2018109).

Data Availability: The raw/processed data required to reproduce these findings cannot be shared at this time as the data also forms part of an ongoing study.

Conflicts of Interest: The authors declare no conflict of interest.

References

1. Mordike, B.L.; Ebert, T. Magnesium: Properties-applications-potential. *Mater. Sci. Eng. A* **2001**, *302*, 37–45. [CrossRef]

2. Yamashita, A.; Horita, Z.; Langdon, T.G. Improving the mechanical properties of magnesium and a magnesium alloy through severe plastic deformation. *Mater. Sci. Eng. A* **2001**, *300*, 142–147. [CrossRef]

3. Song, M.S.; Zeng, R.C.; Ding, Y.F.; Li, R.W.; Easton, M.; Cole, I.; Birbilis, N.; Chen, X.B. Recent advances in biodegradation controls over Mg alloys for bone fracture management: A review. *J. Mater. Sci. Technol.* **2019**, *35*, 535–544. [CrossRef]

4. Wang, B.J.; Wang, S.D.; Xu, D.K.; Han, E.H. Recent progress in the research about fatigue behavior of Mg alloys in air and aqueous medium: A review. *J. Mater. Sci. Technol.* **2017**, *33*, 1075–1086. [CrossRef]

5. Liu, H.; Huang, H.; Sun, J.P.; Wang, C.; Bai, J.; Ma, A.B.; Chen, X.H. Microstructure and mechanical properties of Mg-RE-TM cast alloys containing long period stacking ordered phases: A review. *Acta. Metall. Sin. Engl.* **2019**, *32*, 269–285. [CrossRef]

6. Wang, X.J.; Xu, D.K.; Wu, R.Z.; Chen, X.B.; Peng, Q.M.; Jin, L.; Xin, Y.C.; Zhang, Z.Q.; Liu, Y.; Chen, X.H.; et al. What is going on in magnesium alloys? *J. Mater. Sci. Technol.* **2018**, *34*, 245–247. [CrossRef]

7. Liu, H.; Huang, H.; Wang, C.; Ju, J.; Sun, J.; Wu, Y.; Jiang, J.; Ma, A. Comparative study of two aging treatments on microstructure and mechanical properties of an ultra-fine grained Mg-10Y-6Gd-1.5Zn-0.5Zr alloy. *Metals* **2018**, *8*, 658. [CrossRef]

8. Liu, H.; Huang, H.; Wang, C.; Sun, J.; Bai, J.; Xue, F.; Ma, A.; Chen, X.B. Recent advances in LPSO-containing wrought magnesium alloys: Relationships between processing, microstructure, and mechanical properties. *JOM* **2019**, 1–14. [CrossRef]

9. Zhang, J.H.; Liu, S.J.; Wu, R.Z.; Hou, L.G.; Zhang, M.L. Recent developments in high-strength Mg-RE-based alloys: Focusing on Mg-Gd and Mg-Y systems. *J. Magnes. Alloy* **2018**, *6*, 277–291. [CrossRef]

10. Pan, H.C.; Qin, G.W.; Huang, Y.M.; Ren, Y.P.; Sha, X.C.; Han, X.D.; Liu, Z.Q.; Li, C.F.; Wu, X.L.; Chen, H.W.; et al. Development of low-alloyed and rare-earth-free magnesium alloys having ultra-high strength. *Acta Mater.* **2018**, *149*, 350–363. [CrossRef]

11. Fu, L.; Le, Q.C.; Tang, Y.; Sun, J.Y.; Jia, Y.H.; Song, Z.T. Effect of Ca and RE additions on microstructures and tensile properties of AZ31 alloys. *Mater. Res. Express* **2018**, *5*, 056521. [CrossRef]

12. Cheng, C.; Lan, Q.; Wang, A.; Le, Q.; Yang, F.; Li, X. Effect of Ca additions on ignition temperature and multi-Stage oxidation behavior of AZ80. *Metals* **2018**, *8*, 766. [CrossRef]

13. Kim, S.H.; Lee, J.U.; Kim, Y.J.; Moon, B.G.; You, B.S.; Kim, H.S.; Parka, S.H. Improvement in extrudability and mechanical properties of AZ91 alloy through extrusion with artificial cooling. *Mater. Sci. Eng. A* **2017**, *703*, 1–8. [CrossRef]

14. Kang, J.W.; Sun, X.F.; Deng, K.K.; Xu, F.J.; Zhang, X.; Bai, Y. High strength Mg-9Al serial alloy processed by slow extrusion. *Mater. Sci. Eng. A* **2017**, *697*, 211–216. [CrossRef]

15. Han, L.; Hu, H.; Northwood, D.O. Effect of Ca additions on microstructure and microhardness of an as-cast Mg–5.0 wt. % Al alloy. *Mater. Lett.* **2008**, *62*, 381–384. [CrossRef]

16. Kondori, B.; Mahmudi, R. Effect of Ca additions on the microstructure and creep properties of a cast Mg-Al-Mn magnesium alloy. *Mater. Sci. Eng. A* **2017**, *700*, 438–447. [CrossRef]

17. Xu, S.W.; Oh-ishi, K.; Kamado, S.; Uchida, F.; Homma, T.; Hono, K. High-strength extruded Mg-Al-Ca-Mn alloy. *Scr. Mater.* **2011**, *65*, 269–272. [CrossRef]

18. Li, Z.T.; Zhang, X.D.; Zheng, M.Y.; Qiao, X.G.; Wu, K.; Xu, C.; Kamado, S. Effect of Ca/Al ratio on microstructure and mechanical properties of Mg-Al-Ca-Mn alloys. *Mater. Sci. Eng. A* **2017**, *682*, 423–432. [CrossRef]

19. Li, Z.T.; Qiao, X.G.; Xu, C.; Kamado, S.; Zheng, M.Y.; Luo, A.A. Ultrahigh strength Mg-Al-Ca-Mn extrusion alloys with various aluminum contents. *J. Alloys Compd.* **2019**, *792*, 130–141. [CrossRef]

20. Khorasani, F.; Emamy, M.; Malekan, M.; Mirzadeh, H.; Pourbahari, B.; Krajnák, T.; Minárik, P. Enhancement of the microstructure and elevated temperature mechanical properties of as-cast Mg-Al$_2$Ca-Mg$_2$Ca in-situ composite by hot extrusion. *Mater. Charact.* **2019**, *147*, 155–164. [CrossRef]

21. Wang, C.; Ma, A.; Sun, J.; Liu, H.; Huang, H.; Yang, Z.; Jiang, J. Effect of ECAP process on as-cast and as-homogenized Mg-Al-Ca-Mn alloys with different Mg$_2$Ca morphologies. *J. Alloys Compd.* **2019**, *793*, 259–270. [CrossRef]

22. Watanabe, H.; Yamaguchi, M.; Takigawa, Y.; Higashi, K. Mechanical properties of Mg-Al-Ca alloy processed by hot extrusion. *Mater. Sci. Eng. A* **2007**, *454*, 384–388. [CrossRef]

23. Zhou, D.W.; Liu, J.S.; Peng, P.; Chen, L.; Hu, Y.J. A first-principles study on the structural stability of Al$_2$Ca Al$_4$Ca and Mg$_2$Ca phases. *Mater. Lett.* **2008**, *62*, 206–210. [CrossRef]

24. Jiang, Z.T.; Jiang, B.; Yang, H.; Yang, Q.S.; Dai, J.H.; Pan, F.S. Influence of the Al$_2$Ca phase on microstructure and mechanical properties of Mg-Al-Ca alloys. *J. Alloys Compd.* **2015**, *647*, 357–363. [CrossRef]

25. Su, K.; Deng, K.K.; Xu, F.J.; Nie, K.B.; Zhang, L.; Zhang, X.; Li, W.J. Effect of extrusion temperature on the microstructure and mechanical properties of Mg-5Al-2Ca alloy. *Acta. Metall. Sin. Engl.* **2015**, *28*, 1015–1023. [CrossRef]

26. Homma, T.; Hirawatari, S.; Sunohara, H.; Kamado, S. Room and elevated temperature mechanical properties in the as-extruded Mg-Al-Ca-Mn alloys. *Mater. Sci. Eng. A* **2012**, *539*, 163–169. [CrossRef]

27. Jin, L.; Lin, D.L.; Mao, D.L.; Zeng, X.Q.; Ding, W.J. Mechanical properties and microstructure of AZ31 Mg alloy processed by two-step equal channel angular extrusion. *Mater. Lett.* **2005**, *59*, 2267–2270. [CrossRef]

28. Sun, J.P.; Yang, Z.Q.; Han, J.; Liu, H.; Song, D.; Jiang, J.H.; Ma, A.B. High strength and ductility AZ91 magnesium alloy with multi-heterogenous microstructures prepared by high-temperature ECAP and short-time aging. *Mater. Sci. Eng. A* **2018**, *734*, 485–490. [CrossRef]

29. Garces, G.; Pérez, P.; Barea, R.; Medina, J.; Stark, A.; Schell, N.; Adeva, P. Increase in the mechanical strength of Mg-8Gd-3Y-1Zn alloy containing long-period stacking ordered phases using equal channel angular pressing processing. *Metals* **2019**, *9*, 221. [CrossRef]

30. Chen, B.; Lin, D.L.; Zeng, X.Q.; Lu, C. Microstructure and mechanical properties of ultrafine grained Mg$_{97}$Y$_2$Zn$_1$ alloy processed by equal channel angular pressing. *J. Alloys Compd.* **2007**, *440*, 94–100. [CrossRef]

31. Zhang, J.S.; Zhang, W.B.; Bian, L.P.; Cheng, W.L.; Niu, X.F.; Xu, C.X.; Wu, S.J. Study of Mg-Gd-Zn-Zr alloys with long period stacking ordered structures. *Mater. Sci. Eng. A* **2013**, *585*, 268–276. [CrossRef]

32. Martynenko, N.S.; Lukyanova, E.A.; Serebryany, V.N.; Gorshenkov, M.V.; Shchetinin, I.V.; Raab, G.I.; Dobatkin, S.V.; Estrin, Y. Increasing strength and ductility of magnesium alloy WE43 by equal-channel angular pressing. *Mater. Sci. Eng. A* **2018**, *712*, 625–629. [CrossRef]

33. Ma, A.B.; Nishida, Y.; Suzuki, K.; Shigematsu, I.; Saito, N. Characteristics of plastic deformation by rotary-die equal-channel angular pressing. *Scr. Mater.* **2005**, *52*, 433–437. [CrossRef]

34. Jin, L.; Lin, D.L.; Mao, D.L.; Zeng, X.Q.; Ding, W.J. An electron back-scattered diffraction study on the microstructure evolution of AZ31 Mg alloy during equal channel angular extrusion. *J. Alloys Compd.* **2006**, *426*, 148–154. [CrossRef]

35. Rong, W.; Zhang, Y.; Wu, Y.J.; Chen, Y.L.; Sun, M.; Chen, J.; Peng, L.M. The role of bimodal-grained structure in strengthening tensile strength and decreasing yield asymmetry of Mg-Gd-Zn-Zr alloys. *Mater. Sci. Eng. A* **2019**, *740*, 262–273. [CrossRef]

36. Yang, Z.; Ma, A.; Liu, H.; Sun, J.; Song, D.; Wang, C.; Yuan, Y.; Jiang, J. Multimodal microstructure and mechanical properties of AZ91 Mg alloy prepared by equal channel angular pressing plus aging. *Metals* **2018**, *8*, 763. [CrossRef]

37. Li, B.; Teng, B.G.; Chen, G.X. Microstructure evolution and mechanical properties of Mg-Gd-Y-Zn-Zr alloy during equal channel angular pressing. *Mater. Sci. Eng. A* **2019**, *744*, 396–405. [CrossRef]

38. Akbaripanah, F.; Saniee, F.F.; Mahmudi, R.; Kim, H.K. Microstructural homogeneity, texture, tensile and shear behavior of AM60 magnesium alloy produced by extrusion and equal channel angular pressing. *Mater. Des.* **2013**, *43*, 31–39. [CrossRef]

39. Dumitru, F.D.; Cobos, O.F.H.; Cabrera, J.M. ZK60 alloy processed by ECAP: Microstructural, physical and mechanical characterization. *Mater. Sci. Eng. A* **2014**, *594*, 32–39. [CrossRef]

40. Yu, H.H.; Xin, Y.C.; Wang, M.Y.; Liu, Q. Hall-petch relationship in Mg alloys: A review. *J. Mater. Sci. Technol.* **2018**, *34*, 248–256. [CrossRef]

41. Sun, X.F.; Wang, C.J.; Deng, K.K.; Nie, K.B.; Zhang, X.C.; Xiao, X.Y. High strength SiCp/AZ91 composite assisted by dynamic precipitated Mg$_{17}$Al$_{12}$ phase. *J. Alloys Compd.* **2018**, *732*, 328–335. [CrossRef]

42. Yamasaki, M.; Hashimoto, K.; Hagihara, K.; Kawamura, Y. Effect of multimodal microstructure evolution on mechanical properties of Mg-Zn-Y extruded alloy. *Acta Mater.* **2011**, *59*, 3646–3658. [CrossRef]

Article

The Effects of Severe Plastic Deformation and/or Thermal Treatment on the Mechanical Properties of Biodegradable Mg-Alloys

Andrea Ojdanic [1,2], Jelena Horky [3], Bernhard Mingler [3], Mattia Fanetti [4], Sandra Gardonio [4], Matjaz Valant [4,5], Bartosz Sulkowski [6], Erhard Schafler [1], Dmytro Orlov [4,7] and Michael J. Zehetbauer [1,*]

[1] Physics of Nanostructured Materials, Faculty of Physics, University of Vienna, 1090 Wien, Austria; andrea.ojdanic@univie.ac.at (A.O.); erhard.schafler@univie.ac.at (E.S.)

[2] Faculty of Industrial Engineering, University of Applied Sciences–Technikum Wien, 1200 Wien, Austria

[3] Center for Health & Bioresources, Biomedical Systems, AIT Austrian Institute of Technology GmbH, 2700 Wiener Neustadt, Austria; jelena.horky@ait.ac.at (J.H.); bernhard.mingler@ait.ac.at (B.M.)

[4] Materials Research Laboratory, University of Nova Gorica, SI-5270 Ajdovscina, Slovenia; mattia.fanetti@ung.si (M.F.); sandra.gardonio@ung.si (S.G.); matjaz.valant@ung.si (M.V.); dmytro.orlov@material.lth.se (D.O.)

[5] Institute of Fundamental and Frontier Sciences, University of Electronic Science and Technology of China, Chengdu 610054, China

[6] Department of Material Science and Non-Ferrous Metals Engineering, Faculty of Non-Ferrous Metals, AGH-University of Science and Technology, 30-059 Kraków, Poland; sul5@agh.edu.pl

[7] Division of Materials Engineering, Department Mechanical Engineering, LTH, Lund University, 223-65 Lund, Sweden

* Correspondence: michael.zehetbauer@univie.ac.at; Tel.: +43-6-648-175-540

Received: 22 June 2020; Accepted: 1 August 2020; Published: 6 August 2020

Abstract: In this study, five MgZnCa alloys with low alloy content and high biocorrosion resistance were investigated during thermomechanical processing. As documented by microhardness and tensile tests, high pressure torsion (HPT)-processing and subsequent heat treatments led to strength increases of up to 250%; as much as about 1/3 of this increase was due to the heat treatment. Microstructural analyses by electron microscopy revealed a significant density of precipitates, but estimates of the Orowan strength exhibited values much smaller than the strength increases observed. Calculations using Kirchner's model of vacancy hardening, however, showed that vacancy concentrations of 10^{-5} could have accounted for the extensive hardening observed, at least when they formed vacancy agglomerates with sizes around 50-100 nm. While such an effect has been suggested for a selected Mg-alloy already in a previous paper of the authors, in this study the effect was substantiated by combined quantitative evaluations from differential scanning calorimetry and X-ray line profile analysis. Those exhibited vacancy concentrations of up to about 10^{-3} with a marked percentage being part of vacancy agglomerates, which has been confirmed by evaluations of defect specific activation migration enthalpies. The variations of Young's modulus during HPT-processing and during the subsequent thermal treatments were small. Additionally, the corrosion rate did not markedly change compared to that of the homogenized state.

Keywords: Mg alloy; severe plastic deformation (SPD); intermetallic precipitates; vacancy agglomerates; corrosion

1. Introduction

Biodegradable Mg Alloys

Magnesium alloys, as the lightest structural materials, are becoming increasingly popular for numerous applications, especially for biodegradable implants. When the clinical function of permanent implants is served, they typically must be removed because of allergy problems and/or mechanical instabilities, e.g., the stress shielding effect [1], which arises from different Young's moduli of bone and implant materials. Such a removal can be avoided when using biodegradable materials for implants provided the degradation time can be adjusted to the healing time. The implants dissolve in the human body after fulfilling their purposes by a corrosion process initiated by the body fluids [2].

MgZn-based alloys have been proposed as very suitable biodegradable materials for load-bearing applications due to their low density, comparably high strength and low Young's modulus—coming close to that of bones and thus avoiding stress shielding.

In general, improvements in mechanical properties can be achieved through increased alloy content and/or precipitation formation. In the case of Mg, the first strategy is limited since the solubility of most alloying elements is limited [3,4]. Concerning the formation of precipitates in Mg-Zn systems, Mima and Tanaka [5] identified three important low-temperature ranges for Mg-Zn systems: (i) below 60 °C, the formation of stable Guinier–Preston (GP1) zones; (ii) 60–110 °C, the formation of stable rod-type and basal platelet-type precipitates along with unstable GP1 zones followed by growth of the former at the expense of dissolution of the latter; and (iii) above 110 °C, the formation of stable rod-type and basal platelet-type precipitates, the most stable ones being the rod type [6,7]. In a commercial Mg5.5Zn0.6Zr (wt%) alloy (ZK60), Orlov et al. [8] found intermetallic precipitates similar to the GP1 zones, as a result of special ageing conditions after plastic deformation.

Enhancing solid solution and/or precipitation for the sake of mechanical properties, however, increases the chemical reactivity and finally causes unacceptably large rates of corrosion in most environments [9]. Moreover, in the case of precipitate formation, it may exhibit a markedly enhanced Young's modulus, thereby increasing the shielding effect. For all these reasons, using plastic deformation for the generation of lattice defects acting as barriers to dislocation movement, is an interesting alternative aiming at higher mechanical properties [10]. Importantly, the elastic moduli in texture-free polycrystalline aggregates do not change during plastic deformation.

In comparison to fcc and bcc metals, the critical resolved shear stresses in slip systems of hcp metals have large variations. Therefore, in Mg and its alloys, plastic deformation occurs by slip and/or twinning on a much lower number of systems, which significantly limits the ductility, especially at low temperatures [11]. This problem can be overcome by deformation at elevated temperatures, but then the production of lattice defects becomes increasingly balanced by their thermally activated annihilation, resulting in a decreasing total number of lattice defects. A better way is to process the materials by methods of severe plastic deformation (SPD) [12–16]. Those provide an enhanced hydrostatic pressure that is prevalent during deformation which suppresses the formation of cracks and extends deformability. In many works, equal channel angular pressing (ECAP, [17–23]) was used to deform Mg and Mg alloys for the sake of hardening through refining the microstructure, but still, cracks were formed during deformation at room temperature (RT), and continuous deformation was possible only above 200 °C allowing for grain sizes beyond 0.5 µm. As HPT yields much higher hydrostatic pressures (up to 10 GPa) than ECAP (1.5 GPa), processing can be extended to 10–100 times larger strains, thereby providing grain sizes down to 100 nm at room temperature (RT) processing [24–28]. Therefore, in the present investigation of low-concentration biodegradable Mg alloys, HPT was applied; we were expecting substantial grain refinement through massive dislocation production, by redistribution of solutes and also by a high concentration of vacancies [29–31]. Zehetbauer [32,33] and especially Horky et al. [33] for two selected Mg-alloys reported that these deformation-induced vacancies can form agglomerates that inhibit the dislocation movement and therefore increase the macroscopic strength; however, an extensive study of this effect has not been performed yet.

It was therefore the aim of this work to investigate thoroughly the strengthening capabilities of five biodegradable Mg alloys with low alloy content through the formation of both deformation-induced defects, including vacancy agglomerates, and precipitates via severe plastic deformation and heat treatments. The biodegradable MgZnCa-systems chosen here only included mineral nutrients that are not harmful to the human body. With the alloy constituents Zn and Ca, precipitates were formed, which not only impeded dislocation movement and thereby increased both strength and work hardening, but also stopped grain growth during solidification and thermal treatments [29]. Due to the fact that Ca is less noble than Mg, low MgZnCa systems also provide—according to Hofstetter et al. [9]—desirably slow and homogeneous degradation behavior with the low Mg alloys selected here.

2. Experimental Procedure

2.1. Materials, Samples and Preparations

Five alloys were investigated with compositions Mg5Zn0.3Ca, Mg5Zn0.15Ca, Mg5Zn0.15Ca0.15Zr, Mg5Zn and Mg0.3Ca. The emphasis of analyses was laid on the first three alloys; the last two alloys helped to understand the main findings in-depth. The alloys were cast at the LKR (Leichtmetall-Kompetenzzentrum Ranshofen, Ranshofen, Austria), a subsidiary of AIT (Austrian Institute of Technology, Wien, Austria). The chemical compositions of the alloys are shown in Table 1.

Table 1. Chemical compositions of the investigated alloys.

Alloy	Mg (at%)	Zn (at%)	Ca (at%)	Zr (at%)
Mg5Zn0.3Ca	94.28 ± 0.03	5.44 ± 0.03	0.28 ± 0.03	-
Mg5Zn	94.77 ± 0.03	5.23 ± 0.03	-	-
Mg0.3Ca	99.73 ± 0.03	-	0.27 ± 0.03	-
Mg5Zn0.15Ca	94.90 ± 0.03	5.1 ± 0.03	0.15 ± 0.03	-
Mg5Zn0.15Ca0.15Zr	94.40 ± 0.03	5.6 ± 0.03	0.18 ± 0.03	0.18 ± 0.03

The specimens were homogenized in a resistance furnace in argon protective gas atmosphere in order to avoid strong oxidation, at 450 °C for 24 h for all alloys, except Mg5Zn which was heat treated at 350 °C for 12 h [34]. After that, the samples were either furnace-cooled or quenched to room temperature, and investigated separately.

2.2. HPT-Processing

For HPT-processing, disc-shaped samples with a diameter of 10 mm and a thickness of 0.7 mm were prepared. The hydrostatic pressure applied during HPT was 4 GPa, and the rotation speed was 0.2 rot/min. The samples were torsion strained to shear strains

$$\gamma_T = \frac{2\pi N r}{h} \tag{1}$$

with the von Mises strains ε being by a factor $\sqrt{3}$ smaller [35]. N means the number of rotations, r the sample radius and h the thickness of the sample. HPT-processing was performed at room temperature (RT).

2.3. Heat Treatments

Heat treatments of both the initial homogenized (IS) and the HPT-processed samples were performed over different time periods and at different temperatures between 50 °C and 240 °C in a silicon oil bath with a thermal stability of ±0.5 °C. After the heat treatments, the samples were quenched in RT-water to ensure fast cooling and high accuracy of heat treatment duration [34].

2.4. Characterization of Microstructure

2.4.1. Microhardness Tests

Vickers hardness was chosen as a measure of strength and investigated for all samples by an MHT-4 Micro Indentation Tester (ANTON PAAR, Graz, Austria), by applying a load of 0.5 N for 10 s. After the indentation, the indents' areas were observed with a Light Microscope AXIOPLAN (ZEISS, Jena, Germany) using 50×/0.75 HD DIC and EPIPLAN 100×/0.75 objectives. The diagonals were measured from images taken with a CCD camera being part of the microscope, as shown in Figure 1. Asymmetries of indents did not affect the lengths of diagonals, and no crack formation occurred. When determining the hardness of an HPT-processed sample, the indents were taken at fixed radii from the center of the specimen, in order to ensure comparability of the results. At least 10 indents per sample state were evaluated.

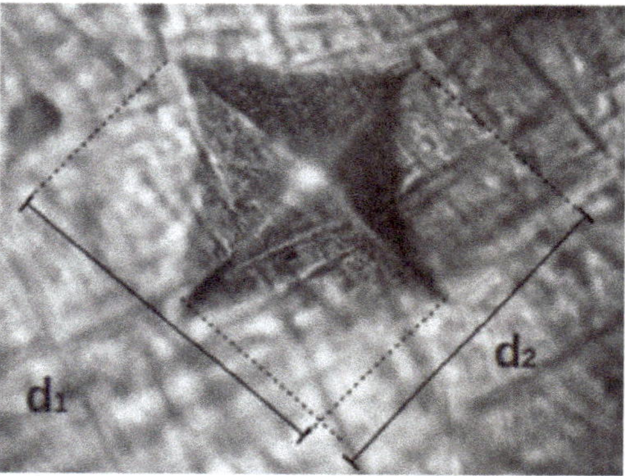

Figure 1. An image of an indentation taken with a CCD camera, with d_1 and d_2 as the measured diagonals of the indentation. They amount to approximately 25 μm.

For the measurements of Young's modulus, a microindentation tester MHT-3 (ANTON PAAR, Graz, Austria) was used.

2.4.2. Tensile Tests

Strength and ductility were determined by tensile testing. Dogbone-shaped specimens with a cross-sectional area of 0.6×0.6 mm^2 and a parallel gauge length of 3.5 mm in average (Figure 2) were cut via spark erosion from the HPT-processed and non-processed discs. The gauge length of the tensile samples was off-centered at a radius of the HPT-disc of approximately 3 mm.

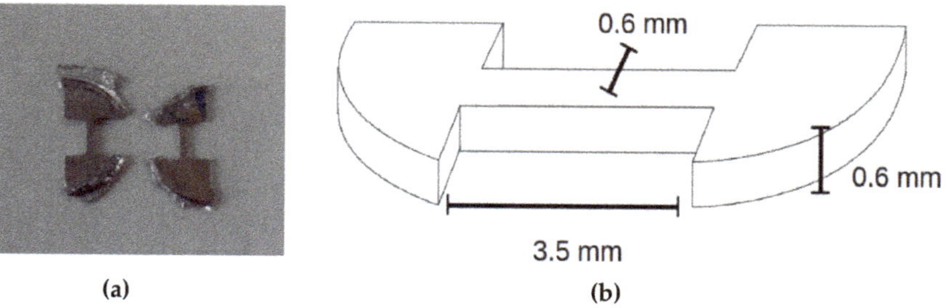

Figure 2. (**a**) Tensile test samples punched from Mg alloys in question, (**b**) dimensions of a punched sample.

A micro-tensile machine produced by MESSPHYSIK/ZWICK-RÖLL (Fürstenfeld, Austria) was used with load capacities in the range of 1–1000 N—suitable for determination of tensile properties and low cycle fatigue tests of small-scaled samples. Loads up to 120 N with a ramp speed of 0.2 mm/min giving the strain rate of approximately 1×10^{-3} s^{-1} were used. To obtain average values for the tensile test results, at least 3 samples were tested for each condition.

2.4.3. Electron Microscopy

In order to follow the relative redistribution of intermetallic precipitates throughout the matrix at the microscale, scanning electron microscopy (SEM) in backscattered-electron (BSE) imaging mode was carried out. For this purpose, a SUPRA 55 VP SEM (ZEISS, Jena, Germany) equipped with an energy-dispersive X-ray spectroscopy (EDS) analysis and imaging system was used. Samples were mechanically ground and polished down to a finish by employing ethanol lubricant and a 0.05 μm Al_2O_3 suspension. Detailed evaluation of the evolution of particles at the nano-scale was carried out by transmission electron microscopy (TEM) and scanning transmission electron microscopy (STEM). For this purpose, a JEM 2100F-UHR microscope (JEOL, Tokyo, Japan) equipped with a high-angle annular dark field (HAADF) detector was used. The same microscope is equipped with an EDS facility (Oxford Instruments, Abingdon, UK), which has been used for elemental analyses of the Mg alloys investigated. The specimens for TEM/STEM analyses were prepared by cutting a 3 mm disk out of the slice, and mechanically polishing with polishing paper (up to 2000 grit) and finishing with 1 μm suspension. During polishing only absolute ethanol has been used as a lubricant. Finally, the specimen was ion milled with grazing Ar+ ion beam, by means of a Precision Ion Polishing System (PIPS-II) (GATAN, Pleasanton, CA, USA) operated at 5 kV at the beginning down to 2 kV for the final step.

2.4.4. Differential Scanning Calorimetry

The differential scanning calorimetry (DSC) measurements were performed using two instruments, a DSC204 (NETZSCH, Selb, Germany) and a DSC8500 (PERKIN-ELMER, Cleveland, OH, USA) using aluminum crucibles. Heating was carried out in a linear way in a temperature range from 25 to 450 °C. The standard heating rate was 10 K/min. During DSC, the annealing of deformation-induced defects can be observed by the occurrence of exothermic peaks. The area of a given peak corresponds to the total energy being stored in all defects of a given type, E_{defect}, from which its density can be derived. For dislocations, their stored energy E_{disl} is related to their density ρ [36]:

$$E_{disl} = Gb^2 \frac{\rho}{4\pi\kappa} \cdot \ln\left(\left(b\sqrt{\rho}\right)^{-1}\right) \tag{2}$$

where G is the shear modulus and b the absolute value of the Burgers vector. κ denotes the arithmetic average of 1 and $(1 - v)$, with $v = 0.343$ as Poisson's ratio, assuming equal parts of edge and screw dislocations. The concentration of vacancies c_v can be evaluated from the stored energy of vacancies E_{vac} divided by the formation enthalpy per vacancy, ΔH (in Mg, $\Delta H = 1.27 \times 10^{-19}$ Joule = 0.79 eV [37]),

$$c_V = \frac{E_{vac}}{\Delta H \times v \times N_a}, \tag{3}$$

with v being the amount of substance and N_a Avogadro's number. Information on the nature of the defects can be obtained from the annealing peak temperature T_{max} and the activation (defect's migration) enthalpy Q. For the latter, the method of Kissinger [38] was applied by evaluating the shift of T_{max} with changing heating rate. By plotting

$$\ln\left(\frac{\phi}{T_{max}^2}\right) = -\frac{Q}{R}\frac{1}{T_{max}} + const \tag{4}$$

for various heating rates Φ and absolute peak temperatures T_{max}, the migration enthalpy Q can be determined; R is the gas constant.

2.4.5. X-ray Diffraction Peak Profile Analysis (XPA)

Selected samples were subjected to X-ray diffraction peak profile analysis (XPA) measurements by means of a self-assembled double diffractometer to measure dislocation density ρ. The X-ray source was a MM9 X-ray rotating anode generator (RIGAKU, Tokyo, Japan) using monochromatic Co-K$_\alpha$ radiation, corresponding to a wavelength $\lambda = 1.79$ nm. The peak profiles were collected by a curved position-sensitive detector of type CPS-590 (INEL, Artenay, France), covering an angular range of 90° between 40° and 130°. For details of the diffraction theory of Bragg peak broadening related to dislocation density and crystallite size, see [39,40]. The evaluation of XPA data was done by the open source software (C) MWP-fit (Version 140518, Budapest, Hungary) [41,42].

2.4.6. Corrosion Tests

Corrosion tests were performed in testing tubes filled with 250 mL simulated body fluid (SBF) with a composition according to SBF27 from [43], a pH value of 7.35. The tests were conducted at a temperature of 37 °C and the SBF was changed every seven days. Two disc-shaped samples with the same diameter of 10 mm and a height between 0.4 mm and 0.7 mm were immersed together in order to have a larger total surface area. The samples were polished with SiC-paper (1200 grid) and cleaned with ethanol directly before the start of the test.

The corrosion of Mg alloys leads to the formation of hydrogen gas, which accumulates at the top of the testing tubes [44,45], according to the reaction

$$Mg + 2H_2O \rightarrow Mg(OH)_2 + H_2$$

The amount of H_2 gas was recorded every 4 h and normalized by the initial sample's surface area. A numerical derivation was conducted to obtain the corrosion rate followed by a moving average algorithm to smoothen the curves. Two tests were performed for each material in each condition. The corrosion test set-up has already been described previously [46,47].

3. Results

3.1. Achieving the Supersaturated Solid-Solution Condition

SEM images of the as-cast alloys reveal a large number of primary precipitates, which are represented by the areas with bright contrast in Figures 3–5. On the microscale, the precipitates are distributed rather homogeneously in every sample condition. The level of contrast in these images signifies compositional changes in the microstructure, as it depends on the atomic numbers of the elements involved. All samples in the as-cast condition showed particles with a size of 5–10 µm, both within grain interiors and along grain boundaries. EDS analyses were done to reveal the chemistry of the primary precipitates. The analyses were achieved on well-defined points on the sample's surface, as shown in Figure 3, revealing a spectrum of the chemical composition of each point investigated (Figure 4). A total of 10 points were used for the primary precipitates and the surrounding matrix to determine the mean value. The chemistry of these precipitates revealed by EDS is shown in Table 2. Further, phase diagrams from literature were used to estimate the compositions of the precipitates. Those of the primary particles should correspond to Mg_2Ca, $Zn_{13}Mg_{12}$ and $Ca_2Mg_5Zn_5$ [48]. A more detailed analysis on primary precipitates could have been done by TEM, but that was not the focus of this study. The total volume fraction of primary precipitates was determined to be about ~2% being estimated from the total area of the particles by means of standard image analysis methods.

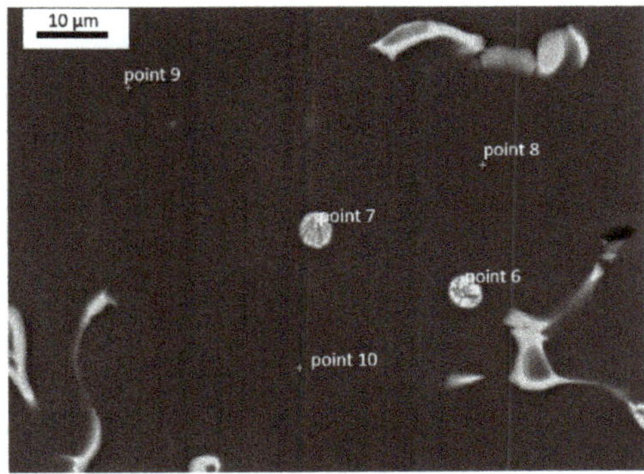

Figure 3. SEM image of Mg5Zn0.3Ca. EDS was carried out at well-defined points (here #6–#10) on the sample's surface. Mainly, the primary precipitates were investigated, but so was the surrounding matrix.

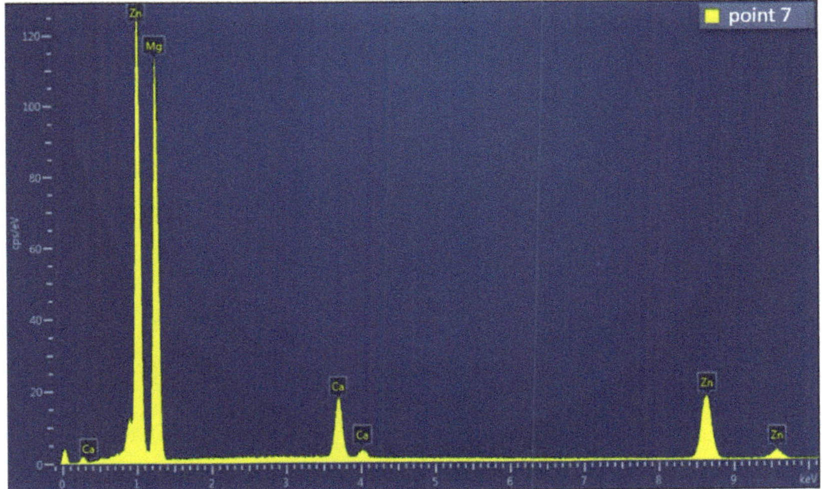

Figure 4. Exemplary EDS analysis of point 7 indicated in Figure 3, for a Mg5Zn0.3Ca sample. Intensity peaks (counts/s) for different energies (eV) correspond to EDS signals from Ca, Zn, Mg, Ca, Ca, Zn, and Zn again (in sequence from left to right).

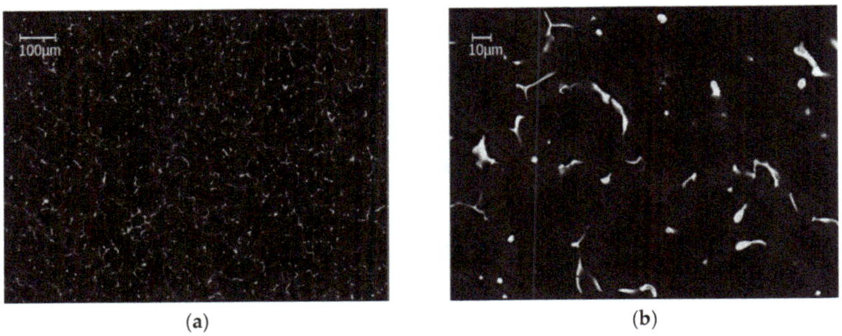

(**a**)   (**b**)

Figure 5. SEM image of the as-cast alloy Mg5Zn0.3Ca, (**a**) with about 100× magnification (**b**) with about 500× magnification.

Table 2. Chemistry of the precipitates revealed by EDS.

Alloy	Mg (at%)	Zn (at%)	Ca (at%)
Mg5Zn0.3Ca	65 ± 5	31 ± 7	4 ± 1
Mg5Zn	76 ± 4	24 ± 3	-
Mg0.3Ca	90 ± 1	-	10 ± 1
Mg5Zn0.15Ca	28 ± 6	70 ± 5	3 ± 1
Mg5Zn0.15Ca0.15Zr	24 ± 5	72 ± 5	4 ± 1

During homogenization, the primary precipitates should have been thermally destroyed. It is not possible, however, to dissolve them completely in the Mg matrix, as its solid solubility is low: Ca has a solubility in Mg of 0.82 at% (1.35 wt%), and Zn has a solubility in Mg of 2.4 at% (6.2 wt%) [49]. Two cooling treatments—quenching and furnace cooling—were performed with all alloys (Figure 6): Quenched samples showed still widely spread precipitates, while the furnace-cooled ones did not, indicating almost a solid-solution state. For the alloys Mg5Zn and Mg5Zn0.3Ca, the volume fraction after quenching could be reduced to ~2% with a precipitation size of less than 5 µm. In case of furnace cooling, the remaining volume fraction of precipitates was less than 1% with a size in the nm-scale. For Mg5Zn0.15Ca and Mg5Zn0.15Ca0.15Zr both cooling treatments could reduce the volume fraction of primary precipitates to below 1%. As a common name for both states, "initial state (IS)" is used in all the following parts of this paper.

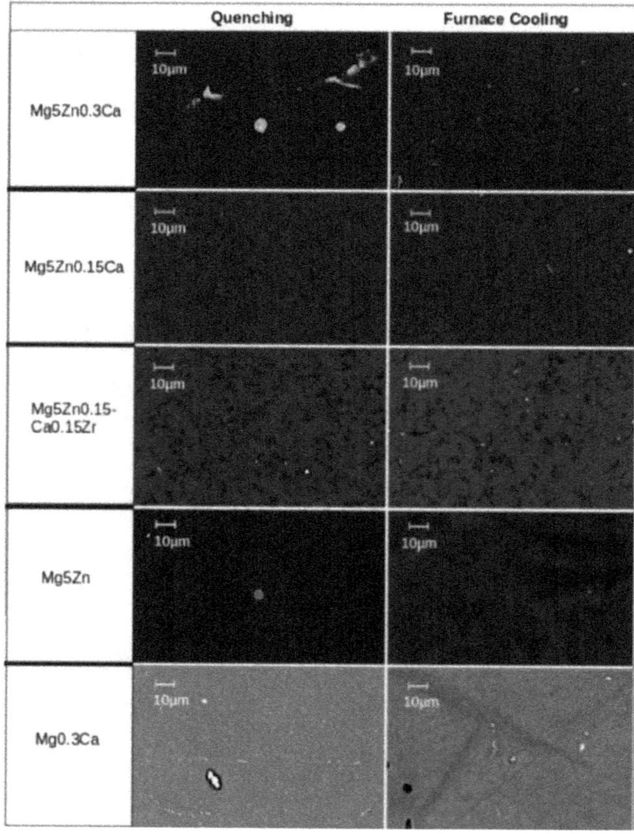

Figure 6. SEM images of all alloys after best-condition annealing, and quenching or furnace-cooling.

The results of microhardness (HV0.05) measurements of as-cast Mg5Zn0.3Ca, Mg5Zn, Mg0.3Ca, Mg5Zn0.15Ca and Mg5Zn0.15Ca0.15Zr samples, and in their furnace-cooled or quenched states, are listed in Table 3.

Table 3. Microhardness results of Mg5Zn0.3Ca, Mg5Zn0.15Ca, Mg5Zn0.15Ca0.15Zr and the two alloys Mg5Zn and Mg0.3Ca in the as-cast, furnace-cooled and quenched states.

Alloy	HV0.05 as-Cast	HV0.05 Furnace-Cooled	HV0.05 Quenched
Mg5Zn0.3Ca	65 ± 6	58 ± 1	70 ± 2
Mg5Zn	70 ± 9	55 ± 2	76 ± 4
Mg0.3Ca	50 ± 3	47 ± 1	54 ± 3
Mg5Zn0.15Ca	77 ± 2	78 ± 3	83 ± 5
Mg5Zn0.15Ca0.15Zr	78 ± 4	74 ± 3	76 ± 3

3.2. The Effect of Severe Plastic Deformation

After reaching the homogenized (initial) state, the samples were HPT processed at a hydrostatic pressure of 4 GPa by 0.5 and 2 rotations, reaching torsional strains up to $\gamma_T \sim 100$. After that, microhardness (HV0.05) measurements were carried out while distributing them over the whole cross section. In particular, microhardness was measured along entire radii of the disc-shaped sample. In Figure 7, the microhardness values are presented as a function of torsional shear strain γ_T calculated according to Equation (1). After HPT of the furnace-cooled samples, the microhardness can be said to increase with increasing γ_T, and finally reaches saturation at around HV0.05 = 160 for all four alloys, while for Mg0.3Ca the saturation is reached already at around HV0.05 = 72 (Figure 7).

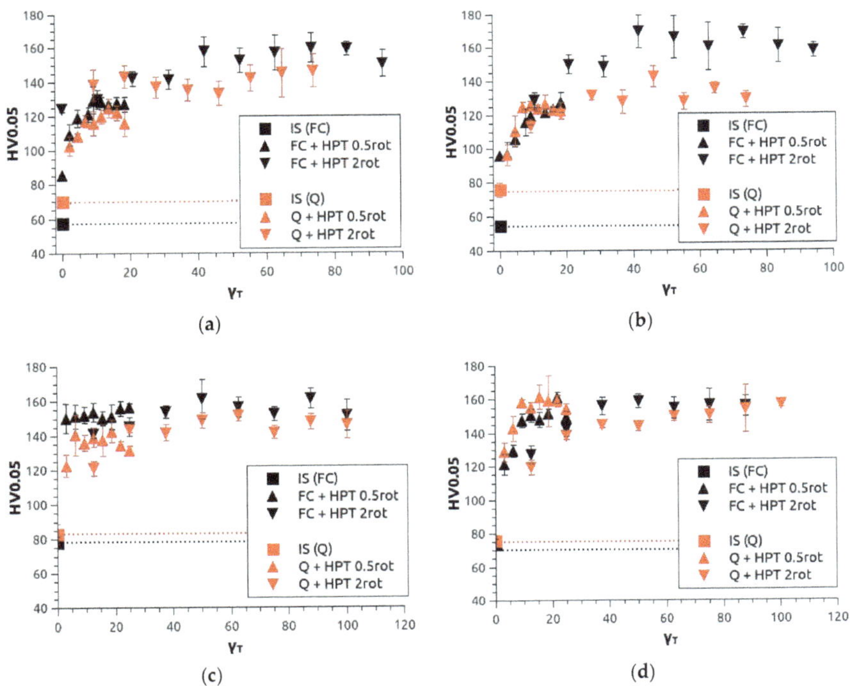

Figure 7. *Cont.*

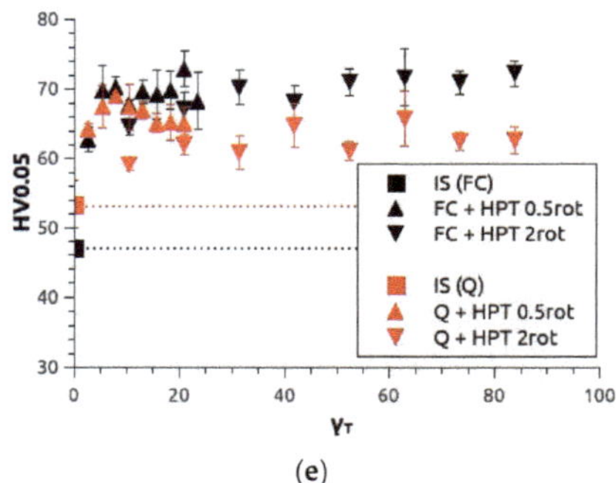

Figure 7. Microhardness results for (**a**) Mg5Zn0.3Ca, (**b**) Mg5Zn, (**c**) Mg5Zn0.15Ca, (**d**) Mg5Zn0.15Ca0.15Zr and (**e**) Mg0.3Ca in the initial state (IS–dotted line) after furnace-cooling (FC, black) and quenching (Q, red) and additional HPT deformation at room temperature at 4 GPa, 0.5 and 2 rotations giving torsional strains up to $\gamma_T{\sim}100$.

However, starting with the quenched IS condition, which showed a higher microhardness than the furnace-cooled initial state, microhardness after HPT processing saturated at lower values, for Mg5Zn0.3Ca and Mg5Zn at around HV0.05 = 140 and for Mg5Zn0.15Ca and Mg5Zn0.15Ca0.15Zr at around HV0.05 = 150 for the highest γ_T. For Mg0.3Ca, the saturation hardness was around HV0.05 = 64 only. The microhardness of the alloy with 0.15% Zr showed a slight tendency to decrease at higher strains (30-100) after a peak value at lower strains of approximately $\gamma_T = 20$. In terms of relative effects after HPT processing, the samples in the IS (furnace-cooled) showed microhardness increases of 175% for Mg5Zn0.3Ca, 190% for Mg5Zn, 135% for Mg5Zn0.15Ca0.15Zr, 85% for Mg5Zn0.15Ca and 53% for Mg0.3Ca. Even in relative numbers, the HPT-processed IS (quenched) samples showed smaller increases, 100% for Mg5Zn0.3Ca, 87% for Mg5Zn, 77% for Mg5Zn0.15Ca, 113% for Mg5Zn0.15Ca0.15Zr and 19% for Mg0.3Ca.

3.3. The Effects of Isothermal Heat Treatments

Microhardnesses of heat-treated alloys in the initial state and HPT-processed state have been reported in [34]. After homogenization, followed by furnace-cooling or quenching, Mg alloy samples were not only HPT-processed to $\varepsilon{\sim}5$ and $\varepsilon{\sim}20$, but also heat-treated for up to 29 h, at a constant temperature between 50–220 °C in intervals of 20 °C. After each isothermal heat treatment, microhardness was measured. Similarly to the results reported by Horky et al. [33] (see also Sections 4.3 and 4.4), further hardness increases were observed in all alloys studied in this work except Mg0.3Ca, as discussed later. The highest hardness was observed after a heat treatment at 100 °C (see Figure 8a–e) and at an annealing time of 1 h (Figure 8).

The increases in the IS (furnace-cooled) samples were larger than in the IS (quenched); i.e., heat treatments of the furnace-cooled and HPT-processed samples gave a total increase of microhardness by 40%, while the heat treatments of the quenched and HPT-processed samples gave 15% only.

The binary alloy Mg0.3Ca had its hardness maxima after heat treatments at 75 °C and between 150 and 160 °C. When exceeding the annealing temperature of 100 °C (or 160 °C in the case of Mg0.3Ca), microhardness decreased significantly for all alloys; and at annealing temperatures near 220 °C, it reached original values before heat treatment.

It is interesting to note that the samples without any HPT-processing also showed a thermally-induced hardness maximum at around the same annealing temperatures, i.e., 100 °C and 160 °C, although the values remained substantially below those of the HPT-processed samples (Figure 8a–e).

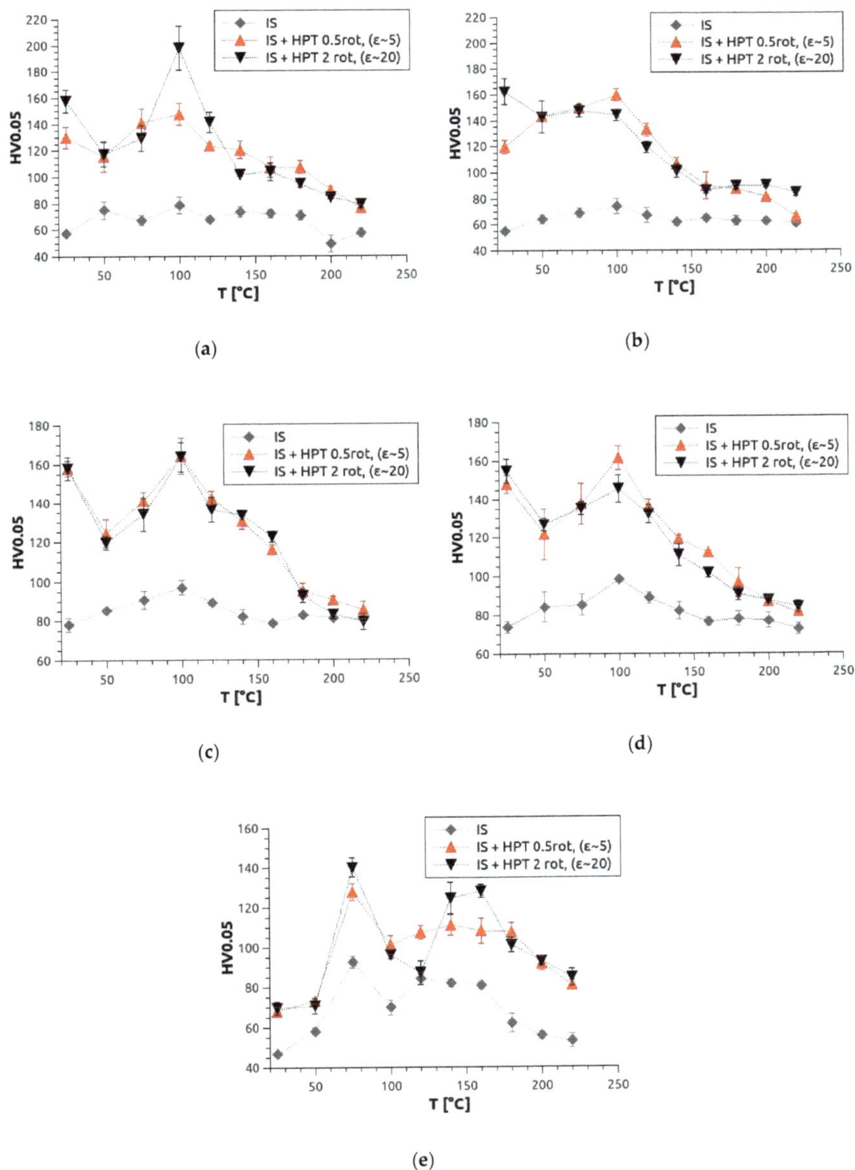

Figure 8. Microhardness after heat treatments for 1 h at various temperatures for (**a**) Mg5Zn0.3Ca (**b**) Mg5Zn (**c**) Mg5Zn0.5Ca (**d**) Mg5Zn0.15Ca0.15Zr and (**e**) Mg0.3Ca in the IS (furnace-cooled) (grey), at ε~5 (red) and ε~20 (black).

The peak temperature of 100 °C was chosen to investigate how the extension of heat treating time up to 29 h affects hardness of Mg5Zn0.3Ca, Mg5Zn and Mg0.3Ca alloys HPT-processed to ε~5 and ε~20; see Figure 9. For almost all materials, hardness increased after one hour but did not increase further within the next 23 h of heat treatment. Only after longer heat treatment time, substantial decreases in hardness were observed. Figure 10 shows the results for a fixed heat treatment time of 24 h, and varying temperature between 25 and 150 °C. For HPT-processed samples, the highest hardness values were observed, most of them again at T = 100 °C. Significantly smaller or no hardness maxima at all were found in the case of IS samples heat treated at T = 100 °C.

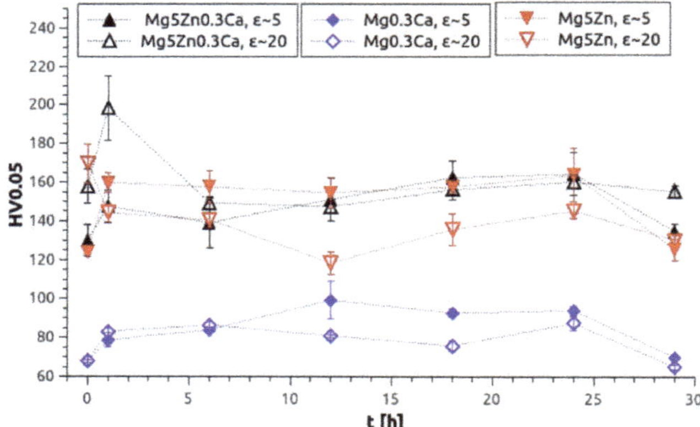

Figure 9. Microhardness after heat treatments at the peak temperature of 100 °C with varying annealing times up to 29 h for Mg5Zn0.3Ca (black), Mg5Zn (red) and Mg0.3Ca (blue) processed at different strains.

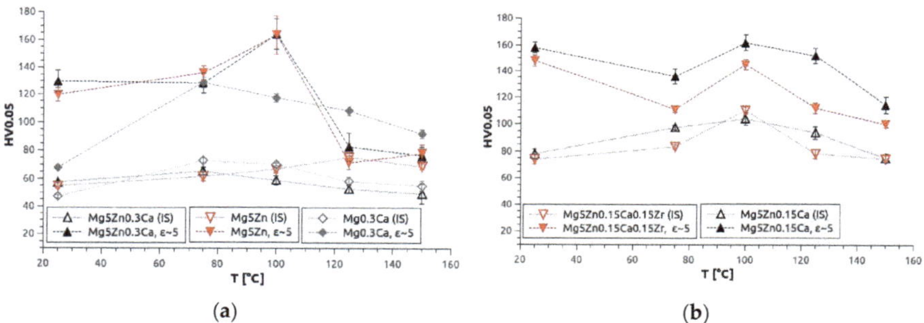

Figure 10. Results of microhardness measurements of Mg alloys in the IS (furnace-cooled) and after HPT-processing at 4 GPa and 0.5 rotations, and additional heat treatment for 24 h at varying temperatures. (**a**) Mg5Zn0.3Ca (black), Mg5Zn (red) and Mg0.3Ca (grey); and (**b**) Mg5Zn0.15Ca0.15Zr (red) and Mg5Zn0.15Ca (black).

3.4. Total Hardness Increase after Processing

The total hardness increase along the processing history of Mg5Zn0.3Ca and Mg5Zn in the IS (quenched and furnace-cooled) is summarized in Figure 11. A percentage in a bar shows the hardness increase from one processing stage to the next one. A percentage above a bar shows the total hardness increase from the first processing stage to the last one. A very high total hardness increase of up to 200% for Mg5Zn could be reached for the furnace-cooled samples, while the quenched ones showed a total hardness increase of only 100%.

The maximum hardness increase for furnace-cooled samples of about 200% could only be reached for Mg5Zn and Mg5Zn0.3Ca. Mg0.3Ca (Figure 12) showed a surprisingly high hardness increase of 150%, whereas Mg5Zn0.15Ca and Mg5Zn0.15Ca0.15Zr showed only an increase of 100%. The last two alloys did not show a reaction to heat treatments at all.

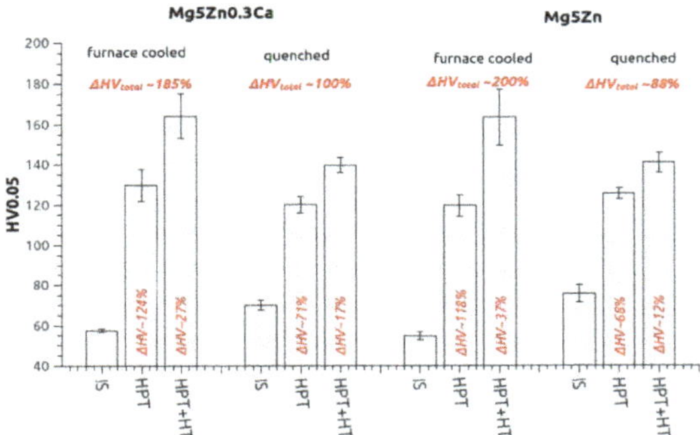

Figure 11. Total microhardness increases of Mg5Zn0.3Ca and Mg5Zn in the IS (furnace-cooled and quenched) and additional HPT deformations after 4 GPa and 0.5 rotations and additional heat treatment at 100 °C for 24 h.

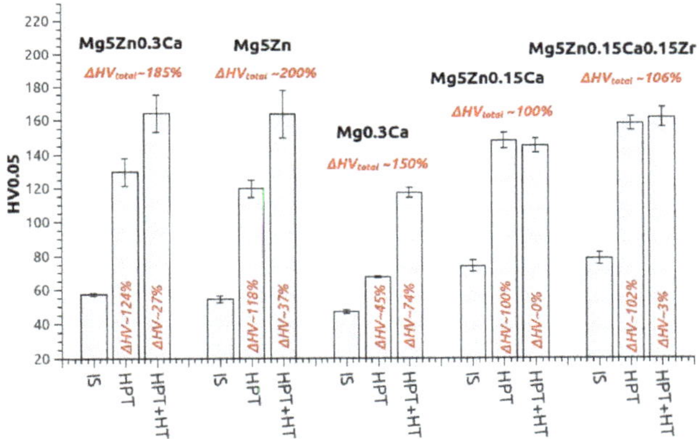

Figure 12. Total hardness increases of Mg5Zn0.3Ca, Mg5Zn, Mg0.3Ca, Mg5Zn0.15Ca and Mg5Zn0.15Ca0.15Zr from the IS (furnace-cooled), HPT processing at 4 GPa and 0.5 rotations; and additionally, heat treatment at the peak temperatures of 100 °C for 24 h, and 75 °C for 24 h for Mg0.3Ca.

3.5. Tensile Strength and Ductility

Tensile tests were conducted for all five alloys in the initial state; HPT-deformed (0.5 and 2 rotations); and additionally, heat treated for 24 h at 100 °C. Representative engineering stress–strain curves obtained by micro-tensile testing are shown in Figure 13. In cases wherein no relative maximum was seen, the ultimate tensile strength (UTS) was derived by determining the maximum load and the initial cross section of the sample. For the determination of yield strength, a constant plastic strain of 1% was chosen. The difference in yield strength and UTS of the IS and HPT-processed samples typically amount to 100 MPa and 200 MPa, respectively. HPT-processing more than doubles the yield strength and also drastically increases the ultimate tensile strength by ~140% for Mg0.3Ca. For Mg5Zn0.3Ca and Mg5Zn, the UTS increased by ~45% and ~75%, respectively. For Mg5Zn0.15Ca and Mg5Zn0.15Ca0.15Zr, HPT processing reduced the maximum elongation to ~5%. This means that the ductility is much lower than that of the material in the initial state. The post-HPT heat treatment (condition of peak hardness) led to a further strong increase in yield strength by up to ~85% and a slight decrease in ultimate tensile strength. However, the elongation to failure still reached 5% only, not showing any response to thermal treatment after HPT-processing. The average values of strength and ductility measured in the tensile tests are summarized in Table 4.

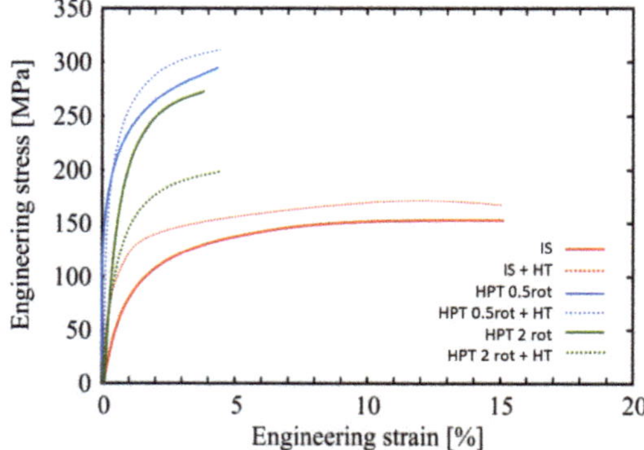

Figure 13. Representative tensile stress–strain curves of IS, IS and heat-treated, HPT-processed (0.5 and 2 rotations) and HPT-processed and subsequently heat-treated Mg5Zn samples. Both heat treatments were done at 100 °C through 24 h.

Table 4. Average values and standard deviations of yield strength σ_{yield}, ultimate tensile strength (UTS) and total engineering strain (ε_{total}) of all alloys in different conditions. HPT deformation was done at 0.5 rotations (0.5) or 2 rotations (2); heat treatments (HT) were done at 100 °C for 24 h.

Alloy	Condition	σ_{yield} (MPa)	UTS (MPa)	ε_{total} (%)
Mg5Zn0.3Ca	IS	70 ± 10	153 ± 10	15 ± 1
	IS + HT	95 ± 15	175 ± 19	14 ± 2
	HPT (0.5)	163 ± 36	221 ± 21	7 ± 2
	HPT (0.5) + HT	303 ± 35	310 ± 55	5 ± 1
	HPT (2)	55 ± 16	55 ± 16	1 ± 0.1
	HPT (2) + HT	209 ± 45	209 ± 45	4 ± 1
Mg5Zn	IS	90 ± 10	174 ± 57	17 ± 3
	IS + HT	155 ± 15	180 ± 28	17 ± 3
	HPT (0.5)	280 ± 10	303 ± 20	5 ± 1
	HPT (0.5) + HT	310 ± 26	329 ± 36	5 ± 1
	HPT (2)	280 ± 22	280 ± 22	6 ± 1
	HPT (2) + HT	185 ± 51	185 ± 51	5 ± 1
Mg0.3Ca	IS	36 ± 20	92 ± 4	18 ± 6
	IS + HT	24 ± 9	57 ± 10	30 ± 10
	HPT (0.5)	210 ± 26	222 ± 28	10 ± 2
	HPT (0.5) + HT	245 ± 5	250 ± 2	5 ± 2
	HPT (2)	202 ± 20	206 ± 17	14 ± 3
	HPT (2) + HT	255 ± 42	258 ± 39	6 ± 1
Mg5Zn0.15Ca	IS	120 ± 5	339 ± 40	17 ± 4
	IS + HT	94 ± 12	180 ± 25	16 ± 2
	HPT (0.5)	246 ± 15	285 ± 19	7 ± 2
	HPT (0.5) + HT	265 ± 5	268 ± 22	6 ± 1
	HPT (2)	306 ± 22	306 ± 22	6 ± 2
	HPT (2) + HT	395 ± 25	418 ± 25	5 ± 1
Mg5Zn0.15Ca0.15Zr	IS	165 ± 15	266 ± 49	15 ± 2
	IS + HT	180 ± 40	204 ± 27	17 ± 4
	HPT (0.5)	230 ± 10	224 ± 82	6 ± 1
	HPT (0.5) + HT	280 ± 10	285 ± 20	5 ± 1
	HPT (2)	303 ± 40	309 ± 24	5 ± 1
	HPT (2) + HT	270 ± 10	292 ± 24	7 ± 2

3.6. Evolution of Young's Modulus

From microhardness indentation tests, values of Young's Modulus E of all materials and treatments can be evaluated too. They lie between 32–50 GPa (Table 5) considering all materials, but within one material they did not change by more than 18% due to the different treatments. Mg5Zn0.3Ca and Mg0.3Ca showed moderate differences of Young's modulus at a strain of γ_T~8. Mg5Zn0.15Ca0.15Zr had a variation of Young's Modulus at a strain of γ_T~2. Pole figures (002) (010) (011) and (102) of the samples with apparent variations of Young's modulus during processing history were determined by XRD, displaying the crystallographic textures of the materials (Figure 14). From the latter, data of Young's Modulus could be simulated (Figure 15 for Mg5Zn0.3Ca): each obtained texture was discretized and represented as a result from 100,000 grains all having the same weight. Based on the orientation of each grain and the compliance constants S_{ij}, the individual Young's modulus of each grain was calculated using Bunge's interdependence [50] of Young's modulus E with the materials' compliances S_{ij}, and with the texture components. Using Voigt's average [51] for the E-values of all grains, an upper limit for the macroscopic E could be calculated. More details of the simulation have been given in publication [10]. The following compliances given in [52] for dilute Mg-alloys were used: $S_{11} = 0.0172738$ GPa^{-1}, $S_{12} = -0.01606$ GPa^{-1}; and $S_{44} = 0.066667$ GPa^{-1}; neglecting influences from constituents Zn and Ca in higher percentages. These influences and the fact of upper-limit calculation may explain some positive, constant offsets of simulated values of E compared to the experimental values.

Table 5. Young's moduli of all alloys in various conditions.

Alloy	Condition	E (GPa)
Mg5Zn0.3Ca	IS	43 ± 1
	IS + HT	42 ± 3
	HPT (0.5)	44 ± 1
	HPT (0.5) + HT	47 ± 2
	HPT (2)	50 ± 5
	HPT (2) + HT	37 ± 3
Mg5Zn	IS	44 ± 4
	IS + HT	46 ± 3
	HPT (0.5)	50 ± 4
	HPT (0.5) + HT	46 ± 2
	HPT (2)	45 ± 3
Mg0.3Ca	IS	32 ± 1
	IS + HT	33 ± 2
	HPT (0.5)	40 ± 2
	HPT (0.5) + HT	43 ± 1
	HPT (2) + HT	33 ± 4
Mg5Zn0.15Ca	IS	44 ± 2
	IS + HT	35 ± 5
	HPT (0.5)	37 ± 3
	HPT (0.5) + HT	32 ± 2
	HPT (2)	45 ± 2
	HPT (2) + HT	46 ± 3
Mg5Zn0.15Ca0.15Zr	IS	40 ± 3
	IS + HT	40 ± 2
	HPT (0.5)	42 ± 2
	HPT (0.5) + HT	37 ± 2
	HPT (2)	44 ± 2
	HPT (2) + HT	45 ± 3

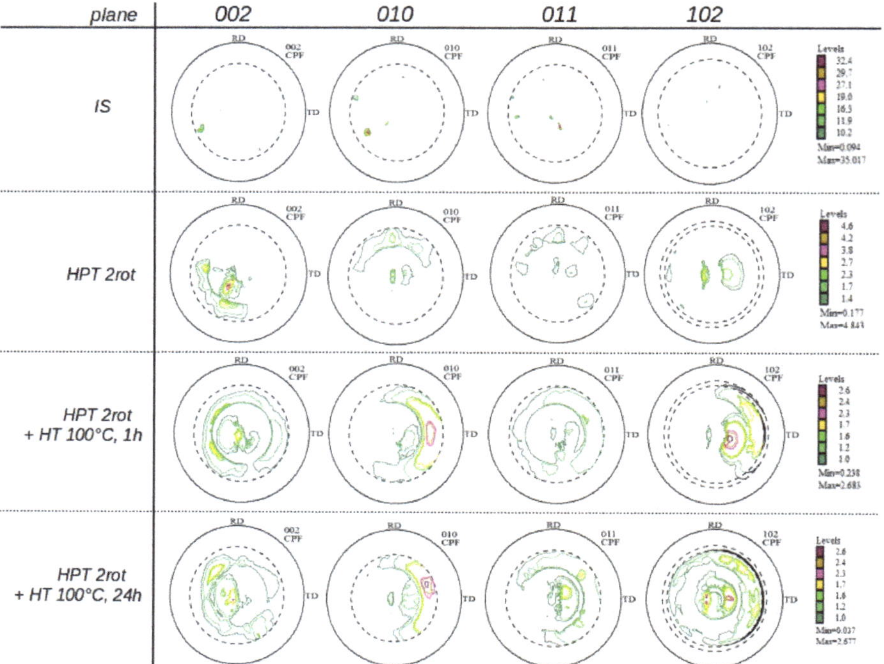

Figure 14. Pole figures displaying crystallographic textures of Mg5Zn0.3Ca samples in the IS (furnace-cooled) and HPT-processed by 2 rotations (torsional strain $\gamma_T \sim 8$) and additional heat treatment at 100 °C for 1 and 24 h.

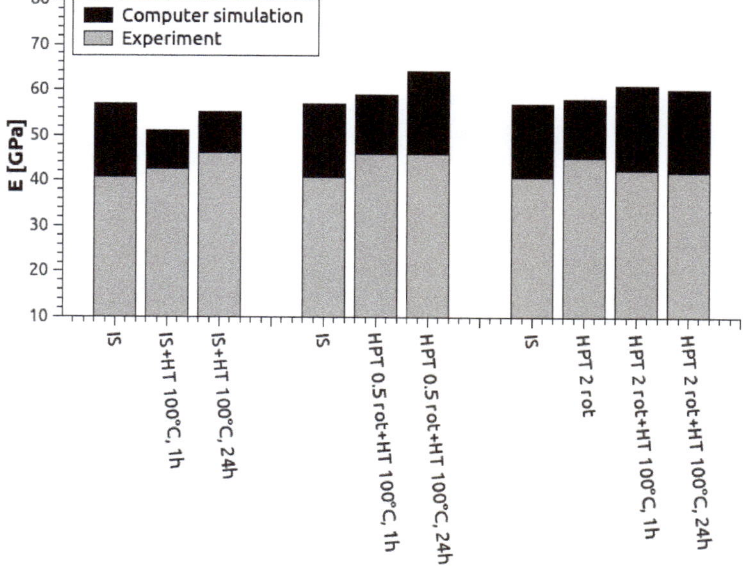

Figure 15. Comparison of measured and simulated values of Young's modulus data obtained from microhardness measurements and texture evaluations for Mg5Zn0.3Ca.

3.7. Electron Microscopy Analysis of Precipitate Structure Evolution

TEM analyses were done in order to understand how the mechanical properties of Mg samples are related to microstructural features, in particular, to different shapes and densities of precipitates that form during heat treatments. The samples were chosen based on the significance of microhardness results shown in Figure 10, for the alloy Mg5Zn0.3Ca, in the following three states:

(1) IS (furnace-cooled) and additionally HPT-deformed at 4 GPa for 0.5 rotations at RT;
(2) IS (furnace-cooled) and HPT-deformed at 4 GPa for 0.5 rotations at RT and heat treated at 100 °C for 24 h;
(3) IS (furnace-cooled) and HPT-deformed at 4 GPa for 0.5 rotations at RT and heat treated at 125 °C for 24 h.

The three precipitate states reveal differences in morphology, size and density of precipitates in the range of 3–100 nm. Primary precipitates, as scattered residuals after annealing, did not show any substantial evolution during the processing history; see Figure 16—first row, SEM 200×.

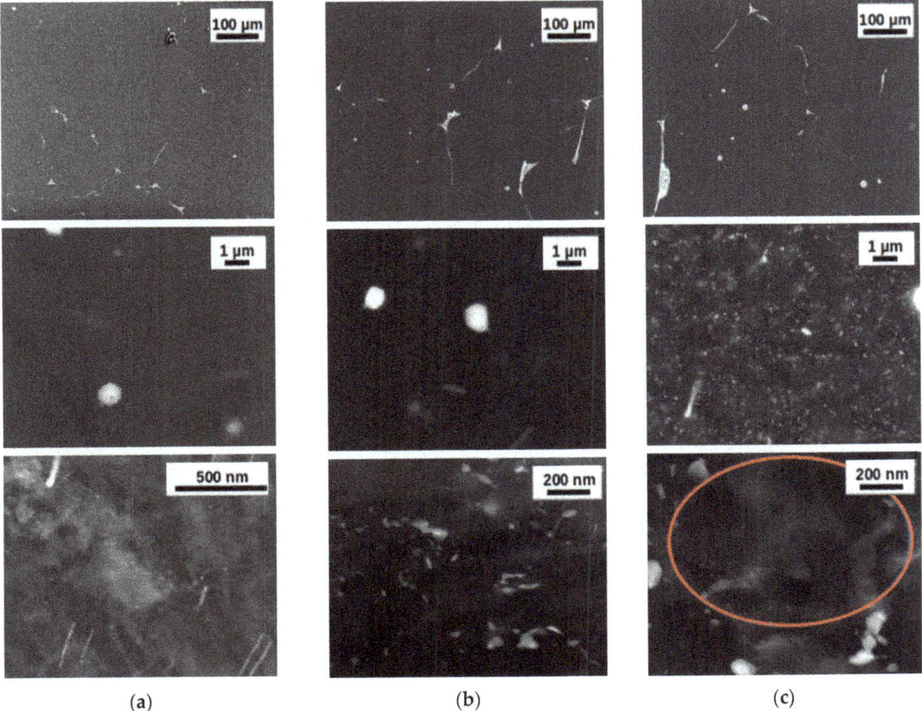

Figure 16. SEM (upper two rows) and STEM (bottom row) images of Mg5Zn0.3Ca samples (**a**) after HPT-processing; (**b**) after HPT and additional heat treatment at 100 °C for 24 h; and (**c**) after HPT and additional heat treatment at 125 °C for 24 h. HPT-processing was carried out by 0.5 rotations under 4 GPa pressure at RT.

Precipitates' structures have been analyzed in all three states (Figure 16—second row, SEM about 10,000×), namely: polyhedral precipitates (P1), elongated precipitates (P2) and thinner phases. The thinner phases are at the boundaries of grains, or follow their contours. P1 and P2 do not change at all during processing, and can be observed in all states (Figure 17). P1 and P2 contain Zn and Ca, as seen in exemplary EDS maps of P1 precipitates (Figure 18).

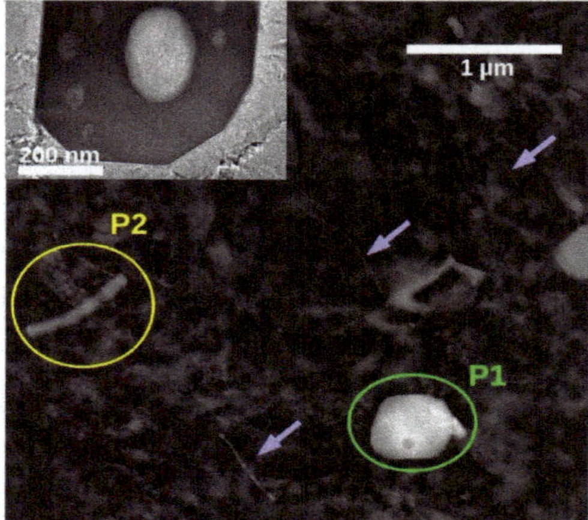

Figure 17. STEM image of Mg5Zn0.3Ca, HPT-processed at 0.5 rotations, showing P1, P2 and thinner precipitates (blue arrows). The TEM image in the top left corner is a close-up of the P1 precipitate. Thinner precipitates are located at the grain boundaries, following their contours.

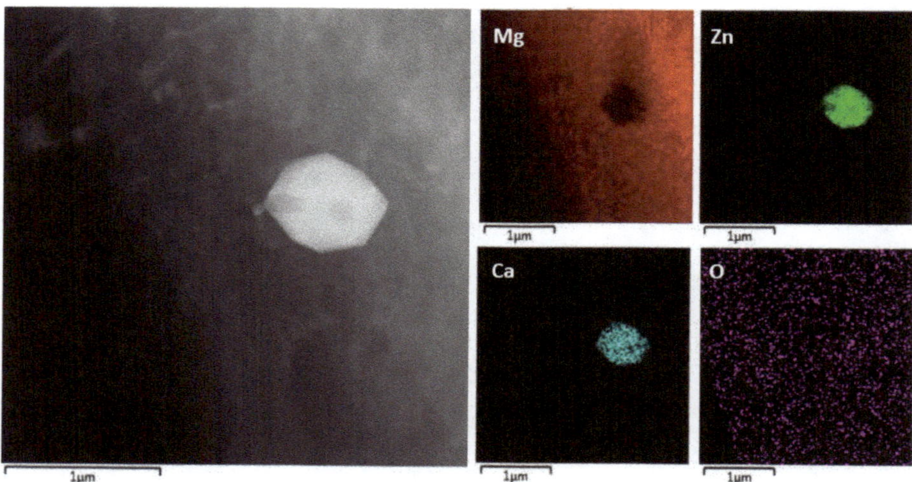

Figure 18. EDS maps of a P1 precipitate in a Mg5Zn0.3Ca sample. These precipitates contain Zn (green) and Ca (blue); Mg (red) was mainly observed in the surrounding matrix. O (pink) appeared in all sample areas; it most likely came from the atmospheric oxidation of prepared TEM foils.

Besides P1, P2 and thinner phases, further precipitates were observed in the heat-treated samples (Figure 16—third row, STEM, about 200,000×). P3-precipitates could be observed in both the heat-treated samples (Figure 19). In the sample heat treated at 100 °C, the P3-precipitates were located at the boundaries between grains and had a size of 20-50 nm. In the sample heat treated at 125 °C, those phases grew in size up to 100 nm. Additionally, ultrafine P4 precipitates (Figure 19) with a size less than 10 nm and up to 50 nm appeared. They were oriented normal to <0001> direction, but further verifications are required. These oriented platelets were not observed in the other two conditions.

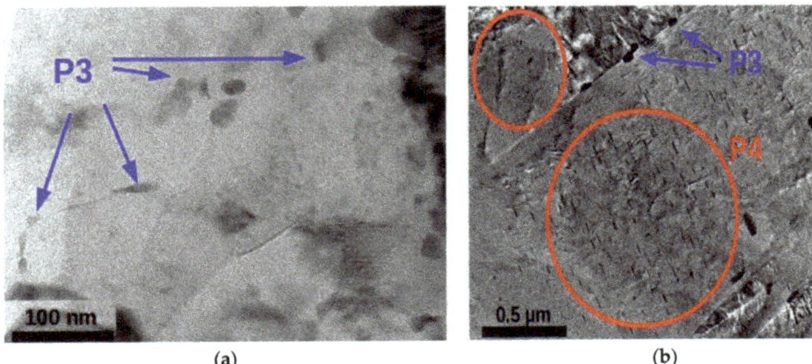

Figure 19. TEM images of the HPT-processed (0.5 rot) and heat-treated Mg5Zn0.3Ca sample (**a**) at 100 °C for 24 h and (**b**) HPT processed and heat treated at 125 °C for 24 h showing P3 and P4 precipitates.

3.8. Determination of SPD-Induced Defect Densities by DSC and XPA

With differential scanning calorimetry (DSC), one can distinguish between lattice defects through selected annealing. Defects always show exothermal peaks, while the endotherm ones indicate phase transformations. Defects with low migration enthalpy Q exhibit their annealing peaks at low temperatures while those with high Q are reflected by peaks at higher temperatures. For the case of defects produced by plastic deformation, it is well established [29,30] that for SPD-processed pure metals, up to three annealing peaks emerge which are dominated by the annealing of (i) single and/or double vacancies at approximately $T = 0.2\ T_m$ (T_m is the melting temperature in K) in the first peak, of (ii) vacancy agglomerates at about $T = 0.3\ T_m$ in the second peak and (iii) of dislocations around $T = 0.3\text{–}0.4\ T_m$ in the third peak, depending on the stress intensity of their arrangement [29]. These peaks can be shifted by up to 100 K to higher temperatures in alloyed metals because of trapping of defects by the alloying atoms.

As a representative for the DSC scans of all Mg alloys investigated, Figure 20 reveals the results for HPT-processed Mg5Zn0.3Ca. Only two peaks emerged in the DSC scans, a peak I between 100–200 °C ($T = 0.4\text{–}0.5\ T_m$), and a peak II around 300–370 °C ($T = 0.6\text{–}0.7\ T_m$). Comparisons with the results mentioned for SPD-processed materials, and especially with those from quenched Mg-alloys only exhibiting vacancy-type peaks [32,33], suggest that peak I indicates the annealing of single and double vacancies, while peak II results from an overlap of the second and third peak indicating the annealing of both vacancy agglomerates and dislocations at very similar temperatures.

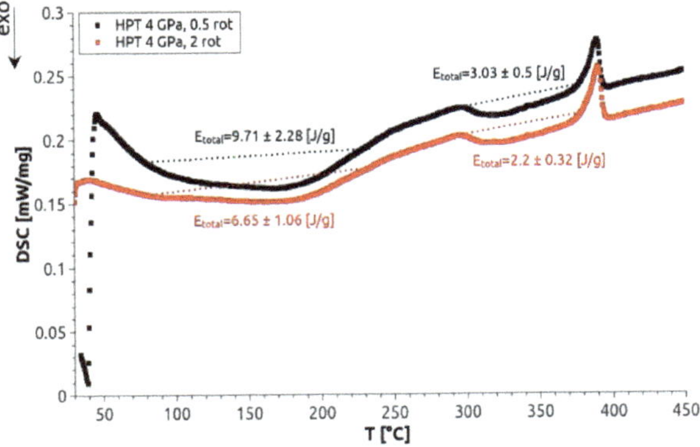

Figure 20. A typical heat flow curve as function of the temperature, for HPT-processed Mg5Zn0.3Ca (black curve—0.5 rotations, red curve—2 rotations) showing two exothermal peaks I and II. The stored energies E_{total} of the HPT-induced defects were evaluated from the areas of the peaks.

For further identification of deformation-induced defects, their migration enthalpies Q (Figure 21) were evaluated by measuring the shift of annealing peak temperatures with different DSC heating rates (Kissinger analysis [38]). Results for activation enthalpies of peak I and peak II were 0.7–1.3 ± 0.1 eV and 1.3–3.8 ± 0.3 eV, respectively, for all alloys Mg5Zn0.3Ca, Mg5Zn, Mg5Zn0.15Ca and Mg5Zn0.15Ca0.15Zr. While peak I remains approximately constant with increasing HPT strain, peak II occasionally does not; for further interpretations, see the discussion section.

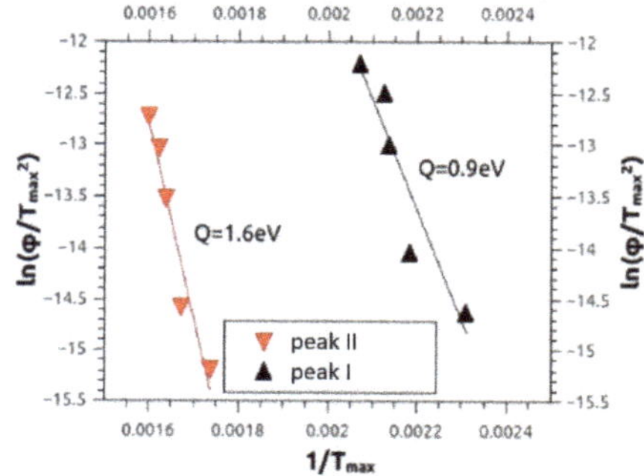

Figure 21. Typical Kissinger plot referring to Equation (4) comprising the peak temperatures T_{max} of peak I and peak II measured by DSC, for HPT-processed Mg5Zn0.15Ca at 4 GPa for 0.5 rotation. The full lines represent the regression to the experimental data; the activation enthalpy Q was calculated from their slopes.

Following the procedure described in papers [30], combined DSC and XPA measurements were done in order to find out the concentrations of vacancies/vacancy agglomerates. The temperature ranges of the two DSC peaks I and II were further investigated by XPA. With constant crystallite size, the broadening of Bragg peak line profiles yields dislocation densities (N) which allow one—by means of Equation (2)—to estimate the energy which is stored in those dislocations (and their arrangements) solely (E_{disl}). The procedure is to subtract E_{disl} from the total stored energy E_{tot} of peak I (or peak II, respectively) in order to obtain E_{vac}, the energy which is to be attributed to the vacancies/vacancy agglomerates. E_{vac} yields the vacancy concentration c_v by applying Equation (3). For all the furnace-cooled and HPT-processed alloys investigated, Table 6 (for peak I) and Table 7 (for peak II) list all these values in sequence.

Table 6. Measured and calculated data for DSC peak I. ρ is the dislocation density, c_v the vacancy concentration.

Sample	E_{total} (J/g)	ρ (m^{-2})	E_{disl} (J/g)	E_{vac} (J/g)	c_v
Mg5Zn0.3Ca	9.7 ± 2	$1.1 \times 10^{14} \pm 10^{13}$	0.06 ± 0.02	9.6 ± 2	$2 \times 10^{3} \pm 1 \times 10^{-4}$
Mg5Zn	5.1 ± 0.5	$2.0 \times 10^{14} \pm 10^{13}$	0.10 ± 0.05	5.1 ± 0.5	$1 \times 10^{-3} \pm 1 \times 10^{-4}$
Mg5Zn0.15Ca	8.2 ± 0.7	$2.0 \times 10^{14} \pm 10^{13}$	0.05 ± 0.01	8.2 ± 0.8	$2 \times 10^{-3} \pm 2 \times 10^{-4}$
Mg5Zn0.15Ca0.15Zr	5.4 ± 0.2	$1.6 \times 10^{14} \pm 10^{13}$	0.02 ± 0.01	5.4 ± 0.3	$1 \times 10^{-3} \pm 9 \times 10^{-5}$
Mg0.3Ca	1.7 ± 0.5	$3.1 \times 10^{14} \pm 10^{13}$	0.05 ± 0.01	1.7 ± 0.5	$5 \times 10^{-4} \pm 1 \times 10^{-4}$

Table 7. Measured and calculated data for DSC peak II. ρ is the dislocation density, c_v the vacancy concentration.

Sample	E_{total} (J/g)	ρ (m^{-2})	E_{disl} (J/g)	E_{vac} (J/g)	c_v
Mg5Zn0.3Ca	3.0 ± 0.5	$4.1 \times 10^{14} \pm 10^{13}$	0.2 ± 0.1	2.7 ± 0.3	$8 \times 10^{-4} \pm 1 \times 10^{-4}$
Mg5Zn	1.5 ± 0.2	$4.4 \times 10^{14} \pm 10^{13}$	0.3 ± 0.1	1.0 ± 0.1	$3 \times 10^{-4} \pm 1 \times 10^{-4}$
Mg5Zn0.15Ca	0.8 ± 0.1	$3.6 \times 10^{14} \pm 10^{13}$	0.4 ± 0.1	0.4 ± 0.1	$1 \times 10^{-4} \pm 9 \times 10^{-5}$
Mg5Zn0.15Ca0.15Zr	1.5 ± 0.01	$3.5 \times 10^{14} \pm 10^{13}$	0.3 ± 0.1	1.1 ± 0.1	$3 \times 10^{-4} \pm 2 \times 10^{-5}$
Mg0.3Ca	0.7 ± 0.1	$2.2 \times 10^{14} \pm 10^{13}$	0.2 ± 0.01	0.5 ± 0.1	$1 \times 10^{-4} \pm 1 \times 10^{-5}$

3.9. Corrosion Tests

The results of corrosion tests are presented in Figure 22. The corrosion rate is plotted as a function of immersion time. It can be seen that the renewal of SBF after 7 and 14 days which removes corrosion products from the liquid and adjust the pH value back to 7.35 resulted in the boosts of hydrogen gas evolution and increases of corrosion rates.

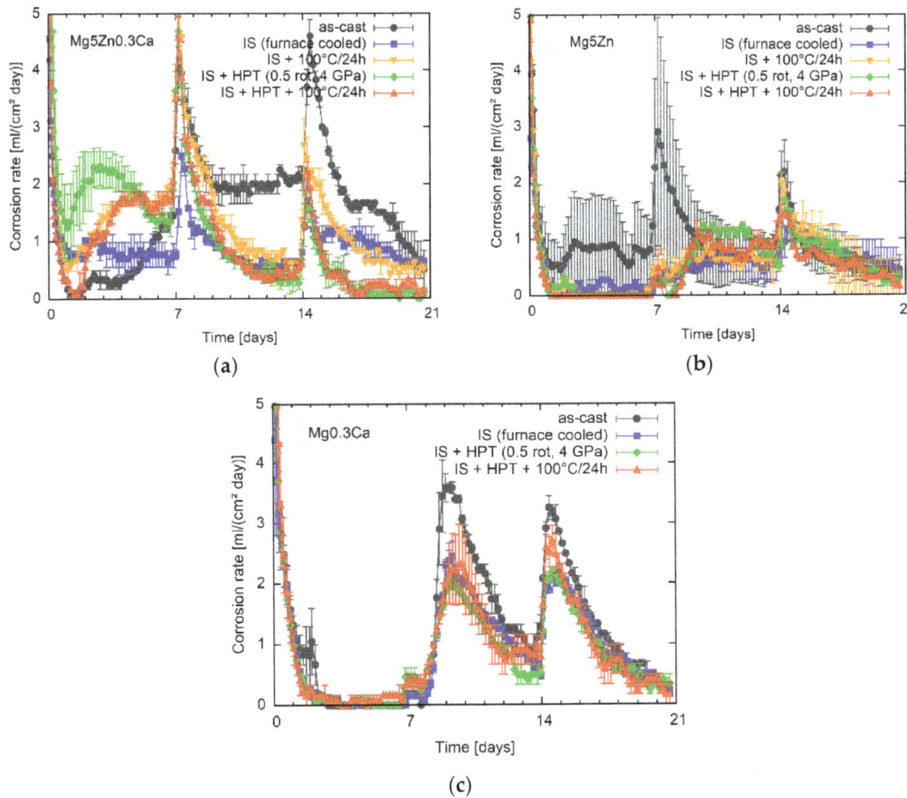

Figure 22. Corrosion rates of (**a**) Mg0.3Ca (**b**) Mg5Zn and (**c**) Mg5Zn0.3Ca in different conditions determined by immersion testing in simulated body fluid at body temperature. The SBF was renewed all 7 days.

It can be concluded from Figure 22 when comparing the curves of the as-cast with those of the IS samples that solid solution treatment decreases the corrosion rate in all the investigated alloys. The difference is least pronounced in the binary Mg-Ca alloy.

HPT-processing increased the corrosion rate of Mg5Zn slightly, while it had no detectable influence on corrosion rate in the case of Mg0.3Ca. In the case of Mg5Zn0.3Ca, HPT resulted in an increased corrosion rate in the first week and a decreased one in the third week.

The heat treatments carried out after solid solution treatment or after HPT-processing have no significant influence on the corrosion rates of Mg0.3Ca and Mg5Zn. In case of the ternary alloy Mg5Zn0.3Ca, the heat treatment after HPT leads to a decrease in the corrosion rate while the same heat treatment in the initial state leads to an increase in the corrosion rate.

The difference between the three alloys is largest in the IS as most of the primary precipitates become dissolved in this condition. It shows that the corrosion rate of Mg5Zn0.3Ca is highest especially in the first week of immersion. Mg5Zn alloy shows the lowest total hydrogen evolution and is less sensitive to the change of SBF. However, microscope images in [47] show that in the case of both Zn containing samples some parts are already missing after three weeks, while the Mg0.3Ca alloy shows a rather homogenous corrosion behavior.

The general appearance of the corrosion behavior is rather independent of the material treatment, meaning that Mg0.3Ca always shows a rather homogenous corrosion behavior and a thick surface layer of degradation products while Mg5Zn shows localized corrosion, and parts of the samples are missing after three weeks. Mg5Zn0.3Ca exhibits a strong surface roughness and some parts of the samples also completely dissolve. The amount of completely dissolved parts in both Zn-containing agrees well with the overall corrosion rates determined by the hydrogen evolution method (Figure 22).

4. Discussion

4.1. The Effect of Solid-Solution Treatment

The initial homogenization treatment of as-cast materials had the intention to re-solutionize primary precipitates, in all the materials considered in this study. Therefore, a homogenization treatment at 450 °C for 24 h was applied for Ca-containing alloys. An exception was made for Mg5Zn, for which a temperature of 350 °C for 12 h was chosen in order to avoid melting. Lower annealing times and temperatures did not remove the primary precipitates, and increasing the annealing time beyond 24 h did not change their density.

In order to stabilize at RT this well-homogenized state and to obtain supersaturated solid-solution condition (which is called the initial state (IS) in the forthcoming text), two types of cooling were used and investigated: furnace-cooling and quenching in ice water. The microhardness values of furnace-cooled alloys Mg5Zn0.3Ca, Mg5Zn, Mg5Zn0.15Ca and Mg5Zn0.15Ca0.15Zr were noticeably lower than those of as-cast samples (Table 3); the microhardness values of the quenched samples, however, were slightly higher.

Our explanation of the latter effect is as follows: As intended, the higher-temperature heating of the materials led to the dissolution of primary precipitates and annihilation of dislocations along with the reduction of the total area of grain boundaries. Nevertheless, the concentration of thermal vacancies was high at the high temperatures. Quenching froze the vacancies, and the higher the cooling rate, the higher was the number of quenched-in vacancies. However, in contrast to the mono-vacancies, di- and multi-vacancies and/or vacancy agglomerates form barriers to the dislocation motion, leading to the so-called quench-hardening [53]. The actually measured rise in microhardness after quenching in comparison with the as-cast state can be thus attributed to the quench-hardening effect.

In contrast, furnace-cooling leads to a decrease in hardness as a result of annealing of thermal vacancies during slow cooling. At the same time, grains start to coarsen and dislocations leave the lattice, both leading to the softening of materials.

SEM images (Figure 6) show that for the alloys Mg5Zn0.15Ca0.15Zr, Mg5Zn0.15Ca and Mg0.3Ca, no significant difference between the IS (furnace-cooled) and IS (quenched) exist in terms of primary precipitates. However, for Mg5Zn0.3Ca, it was not possible to completely dissolve Ca in the Mg matrix. Here, furnace-cooling was more effective than quenching for approaching the solid solution supersaturated state, and fewer residual primary precipitates were found.

However, the negative effect of furnace-cooling is reflected in the possible formation of various complex phases, which may not occur in the quenched samples. The nature and the composition of such phases may affect the microstructure and therefore the properties of the alloy. The densities of these precipitates may be difficult to measure, and the determination of parameters that control the formation of certain phases and/or types of precipitates, is not straight-forward [54].

4.2. The Effect of High-Pressure-Torsion

HPT-processing of a material results in significant increases of densities of dislocations, vacancies and agglomerates during or after HPT-processing. As the dislocations partially arrange into grain boundaries, a decrease in grain size causing an increase in strength and/or hardness takes place.

With HPT, a material workpiece can be exposed to very large torsional strains under hydrostatic pressures up to 10 GPa [55,56]. The large hydrostatic pressure suppresses the annihilation of lattice defects and thus provides grain refinements down to several nanometers and even until an amorphous state. Grain sizes in Mg and other alloys HPT-processed at room temperature can reach around 100 nm [57–62].

As already mentioned, during HPT-deformation of the investigated alloys to 0.5 rotation ($\gamma_T\sim20$) (Figure 11) at room temperature, microhardness increased by up to 130% compared to the furnace-cooled IS, while the increase reached only 80% compared to the quenched IS. When deforming the samples for two rotations and more ($\gamma_T = 20\text{-}100$), microhardness increased further, up to 190% for the furnace-cooled, and up to 100 % for the quenched samples. Again, the samples in the IS (furnace-cooled) showed slightly larger hardness at $\gamma_T > 20$ than the samples in the IS (quenched). This effect may be explained by the fact that in the furnace-cooled samples—in contrast to the quenched ones—some precipitates may exist before HPT, which stimulates the formation of deformation-induced defects, contributing to hardening. Nevertheless, after HPT-processing, both the conditions—furnace-cooled and quenched—reached almost the same hardness level.

4.3. The Effect of Post-HPT Heat Treatments on Strength

The most interesting observation in this work has been the additional substantial strength increase due to a heat treatment, after HPT-processing at RT. Such an effect has been already reported by Horky et al. [33] and Ojdanic et al. [34]. When thermally treating the samples for 1 h (Figure 8), a significant hardness peak was observed for all the furnace-cooled and quenched Mg5Zn0.3Ca and Mg5Zn samples at a temperature of around 100 °C (corresponding to T = 0.4 T_m); only Mg0.3Ca showed a peak at 75 °C. This effect may be attributed to the fact that only in the latter alloy Zn was missing.

In further experiments the peak temperature of 100 °C was set constant, and the annealing time was extended beyond 1 h (Figure 9), which revealed that hardness slightly increases up to 24 h and then decreases. Further heat treatments with the same annealing times of 24 h were done for all alloys for other annealing temperatures too (Figure 10). Again, like in case of only 1 h annealing, the significant hardness peak was found for the annealing treatment at 100 °C.

To sum up, heat treatments at T = 100 °C can increase the hardness of the HPT-processed samples by ~30% (for Mg5Zn0.3Ca and Mg5Zn) and even up to 75% for Mg0.3Ca. The alloys Mg5Zn0.15Ca and Mg5Zn0.15Ca0.15Zr show no response to the heat treatments. For the non-processed samples, hardness increases of up to 50% can be reached, starting from a much lower initial hardness level than in case of all the HPT-processed samples, however.

4.4. The Effect of Precipitates on Strength

In principle, the possible increase of yield strength during heat treatment consists of grain boundary strengthening, vacancy hardening and precipitation hardening, and can be written in a first approximation as

$$\Delta\sigma_{total} = \Delta\sigma_{grains} + \Delta\sigma_{vacancies} + \Delta\sigma_{precipitates} \tag{5}$$

As the grain size was observed to stay constant during the heat treatments considered in this work (maximum 125 °C), changes of $\Delta\sigma_{total}$ could only arise from $\Delta\sigma_{precipitates}$ and/or from $\Delta\sigma_{vacancies}$. For the estimation of contribution from precipitation hardening $\Delta\sigma_{precipitates}$ in Equation (5), we assume that all precipitates are non-shearable and are spherical. By use of a modified Orowan Equation (6) [63], i.e.,

$$\Delta\sigma_{precipitates} = M\frac{0.84Gb}{2\pi(1-v)^{1/2}l}\ln\frac{r}{2b} \tag{6}$$

the increase in yield strength $\Delta\sigma_{precipitates}$ can be estimated. In Equation (6), M means the Taylor factor ($M = 4.2$ for random texture Mg [64]), G the shear modulus ($G = 17$ GPa), b the Burgers vector ($b = 0.32$ nm), v the Poisson's ratio ($v = 0.33$), l the average interparticle spacing and r the average radius; for the following estimation, the values for l and r were taken from STEM images shown in Figure 16b (status after HPT + heat treatment) and Figure 16a (status after HPT only). The estimation leads to a difference of the yield strength due to precipitation hardening of $\Delta\sigma_{HPT + HT} - \Delta\sigma_{HPT} = 115 - 92 = 23$ MPa between the two conditions investigated for the furnace-cooled Mg5Zn0.3Ca samples. The measured difference, however, was 140 MPa; therefore, the result of these calculations is that the difference in the precipitation states cannot explain the extensive increase of the yield strength measured. The precipitates only contribute about 16% to the total increase in yield strength.

4.5. The Effect of Vacancy Agglomerates on Strength

During HPT-processing, vacancies and dislocations are introduced to the sample's lattice. With increasing temperature during heat treatments, the vacancies form agglomerates [65]. Disc-shaped agglomerates form on the close-packed basal planes of the hexagonal Mg lattice [66]. The disc collapses if it is large enough and produces a prismatic dislocation loop. The Burgers vector of such a loop is perpendicular to the plane of the loop, and the loop is therefore immobile. The formed loops are exclusively located on the preferred slip planes of Mg and are therefore strong obstacles for the movement of other dislocations.

Examples of hardening due to the agglomeration of deformation-induced vacancies have already been given some years ago for single and polycrystals of hcp materials for moderate deformation strains [67], and recently even for fcc materials [68,69]. With the latter, the increases in microhardness and yield strength were between 5% and 10% and thereby considerably lower than those observed in this and the recent study of Horky et al. [33]. This may be attributed to the hexagonal lattice of Mg, which makes loop hardening particularly effective because of the coincidence of the loop planes with the preferred dislocation slip plane.

HPT-processed samples provide a significantly higher number of vacancies than non-processed samples in the IS; therefore, the hardness increase during heat treatments is much higher for these samples than for the non-processed ones. The slight hardness increase of the samples in IS is mainly caused by the comparably low number of precipitates (Figure 8) [34].

One interesting fact is that Mg0.3Ca also shows a hardness increase during heat treatments. The peak temperature is around 75 °C and it represents a thermally-induced strength increase of 74% for the HPT-processed sample, which is significantly higher than those measured for Mg5Zn0.3Ca and Mg5Zn. For the other samples, heat treatments contributed ~30% to the increase of hardness. Precipitates probably form in the Mg0.3Ca alloy during heat treatments, but further TEM analyses need to be done.

Previous assumptions were that Zn atoms act as trapping sites for vacancies, but the experimental data on Mg0.3Ca show that Zn alone cannot be responsible for the hardness increase due to trapping-induced vacancy agglomeration. Samples with Ca content of 0.15% (Figure 10) showed almost no increase in hardness during heat treatments when furnace-cooled. In this case the vacancies, induced by HPT deformation, may have stayed single and did not agglomerate, even though there was a Zn content of 5%; this is indicated by the vacancy concentrations c_v for the DSC peaks I and II shown in Tables 6 and 7, respectively. Obviously, the presence of Ca favors the formation of vacancy agglomerates.

As already reported in the previous Section 3, DSC and XPA measurements of furnace-cooled/HPT-processed/heat-treated samples show very large concentrations of vacancies up to ~10^{-3}; (Tables 6 and 7). These start to migrate/agglomerate/anneal at the very temperature (70–100 °C) where the largest values of microhardness appear. The present Kissinger analyses confirm this conclusion as they exhibit an enthalpy between $Q(I)$ = 0.7–1.3 eV, which agrees well with literature values of vacancy migration enthalpies for Mg and Mg alloys, being 0.8–1 eV [37].

Moreover, it is well known from literature [30,68] that temperatures of peaks representing the annealing of single or double vacancies do not shift with the deformation degree applied. In contrast to that, the peak of dislocation annealing has been often found to shift to lower annealing temperatures the higher the deformation is. This effect can be understood as stress-assisted annihilation of dislocations because of their increasing stress field intensity due to increasing dislocation density at growing plastic deformation (see, e.g., [70]). In the present cases, variations of $Q(II)$ were indeed found (Figure 23a,c) although occasionally at higher annealing temperatures as well. From this point of view, it is not surprising that the span of measured $Q(II)$ = 1.3–3.8 ± 0.3 eV for dislocation annealing was much larger than that for vacancy annealing, $Q(I)$. To substantiate these findings, a comparison with the dependence of annealing peak temperature on the applied deformation strain is helpful, as those values are known to be more accurate than those of activation enthalpies. Here, no such variations could be observed (Figure 23b,d), indicating that there may have been defects other than dislocations, too. When considering the evaluation of defect densities in this peak, vacancy concentrations of the order 10^{-4} could still be found, representing a substantial proportion of defects found in this peak II [30].

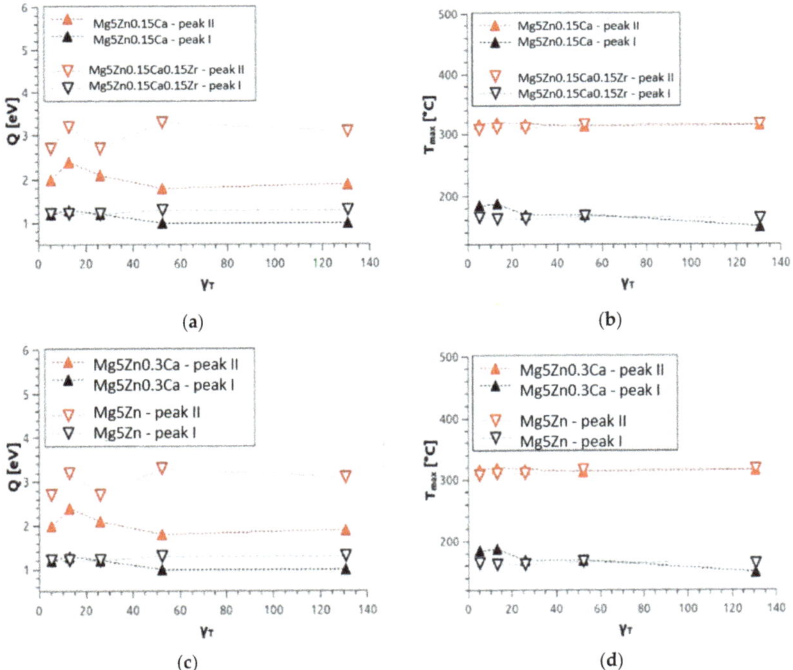

Figure 23. (**a**) Activation enthalpies Q as evaluated from DSC scans and peak temperatures T_{max} measured by DSC for different HPT-processed alloys: (**a,b**) Mg5Zn0.15Ca (full symbols) and Mg5Zn0.15Ca0.15Zr (open symbols), and (**c,d**) Mg5Zn0.3Ca (full symbols) and Mg5Zn (open symbols), all alloys as a function of torsional strain γ_T. The red symbols represent the migration enthalpies of dislocation-type and agglomerate defects (peak II) and the black ones those of single/double vacancies (peak I). Errors are about 0.1 eV for Q, and 5 °C for T_{max}.

A theoretical description of vacancy loop hardening (increase in yield strength $\Delta\sigma_{loops}$) of hexagonal metals is given by Kirchner [71]:

$$\Delta\sigma_{loops} = \frac{Gb}{k}N^a d^{3a-1} \tag{7}$$

with N being the loop density (number of loops/m^3); d the average loop diameter; and a and k constants that depend on the ratio of loop distance ($\lambda = N^{-1/3}$) to diameter d. G and b represent the shear modulus and the Burgers vector, respectively. For a ratio $\lambda/d > 10$, the constants amount to $a = 1/2$ and $k = 0.122$; otherwise, $a = 4/3$ and $k = 0.001$. According to Equation (7), the strengthening potential of loops is higher if the loop density is larger.

Although we do not know the number and the size of the loops in our HPT-processed and heat-treated samples, we can estimate the amount of loop hardening via Kirchner's equation (Equation (7)) by inserting the previously determined values for the vacancy concentration of HPT-processed materials and by an assumption for d. The number of vacancies per loop (vac_{loop}) and the loop density N when assuming circular loops are given by

$$vac_{loop} = \frac{d^2\pi}{4b^2} \tag{8}$$

$$N = \frac{c_V}{vac_{loop}} \tag{9}$$

where c_v is the concentration of vacancies assuming that all of them form loops. For simulations of the yield strength using Equation (7), the determined vacancy concentrations (Tables 6 and 7) were used for each alloy, and average loop diameters of 10–100 nm were assumed. Figure 24 shows the calculated dependence of theoretical yield strength on measured vacancy concentration for all alloys, samples in the IS (furnace-cooled) and HPT-processed at room temperature.

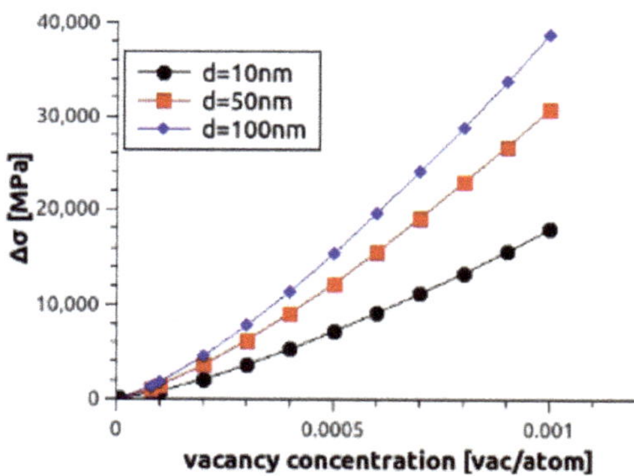

Figure 24. Illustration of the increase of yield strength as a function of vacancy concentration for the case of Mg5Zn alloy, samples of which were in the (IS (furnace-cooled) + HPT-processed) state.

The theoretical model predicts tremendously high increases in yield strength of up to 98 GPa (at $c_v = 2 \times 10^{-3}$ being the highest measured vacancy concentration), for the furnace-cooled materials Mg5Zn0.3Ca and Mg5Zn0.15Ca. The lowest increase was found for the Mg0.3Ca alloy with around 16 GPa. Mg5Zn0.15Ca0.15Zr and Mg5Zn showed increases of around 39 GPa.

From a first perspective, those strength increases predicted by the model are far higher than the experimentally measured ones: On the Mg5Zn0.3Ca alloy processed by HPT at low temperatures, one can conclude that from the experimentally determined thermally-induced increase of yield strength of 115 MPa, around 23 MPa were caused by precipitates. This means that about 92 MPa represents hardening from vacancy agglomerates.

Further simulations of strength increases due to loop hardening along Kirchner's model were done assuming fixed loop diameters at various fixed vacancy concentrations (Figure 25). The shaded area represents the field which spans all strength increases measured by tensile tests after HPT and heat treatment, for all materials investigated. The lines and full points represent fixed values of vacancy concentrations (lines) and fixed values of loop diameters (points). In addition to the estimations reported above, it is here even more evident that vacancy concentrations of the order of 10^{-5} at maximum lead to the increased values of tensile strength measured. It means that the vacancy concentrations 10^{-3} measured by DSC and XPA are far too high to account for these strength increases. Obviously, a major part of the vacancies does not contribute to hardening as they were not part of agglomeration and still stay single.

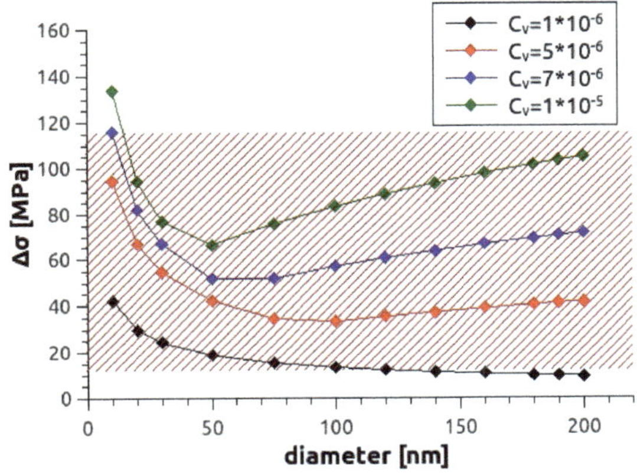

Figure 25. Increase of strength as a function of loop diameter at various vacancy concentrations. The red shaded area represents the values all measured vacancy-induced strength increases found in all materials which were investigated in this work.

4.6. The Effects of Processing Routes on Corrosion Behavior

Comparing now the corrosion properties of all alloys with all their microstructures, including HPT and thermal treatments at 100 °C, it turned out that none of the treatments that produced vacancies/vacancy agglomerates, dislocations and grain boundaries, had a significant impact on corrosion properties. Even a 24 h heat treatment following the HPT-processing did not change the corrosion rate or led to its increase. It can be concluded that the corrosion rate is much more affected by the composition of the alloy. As a side result, even the implementation of IS state out of the as-cast state *decreased* the corrosion rates of all materials studied here. It is expected that the dissolution of primary precipitates decreases the corrosion rate by reducing micro-galvanic corrosion effects.

4.7. The Effects of Processing Routes on the Evolutions of Texture and Young's Modulus

In general, the microindentation measurements of Young's modulus E showed moderate changes of E between the IS and all further treatments, being 18 GPa between all materials, and 14 GPa at maximum within the same material. Apart from a general positive offset of the simulated data compared to the measured ones because of an upper-limit calculation, the majority of simulations followed the measured values of E, at least in all cases of HPT-deformation: E slightly increased with increasing torsional strain γ_T, which increased the intensity of shear texture; the latter (and accordingly also E) reached saturation at highest γ_T where steady state deformation sets in. Thermal treatment, however, should not affect the texture and/or E as long as only recovery processes occur. On the other hand, recrystallization processes must be involved when E changes during thermal treatment, and when this change is also reflected in the simulation, like in the cases of (IS and heat-treated) samples shown in Figure 15. When the simulation does not follow the measured change of E during thermal treatment (e.g., as seen during long-time thermal treatment of HPT 0.5 rot sample) we must suppose the thermally-induced formation of a new phase which is perhaps not registered by XRD due to its small volume fraction. At this point, we conclude the discussion because, as already mentioned, the measured variations in E did not exceed 50 GPa which is still close to that of bone (E = 10–30 GPa) [72]; thus the requirements to avoid the stress-shielding effect [1] in implant applications were still fulfilled.

4.8. Tensile tests

When comparing the stress–strain curves of the alloys in the IS (furnace-cooled)/low-temperature HPT-processing for 0.5 rotations (Figure 13), it can be seen that ductility was markedly reduced in all the HPT-processed samples. After two or more rotations, the material became very brittle, as the failure occurred right after the elastic limit. After additional heat treatment, the samples remained very brittle. Since thermal treatments primarily launched the formation of vacancy agglomerates, it is evident that those were responsible for this brittleness observed.

The values of ultimate tensile strength (*UTS*) derived from the tensile tests suggest comparing them with the Vickers microhardness values (*HV*) as, according to literature [70,73], the *UTS* can be related to *HV* as

$$HV = m \times UTS \tag{10}$$

with *m* as the so-called "Tabor factor." For continuum materials or at least highly isotropic ones, *m* = 3, as the result of the theory of stress distribution of a force on a semi-infinite body [73]. Inserting now our measured data for *UTS* and *HV*, *m* turns out to be *m* = 4.2 with a relatively small standard deviation of ± 0.5 confirming the validity of (11). The strong deviation to the continuum value *m* = 3 can be explained by the fact that Mg alloys are typical examples of highly anisotropic materials. Therefore, another constant value *m* of Equation (10) is expected at least unless the materials exhibit strong textures [74].

5. Summary and Conclusions

The present study showed methods for the optimization of various MgZnCa alloys concerning their strength and corrosion properties. Several Mg alloys (Mg5Zn0.3Ca, Mg5Zn0.15Ca, Mg5Zn0.15Ca0.15Zr) have been investigated with respect to Vickers hardness, Young's modulus, ductility and corrosion properties, and the results were compared to investigations of two binary alloys, Mg5Zn and Mg0.3Ca. The combined application of severe plastic deformation by HPT and heat treatments at low temperatures provided the production of both special intermetallic precipitates and vacancy agglomerates and thus tremendous increases of the alloy's strength. Investigations were done using electron microscopes, microhardness and nanoindentation testing, a microtensile testing facility, XRD and DSC.

Homogenization of all materials was followed by furnace cooling; the fraction of primary precipitates could be reduced to below 1%, their size being close to 1 nm. The main results of our investigations are the following:

(1)   Microhardness increases of up to 250% after furnace cooling, further processing by HPT and/or heat treatment could be reached.

(2)   After those treatments, SEM and STEM investigations revealed complex precipitates, which contribute only ~16% to the hardness increase; this is the result of estimations applying the Orowan equation for precipitation hardening.

(3)   Trying for the first time quantitative analyses of HPT-induced defects, DSC and XPA measurements were undertaken which showed very large concentrations of vacancies (up to ~10−3;) in the furnace-cooled/HPT-processed/heat-treated samples. These started to migrate/agglomerate/anneal at the very temperature (70–100 °C) at which the largest microhardness appeared. Kissinger analyses confirmed that conclusion, as they exhibit a vacancy migration enthalpy between Q = 0.7–1.3 eV which agrees well with literature values for Mg and Mg alloys.

(4)   Theoretical calculations using Kirchner's model indicated that about 1% of the HPT-induced vacancies formed vacancy agglomerates which could account for the significant thermally-induced hardness increases.

(5)   Tensile tests showed that the samples were rather brittle due to the high number of vacancies after HPT deformation and heat treatment. Elongations did not exceed 5%.

(6)   The Young's modulus varied slightly during the processing history because of deformation, thermal treatment and second phase formation due to the evolutions of deformation textures and precipitates, but still remained too small to cause stress shielding; its maximum increase with regard to the homogenized state amounted to 15%.

(7)   Corrosion tests showed that neither the formation of vacancy agglomerates, dislocations and grain boundaries nor that of precipitates has a significant effect on corrosion rate. Mainly, the composition of biomedical binary or ternary Mg-alloys controls the corrosion rate.

Author Contributions: Conceptualization, M.V.; formal analysis, A.O.; investigation, A.O., J.H., M.F. and B.S.; data curation, A.O., J.H., M.J.Z., E.S., D.O. and M.F.; writing—original draft preparation, A.O.; writing—review and editing, M.J.Z., D.O., J.H., B.M. and M.F.; supervision, E.S., M.J.Z. and D.O.; project administration, D.O., M.J.Z., M.V. and S.G.; funding acquisition, D.O., M.J.Z., M.V. and S.G. All authors have read and agreed to the published version of the manuscript.

Funding: This research was funded by the Austrian Science Fund FWF (Fonds zur Förderung der Wissenschaftlichen Forschung Österreich), grant number I2815-N36, and the Slovenian Research Agency (Agencija Raziskovalno Republike Slovenije ARRS), grant number J2-7157.

Acknowledgments: The authors gratefully appreciate financial support from the projects J2-7157 and research program P2-0412 of the Slovenian Research Agency ARRS, and I2815-N36 of the Austrian Science Fund FWF. B.S., A.O. and M.Z. thank the Austrian and Slovenian Exchange Services for mutual visits within projects PL 14/2017 and PL 14/2019.

References

1.   Ridzwan, M.I.Z.; Shuib, S.; Hassan, A.Y.; Shokri, A.A.; Ibrahim, M.M. Problem of stress shielding and improvement to the hip implant designs: A review. *J. Med. Sci.* **2007**, *7*, 460–467.

2.   Atrens, A.; Liu, M.; Zainal Abidin, N.I. Corrosion mechanism applicable to biodegradable magnesium implants. *Mater. Sci. Eng. B* **2011**, *176*, 1609–1636. [CrossRef]

3.   Mordike, B.L.; Ebert, T. Magnesium. *Mater. Sci. Eng. A* **2001**, *302*, 37–45. [CrossRef]

4.   Agnew, S.R.; Nie, J.F. Preface to the viewpoint set on: The current state of magnesium alloy science and technology. *Scr. Mater.* **2010**, *63*, 671–673. [CrossRef]

5.   Mima, G.; Tanaka, Y. Main factors affecting the aging of magnesium-zinc alloys. *Trans. JIM* **1971**, *12*, 76–81. [CrossRef]

6.   Mima, G.; Tanaka, Y. Aging characteristics of magnesium-4 wt percent zinc alloy. *Trans. JIM* **1971**, *12*, 71–75. [CrossRef]

7.   Clark, J.B. Transmission electron microscopy study of age hardening in a Mg-5 wt.% Zn alloy. *Acta Metall.* **1965**, *13*, 1281–1289. [CrossRef]

8.   Orlov, D.; Pelliccia, D.; Fang, X.; Bourgeois, L.; Kirby, N.; Nikulin, A.Y.; Ameyama, K.; Estrin, Y. Particle evolution in Mg–Zn–Zr alloy processed by integrated extrusion and equal channel angular pressing: Evaluation by electron microscopy and synchrotron small-angle X-ray scattering. *Acta Mater.* **2014**, *72*, 110–124. [CrossRef]

9.   Hofstetter, J.; Becker, M.; Martinelli, E.; Weinberg, A.M.; Mingler, B.; Kilian, H.; Pogatscher, S.; Uggowitzer, P.J.; Löffler, J.F. High-Strength Low-Alloy (HSLA) Mg–Zn–Ca alloys with excellent biodegradation performance. *JOM* **2014**, *66*, 566–572. [CrossRef]

10.   Panigrahi, A.; Sulkowski, B.; Waitz, T.; Ozaltin, K.; Chrominski, W.; Pukenas, A.; Horky, J.; Lewandowska, M.; Skrotzki, W.; Zehetbauer, M. Mechanical properties, structural and texture evolution of biocompatible Ti-45Nb alloy processed by severe plastic deformation. *J. Mech. Behav. Biomed. Mater.* **2016**, *62*, 93–105. [CrossRef] [PubMed]

11. Agnew, S.R.; Duygulu, O. A mechanistic understanding of the formability of magnesium: Examining the role of temperature on the deformation mechanisms. *MSF* **2003**, *419*, 177–188. [CrossRef]

12. Alexander, D.J. New methods for severe plastic deformation processing. *J. Mater. Eng. Perform.* **2007**, *16*, 360–374. [CrossRef]

13. Estrin, Y.; Vinogradov, A. Extreme grain refinement by severe plastic deformation: A wealth of challenging science. *Acta Mater.* **2013**, *61*, 782–817. [CrossRef]

14. Valiev, R. Nanostructuring of metals by severe plastic deformation for advanced properties. *Nat. Mater.* **2004**, *3*, 511–516. [CrossRef] [PubMed]

15. Somekawa, H.; Singh, A.; Mukai, T. Synergetic effect of grain refinement and spherical shaped precipitate dispersions in fracture toughness of a Mg-Zn-Zr Alloy. *Mater. Trans.* **2007**, *48*, 1422–1426. [CrossRef]

16. Langdon, T.G. Twenty-five years of ultrafine-grained materials: Achieving exceptional properties through grain refinement. *Acta Mater.* **2013**, *61*, 7035–7059. [CrossRef]

17. Fintová, S.; Kunz, L. Fatigue properties of magnesium alloy AZ91 processed by severe plastic deformation. *J.Mech. Behav. Biomed. Mater.* **2015**, *42*, 219–228. [CrossRef]

18. Dumitru, F.-D.; Higuera-Cobos, O.F.; Cabrera, J.M. ZK60 alloy processed by ECAP: Microstructural, physical and mechanical characterization. *Mater. Sci. Eng. A* **2014**, *594*, 32–39. [CrossRef]

19. Jiang, J.; Wang, Y.; Qu, J. Microstructure and mechanical properties of AZ61 alloys with large cross-sectional size fabricated by multi-pass ECAP. *Mater. Sci. Eng. A* **2013**, *560*, 473–480. [CrossRef]

20. Yuan, Y.; Ma, A.; Jiang, J.; Lu, F.; Jian, W.; Song, D.; Zhu, Y.T. Optimizing the strength and ductility of AZ91 Mg alloy by ECAP and subsequent aging. *Mater. Sci. Eng. A* **2013**, *588*, 329–334. [CrossRef]

21. Kulyasova, O.; Islamgaliev, R.; Mingler, B.; Zehetbauer, M. Microstructure and fatigue properties of the ultrafine-grained AM60 magnesium alloy processed by equal-channel angular pressing. *Mater. Sci. Eng. A* **2009**, *503*, 176–180. [CrossRef]

22. Bryła, K.; Horky, J.; Krystian, M.; Lityńska-Dobrzyńska, L.; Mingler, B. Microstructure, mechanical properties, and degradation of Mg-Ag alloy after equal-channel angular pressing. *Mater. Sci. Eng. C* **2020**, *109*, 110543. [CrossRef] [PubMed]

23. Bryła, K.; Krystian, M.; Horky, J.; Mingler, B.; Mroczka, K.; Kurtyka, P.; Lityńska-Dobrzyńska, L. Improvement of strength and ductility of an EZ magnesium alloy by applying two different ECAP concepts to processable initial states. *Mater. Sci. Eng. A* **2018**, *737*, 318–327. [CrossRef]

24. Qiao, X.G.; Zhao, Y.W.; Gan, W.M.; Chen, Y.; Zheng, M.Y.; Wu, K.; Gao, N.; Starink, M.J. Hardening mechanism of commercially pure Mg processed by high pressure torsion at room temperature. *Mater. Sci. Eng. A* **2014**, *619*, 95–106. [CrossRef]

25. Bonarski, B.J.; Schafler, E.; Mingler, B.; Skrotzki, W.; Mikulowski, B.; Zehetbauer, M.J. Texture evolution of Mg during high-pressure torsion. *J. Mater. Sci.* **2008**, *43*, 7513–7518. [CrossRef]

26. Edalati, K.; Yamamoto, A.; Horita, Z.; Ishihara, T. High-pressure torsion of pure magnesium: Evolution of mechanical properties, microstructures and hydrogen storage capacity with equivalent strain. *Scr. Mater.* **2011**, *64*, 880–883. [CrossRef]

27. Čížek, J.; Procházka, I.; Smola, B.; Stulíková, I.; Kužel, R.; Matěj, Z.; Cherkaska, V.; Islamgaliev, R.K.; Kulyasova, O. Microstructure and thermal stability of ultra fine grained Mg-based alloys prepared by high-pressure torsion. *Mater. Sci. Eng. A* **2007**, *462*, 121–126. [CrossRef]

28. Čížek, J.; Procházka, I.; Smola, B.; Stulíková, I.; Očenášek, V.; Islamgaliev, R.K.; Kulyasova, O.B. The enhanced kinetics of precipitation effects in ultra fine grained mg alloys prepared by high pressure torsion. *Def. Diff. Forum* **2008**, *273–276*, 75–80. [CrossRef]

29. Korznikova, E.; Schafler, E.; Steiner, G.; Zehetbauer, M.J. *Measurements of Vacancy Type Defects in SPD Deformed Ni*; Zhu, Y.T., Langdon, T.G., Horita, Z., Zehetbauer, M.J., Semiatin, S.L., Lowe, T.C., Eds.; The Minerals, Metals & Materials Society (TMS): Warrendate, PA, USA, 2006; pp. 97–102.

30. Setman, D.; Schafler, E.; Korznikova, E.; Zehetbauer, M.J. The presence and nature of vacancy type defects in nanometals detained by severe plastic deformation. *Mater. Sci. Eng. A* **2008**, *493*, 116–122. [CrossRef]

31. Schafler, E.; Steiner, G.; Korznikova, E.; Kerber, M.; Zehetbauer, M.J. Lattice defect investigation of ECAP-Cu by means of X-ray line profile analysis, calorimetry and electrical resistometry. *Mater. Sci. Eng. A* **2005**, *410–411*, 169–173. [CrossRef]

32. Zehetbauer, M. Effects of non-equilibrium vacancies on strengthening. *KEM* **1995**, *97*, 287–306. [CrossRef]

33. Horky, J.; Ghaffar, A.; Werbach, K.; Mingler, B.; Pogatscher, S.; Schäublin, R.; Setman, D.; Uggowitzer, P.J.; Löffler, J.F.; Zehetbauer, M.J. Exceptional strengthening of biodegradable Mg-Zn-Ca alloys through high pressure torsion and subsequent heat treatment. *Materials* **2019**, *12*, 2460. [CrossRef] [PubMed]

34. Ojdanic, A.; Schafler, E.; Horky, J.; Orlov, D.; Zehetbauer, M. Strengthening of a biodegradable Mg–Zn–Ca alloy ZX50 after processing by HPT and heat treatment. In *Magnesium Technology 2018*; Orlov, D., Joshi, V., Solanki, K., Eds.; Springer International Publishing: Cham, Switzerland, 2018; pp. 277–282. ISBN 978-3-319-72332-7.

35. Pippan, R.; Scheriau, S.; Hohenwarter, A.; Hafok, M. Advantages and limitations of HPT: A Review. *MSF* **2008**, *584*, 16–21. [CrossRef]

36. Bever, M.B.; Holt, D.L.; Titchener, A.L. The stored energy of cold work. *Prog. Mater. Sci.* **1973**, *17*, 5–177. [CrossRef]

37. Tzanetakis, P.; Hillairet, J.; Revel, G. The formation energy of vacancies in aluminium and magnesium. *Phys. Status Solidi* **1976**, *75*, 433–439. [CrossRef]

38. Kissinger, H.E. Reaction kinetics in differential thermal analysis. *Anal. Chem.* **1957**, *29*, 1702–1706. [CrossRef]

39. Ungár, T.; Tichy, G.; Gubicza, J.; Hellmig, R.J. Correlation between subgrains and coherently scattering domains. *Powder Diffr.* **2005**, *20*, 366–375. [CrossRef]

40. Schafler, E.; Zehetbauer, M. Characterization of nanostructured materials by X-ray line profile analysis. *Rev. Adv. Mater. Sci.* **2005**, *10*, 28–33. [CrossRef]

41. Ribárik, G.; Gubicza, J.; Ungár, T. Correlation between strength and microstructure of ball-milled Al–Mg alloys determined by X-ray diffraction. *Mater. Sci. Eng. A* **2004**, *387–389*, 343–347. [CrossRef]

42. Convolutional Multiple Whole Profile fitting Main Page. Available online: http://csendes.elte.hu/cmwp/ (accessed on 27 March 2020).

43. Müller, L.; Müller, F.A. Preparation of SBF with different HCO3- content and its influence on the composition of biomimetic apatites. *Acta Biomater.* **2006**, *2*, 181–189. [CrossRef]

44. Kirkland, N.T.; Birbilis, N.; Staiger, M.P. Assessing the corrosion of biodegradable magnesium implants: A critical review of current methodologies and their limitations. *Acta Biomater.* **2012**, *8*, 925–936. [CrossRef] [PubMed]

45. Thomas, S.; Medhekar, N.V.; Frankel, G.S.; Birbilis, N. Corrosion mechanism and hydrogen evolution on Mg. *Curr. Opin. Solid State Mater. Sci.* **2015**, *19*, 85–94. [CrossRef]

46. Orlov, D.; Reinwalt, B.; Tayeb-Bey, I.; Wadsö, L.; Horky, J.; Ojdanic, A.; Schafler, E.; Zehetbauer, M. Advanced immersion testing of model mg-alloys for biomedical applications. In *Magnesium Technology 2020*; Jordon, J.B., Miller, V., Joshi, V.V., Neelameggham, N.R., Eds.; Springer Nature: Berlin/Heidelberg, Germany, 2020; pp. 235–242. ISBN 978-3-030-36646-9.

47. Steiner Petrovič, D.; Mandrino, D.; Šarler, B.; Horky, J.; Ojdanic, A.; Zehetbauer, M.J.; Orlov, D. Surface analysis of biodegradable mg-alloys after immersion in simulated body fluid. *Materials* **2020**, *13*, 1740. [CrossRef]

48. Cao, J.D.; Weber, T.; Schäublin, R.; Löffler, J.F. Equilibrium ternary intermetallic phase in the Mg–Zn–Ca system. *J. Mater. Res.* **2016**, *31*, 2147–2155. [CrossRef]

49. Kirkland, N.T.; Birbilis, N. *Magnesium Biomaterials: Design, Testing and Best Practice*; Springer: Berlin/Heidelberg, Germany, 2014.

50. Bunge, H.-J. Über die elastischen konstanten kubischer materialien mit beliebiger textur. *Kristall und Technik* **1968**, *3*, 431–438. [CrossRef]

51. Voigt, W. Theoretische studien über die elastizitätsverhältnisse der kristalle. *Abhandlungen der Königlichen Gesellschaft der Wissenschaften in Göttingen* **1887**, *34*, 1–155.

52. Long, T.R.; Smith, C.S. Single-crystal elastic constants of magnesium and magnesium alloys. *Acta Metall.* **1957**, *5*, 200–207. [CrossRef]

53. Kimura, H.; Maddin, R. *Quenching Hardening in Metals*; American Elsevier Publishing Co., Inc.: New York, NY, USA, 1971.

54. Sundman, B.O.; Ansara, I. The Gulliver–Scheil method for the calculation of solidification paths. In *The SGTE Casebook: Thermodynamics at Work*, 2nd ed.; Hack, K., Ed.; Woodhead Pub: Cambridge, UK; Boca Raton, FL, USA, 2008; pp. 343–346. ISBN 978-1-84569-215-5.

55. Pippan, R. *Bulk Nanostructured Materials*; Zehetbauer, M.J., Zhu, Y.T., Eds.; Wiley-VCH: Weinheim, Germany, 2009; p. 217.

56. Zehetbauer, M.; Grössinger, R.; Krenn, H.; Krystian, M.; Pippan, R.; Rogl, P.; Waitz, T.; Würschum, R. Bulk nanostructured functional materials by severe plastic deformation. *Adv. Eng. Mater.* **2010**, *12*, 692–700. [CrossRef]

57. Zhilyaev, A.; Langdon, T.G. Using high-pressure torsion for metal processing: Fundamentals and applications. *Prog. Mater. Sci.* **2008**, *53*, 893–979. [CrossRef]

58. Lee, S.; Edalati, K.; Horita, Z. Microstructures and mechanical properties of pure V and Mo processed by high-pressure torsion. *Mater. Trans.* **2010**, *51*, 1072–1079. [CrossRef]

59. Lee, S.; Horita, Z.; Hirosawa, S.; Matsuda, K. Age-hardening of an Al–Li–Cu–Mg alloy (2091) processed by high-pressure torsion. *Mater. Sci. Eng. A* **2012**, *546*, 82–89. [CrossRef]

60. Huang, Y.; Maury, N.; Zhang, N.X.; Langdon, T.G. Microstructures and mechanical properties of pure tantalum processed by high-pressure torsion. In *IOP Conference Series: Materials Science and Engineering*; IOP Science: Bristol, UK, 2014; Volume 63, p. 012100. [CrossRef]

61. Wei, Q.; Zang, H.; Schuster, B.E.; Ramesh, K.T.; Valiev, R.Z.; Kecskes, L.J.; Dowding, R.J.; Magness, L.; Cho, K. Microstructure and mechanical properties of super-strong nanocrystalline tungsten processed by high-pressure torsion. *Acta Mater.* **2006**, *54*, 4079–4089. [CrossRef]

62. Furuta, T.; Kuramoto, S.; Horibuchi, K.; Ohsuna, T.; Horita, Z. Ultrahigh strength of nanocrystalline iron-based alloys produced by high-pressure torsion. *J. Mater. Sci.* **2010**, *45*, 4745–4753. [CrossRef]

63. Embury, J.D. Plastic flow in dispersion hardened materials. *Metall. Mater. Trans. A* **1985**, *16*, 2191–2200. [CrossRef]

64. Shen, J.H.; Li, Y.L.; Wei, Q. Statistic derivation of Taylor factors for polycrystalline metals with application to pure magnesium. *Mater. Sci. Eng. A* **2013**, *582*, 270–275. [CrossRef]

65. Hull, D.; Bacon, D.J. *Introduction to Dislocations*; Butterworth-Heinemann: Oxford, UK, 2001.

66. Hampshire, J.M.; Hardie, D. Hardening of pure magnesium by lattice defects. *Acta Metall.* **1974**, *22*, 657–663. [CrossRef]

67. Zehetbauer, M.; Seumer, V. Cold work hardening in stages IV and V of F.C.C. metals—I. Experiments and interpretation. *Acta Metall. Mater.* **1993**, *41*, 577–588. [CrossRef]

68. Cengeri, P.; Kerber, M.B.; Schafler, E.; Zehetbauer, M.J.; Setman, D. Strengthening during heat treatment of HPT processed copper and nickel. *Mater. Sci. Eng. A* **2019**, *742*, 124–131. [CrossRef]

69. Su, L.H.; Lu, C.; Tieu, A.K.; He, L.Z.; Zhang, Y.; Wexler, D. Vacancy-assisted hardening in nanostructured metals. *Mater. Lett.* **2011**, *65*, 514–516. [CrossRef]

70. Zhang, P.; Li, S.X.; Zhang, Z.F. General relationship between strength and hardness. *Mater. Sci. Eng. A* **2011**, *529*, 62–73. [CrossRef]

71. Kirchner, H. Loop hardening of hexagonal metals. *Z. Met.* **1976**, *67*, 525–532.

72. Radha, R.; Sreekanth, D. Insight of magnesium alloys and composites for orthopedic implant applications—A review. *J. Magnes. Alloys* **2017**, *5*, 286–312. [CrossRef]

73. Tabor, D. *The Hardness of Metals*; Oxford University Press: Oxford, UK, 2000.

74. Khodabakhshi, F.; Haghshenas, M.; Eskandari, H.; Koohbor, B. Hardness–strength relationships in fine and ultra-fine grained metals processed through constrained groove pressing. *Mater. Sci. Eng. A* **2015**, *636*, 331–339. [CrossRef]

Article

Mechanical Properties of Ti-15Mo Alloy Prepared by Cryogenic Milling and Spark Plasma Sintering

Anna Veverková [1,*], Jiří Kozlík [1], Kristína Bartha [1], Tomáš Chráska [2], Cinthia Antunes Corrêa [1] and Josef Stráský [1]

1 Department of Physics of Materials, Charles University, 121 16 Prague, Czech Republic;
Jiri.Kozlik@seznam.cz (J.K.); bartha.kristina@gmail.com (K.B.); cinthiacac@gmail.com (C.A.C.);
josef.strasky@gmail.com (J.S.)
2 Institute of Plasma Physics, Czech Academy of Sciences, 182 00 Prague, Czech Republic; tchraska@ipp.cas.cz
* Correspondence: annaveverkova4@gmail.com; Tel.: +420-95155-1623

Received: 18 October 2019; Accepted: 26 November 2019; Published: 28 November 2019

Abstract: Metastable β-Ti alloy Ti-15Mo was prepared by cryogenic ball milling in a slurry of liquid argon. Material remained ductile even at low temperatures, which suppressed particle refinement, but promoted intensive plastic deformation of individual powder particles. Repetitive deformation of powder particles is similar to the multidirectional rolling and resembles bulk severe plastic deformation (SPD) methods. Initial and milled powders were compacted by spark plasma sintering. Sintered milled powder exhibited a refined microstructure with small β-grains and submicrometer sized α-phase precipitates. The microhardness and the yield tensile strength of the milled powder after sintering at 850 °C attained 350 HV and 1200 MPa, respectively. Low ductility of the material can be attributed to high oxygen content originating from the cryogenic milling. This pioneering work shows that cryogenic milling followed by spark plasma sintering is able to produce two-phase β-Ti alloys with refined microstructure and very high strength levels.

Keywords: metastable β-Ti alloys; powder metallurgy; cryogenic milling; spark plasma sintering

---

1. Introduction

Metastable β-alloys constitute a specific group of Ti alloys in which none of the α-phase, α′-phase or α″-phase form after quenching from a temperature above the temperature of β-transus (774 °C for Ti-15Mo [1]). β-Ti alloys are widely used in the aircraft industry due to their high specific strength [2]. Utilization of these alloys in biomedicine is also expected [3]. Currently, Ti-6Al-4V is a commonly used alloy for biomedical implants manufacturing, despite vanadium being considered as a toxic element [4]. Therefore, Ti-15Mo, which contains only biocompatible elements, is a perspective biocompatible alternative. Furthermore, β-Ti alloys in β-solution treated condition exhibit lower modulus of elasticity, which is closer to that of a human bone [5].

Mechanical properties of metastable β-Ti alloys can be further improved by thermomechanical treatment. Both microstructure refinement and precipitation of α-phase increase the strength of the alloy. α-phase particles precipitate typically at temperatures in the range of 500–750 °C. During this phase transformation, so-called element partitioning occurs. β-stabilizing elements (molybdenum (Mo), niobium (Nb), vanadium (V), etc.) diffuse from the growing α-phase particles to the surrounding β-matrix [6,7]. In addition to α and β-phase, metastable ω-phase can also occur in some metastable β-alloys including Ti-15Mo alloy. During quenching, so-called ω_{ath} (athermal) forms by a displacive transformation. When annealing at temperatures in range 250–450 °C, ω_{iso} (isothermal) forms by a diffusional transformation [8]. ω-phase also strengthens the material.

Powder metallurgy is an alternative and often more suitable way of producing titanium alloys [9,10]. Due to the possibility of near-net-shape processing, the material waste and costs associated with the material processing are reduced. Spark plasma sintering (SPS) was used as a compaction method in this study. During SPS, the powder is compacted by pulse electric current and the powder particles are joined together by the Joule heat. The sintering therefore occurs primarily at the point of contact of powder particles [11]. Therefore, lower temperatures and shorter times are sufficient for compaction in comparison to other compaction methods. Utilization of SPS for compaction helps therefore to preserve the fine microstructure and restrict the grain growth [11,12].

Elemental powders, master alloys or pre-alloyed powder can be used for sintering. Sintering of elemental powders or master alloys must provide alloying (homogenization) of the material and therefore high sintering temperatures (1200–1700 °C) must be used for processing of metastable β-Ti alloys due to high melting points of β-stabilizing elements and their low diffusivity [2,13]. On the other hand, successful compaction of pre-alloyed powder can be achieved at temperatures comparable to the temperature of β-transus of an alloy [14].

Besides the parameters of sintering, the final bulk material is also affected by the shape, the size and the microstructure of powder particles [14]. Mechanical milling is commonly used to fragment powder particles [15]. In this study, intensive ball milling at cryogenic temperatures was used. Cryogenic temperatures are often used for milling of organic materials to ensure their brittleness [16]. In metallic materials, cryogenic temperatures help to suppress dynamic recovery and recrystallization and to achieve very refined microstructure [17].

Cryogenic milling employing the Szegvari type attritor has been previously applied to commercially pure titanium [18–21]. It was found that titanium remains ductile even at cryogenic temperatures (unless contaminated by nitrogen [21]) and powder particles are not significantly refined [22]. On the other hand, powder particles are significantly deformed by repetitive plastic deformation [15,22,23]. In our previous study we proved that deformation by ball milling resembles the multi-directional forging and causes grain refinement [24]. In this respect, mechanical milling can be regarded as a method of severe plastic deformation (SPD). Various metastable β-Ti alloys have been prepared by SPD methods and improvement of strength was reported [25,26]. Recently the Ti-15Mo alloy was prepared by high pressure torsion (HPT), which caused the grain refinement and an increase of dislocation density [27,28]. On the other hand, HPT samples are very small and thermal stability of ultra-fine grained Ti alloys prepared by SPD methods is limited [29,30]. Similarly to bulk SPD methods, grain size after cryogenic milling can be reduced to tens of nanometers as shown for commercially pure Ti (CP Ti) in [22]. However, the grain size in CP Ti significantly increases after sintering due to recrystallization [24].

Ti-15Mo pre-alloyed powder was milled in liquid argon slurry. Note that liquid nitrogen cannot be used, because N atoms diffuse into the powder during milling causing embrittlement [21,31]. In order to prevent cold welding of the powder to the milling tank, shaft and balls, a process control agent such as stearic acid must be added to it [15,22].

Milled powder was subsequently sintered at temperatures in the range of 750–850 °C for 3 min. For comparison, initial gas atomized powder was sintered using the same parameters. Due to the fact that titanium is a strong gatherer of nitrogen and oxygen, some contamination is unavoidable during powder metallurgy processing [22,32]. Therefore, contamination by oxygen, nitrogen (and possibly by carbon and hydrogen) must be always monitored in order to assess strengthening mechanisms appropriately.

2. Materials and Methods

Ti-15Mo powder was prepared by gas atomization and supplied by TLS Technik GmbH and Co. Spezialpulver KG, Bitterfeld-Wolfen, Germany.

The milling was performed for 4 h at 700 revolutions per minute (RPM) milling speed in Union Process 01-HD (Union Process, Akron, OH, USA) attritor (1400 cm^3) device in liquid argon. Stainless steel balls of a diameter 6.35 mm and ball to powder ratio (BPR) 16:1 were used. 1.8 g of stearic acid was added into 180 g of Ti-15Mo powder as a process control agent to prevent cold welding.

Both gas atomized (initial) and cryo-milled (referred to as milled) powders were compacted by a spark plasma sintering (SPS) device by FCT Systeme GmbH (Rauenstein, Germany). The sintering was performed for 3 min in the temperature range from 750 °C to 850 °C. The temperature was measured by a thermocouple inserted into the graphic die 4 mm from the sintered sample. The powder was compressed by a pressure of 80 GPa and heated 50 °C below the desired temperature always in one minute (heating rates were from 700 °C/min to 800 °C/min depending on the sintering temperature). The sample was subsequently heated up to the desired sintering temperature with the heating rate of 100 °C/min. These two steps were designed to avoid temperature overshooting. The example of the temperature and pressure evolution during sintering at the temperature of 750 °C is shown in Figure 1. The cooling was not actively controlled.

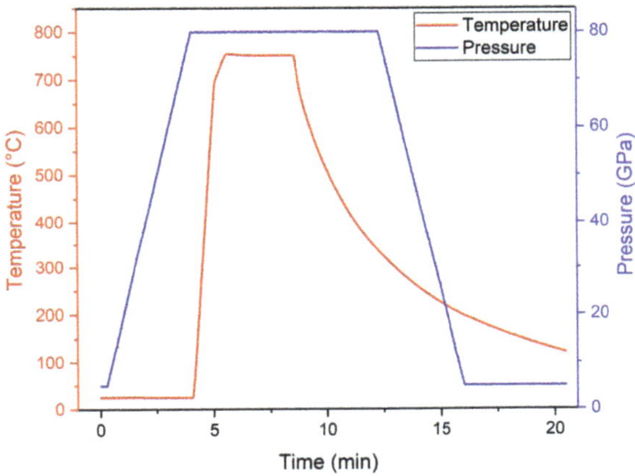

Figure 1. An example of the temperature and pressure evolution during the sintering process.

Scanning electron microscope (SEM) observations were performed with the scanning electron microscope FEI Quanta 200F (FEI, Hillsboro, OR, USA). For this purpose, both initial and milled powders were simply stuck on a conductive foil. Bulk samples for SEM microstructural observations were prepared by standard mechanical grinding and polishing followed by the three step vibratory polishing. Fracture surface after tensile tests were also observed.

The fraction of the α-phase and porosity of samples were determined by image analysis in ImageJ (version 1.52r, Wayne Rasband, Research Services Branch, National Institute of Mental Health, Bethesda, MD, USA). For this purpose, 10 micrographs of size of about 1000 μm^2 from each sample were made by a scanning electron microscope FEI Quanta 200F operated at 10 kV and at 5 kV in case of sintered initial powder and sintered milled powder, respectively. The fraction of the α-phase was determined from electron back-scatter (BSE) micrographs. Porosity of samples sintered from initial powder was determined by image analysis from secondary electron (SE) micrographs. The very low overall porosity of sintered milled powder disallowed its determination by image analysis. The Archimedes method was used instead. The error of this measurement is given only by the precision of the sample weighing, which is 0.5 mg. For comparison, the porosity of the initial powder sintered at 800 °C was measured by both methods. It was found that image analysis underestimates the overall porosity and Archimedes method proved to be more reliable. Porosity of sintered initial powder was therefore calibrated according to the porosity of the initial powder sintered at 800 °C.

Contamination of material by oxygen was determined in powders as well as in selected sintered samples by carrier gas hot extraction (CGHE). The microhardness was measured by the Vickers method (0.5 kgf load, 30 indents per sample) using Qness Q10a (Qness, Golling, Austria) instrument with automatic evaluation of the measurement.

X-ray diffraction (XRD) measurements were performed on a Bruker D8 Advanced diffractometer (Bruker AXS, Karlsruhe, Germany) in Bragg–Brentano geometry using Cu K$_\alpha$ radiation, $\lambda = 1.54051$ Å, variable divergence slits and a Sol-X detector. Bragg–Brentano geometry is an arrangement in which the incident and diffracted beams are focused on a circle with the measured sample in the middle. Diffraction patterns were collected at room temperature in the 2 range from 30° to 130° with a step size of 0.02° and an exposure time of 5 s/step. The patterns were fitted and refined employing le Bail algorithm with a pseudo-Voigt profile using program Jana2006 (V. Petříček, M. Dušek and L. Palatinus, Institute of Physics Academy of Sciences, Prague, Czech Republic).

The position of the flat dog-bone shaped tensile sample in the sintered tablet is shown in Figure 2. The sample was 20 mm long and its gauge length was 5 mm. The total width was 7 mm while the gauge length and thickness were 1.2 mm and 1 mm, respectively. The diameter of the hole for the pin was 2.3 mm. The "bricks" in Figure 2 symbolize arrangement of flat milled powder particles in the tensile sample caused by their stacking in the sintering die [24]. Tensile tests were performed at Instron 5882 machine. The constant crosshead velocity was 0.03 mm/s, resulting in the initial strain rate of $\dot{\varepsilon} = 10^{-4}\text{s}^{-1}$.

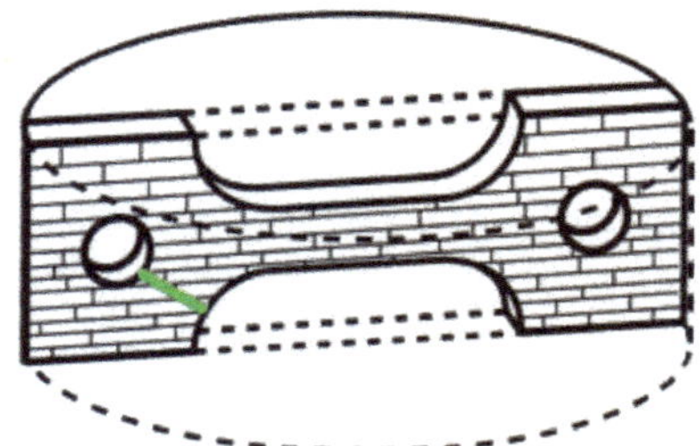

Figure 2. A position of tensile sample in a sintered tablet/cylinder (the green line is described in a discussion of tensile tests below).

3. Results and Discussion

3.1. Cryogenic Milling

The SEM micrograph of gas atomized Ti-15Mo powder is shown in Figure 3a. The initial gas-atomized powder particles are ball-shaped with the size from several µm up to 30 µm. Powder particles after cryomilling are shown in Figure 3b. Powder particles were not refined by cryomilling due to high ductility of titanium and its alloys even at low temperatures [22]. Powder particles were, however, plastically deformed and changed their shape from balls to disks. The size of disk-shaped powder particles is up to around 50 µm.

The contamination by oxygen (shown in Table 1) increased during milling from 0.20 wt. % to 0.78 wt. % despite milling in inert atmosphere of liquid argon. It was previously shown that handling powders in a glove box does not reduce the contamination significantly and therefore it is assumed that contamination origins from the milling [16]. Contamination during milling can be caused both by addition of the process control agent—the stearic acid—and by contamination from the used liquid argon, which may contain some trace amount of oxygen. It has been calculated that adding of 1 wt. % of stearic acid can cause contamination by oxygen by 0.11 wt. %. On the other hand, almost 100 L of 99.999% purity argon was used to mill 180 g of powder. The total content of oxygen in the liquid argon (Ar) corresponds to about 0.5 wt. % of oxygen in the milled powder. Some additional oxygen may be also absorbed.

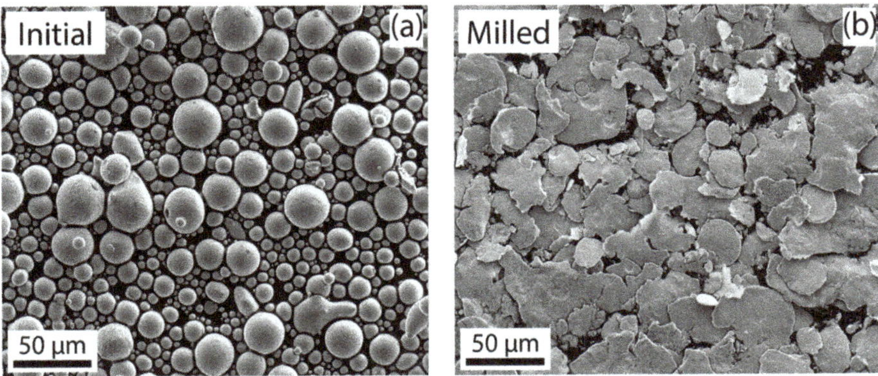

Figure 3. Scanning electron microscopy (SEM) micrographs of (**a**) initial powder and (**b**) milled powder.

Table 1. Contamination of initial and milled powder by oxygen.

Sample	Oxygen (wt. %)
Initial powder	0.20
Milled powder	0.78

3.2. Spark Plasma Sintering

Contamination of samples sintered at 750 °C and 850 °C is shown in Table 2. The content of oxygen in the initial powder slightly increased after sintering, while in the milled powder, it decreased due to the decomposition of stearic acid at high temperatures. The sintering temperature did not have a significant effect on contamination by oxygen for both initial and milled powders.

Table 2. Contamination of sintered samples by oxygen.

Sample	Oxygen (wt. %)
Initial sintered at 750 °C for 3 min	0.267
Initial sintered at 850 °C for 3 min	0.246
Milled sintered at 750 °C for 3 min	0.572
Milled sintered at 850 °C for 3 min	0.601

Figure 4 shows BSE micrographs of sintered powders. Black areas in these BSE figures are α-phase, pores and impurities from polishing. In order to analyze images, these three phenomena were distinguished using SE observation. SE micrographs of the same areas as in Figure 4 are shown in Figure 5. In SE signal, pores are black, impurities from polishing are white and the α-phase is grey.

SEM micrographs of initial powder sintered at 750 °C and at 850 °C are shown in Figures 4 and 5. Initial powder sintered at 750 °C (Figures 4a and 5a) contains a significant amount of α-phase while initial powder sintered at 850 °C (cf. Figures 4b and 5b) contains only a small amount of grain boundary α-phase and black areas in BSE micrographs are mostly impurities from polishing. BSE micrographs of milled powder sintered at 750 °C and at 850 °C are shown in Figure 4c,d, respectively and in Figure 5c,d, respectively. Sintered milled powder has finer microstructure and contains higher amount of α-phase in comparison with the initial one. With increasing sintering temperature the microstructure coarsens and the fraction of α-phase decreases.

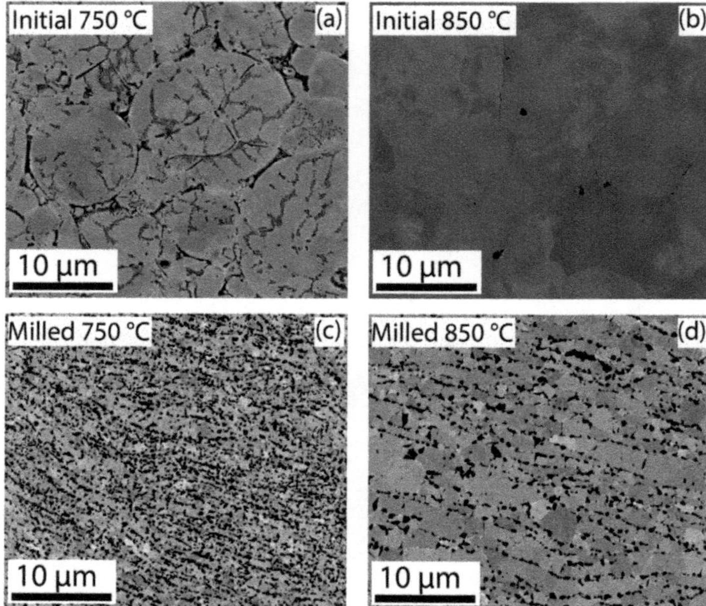

Figure 4. Examples of back-scattered electrons (BSE) micrographs (**a**) initial powder sintered at 750 °C, (**b**) initial powder sintered at 850 °C, (**c**) milled powder sintered at 750 °C and (**d**) milled powder sintered at 850 °C.

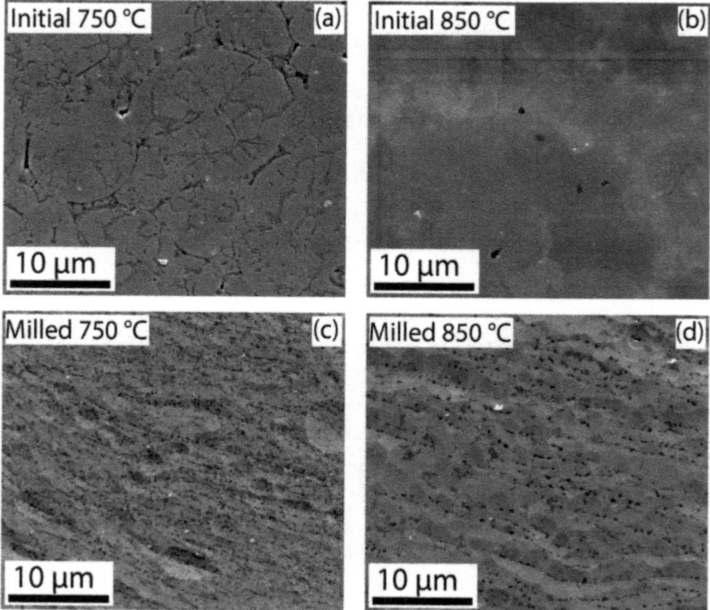

Figure 5. Examples of secondary electrons (SE) micrographs (**a**) initial powder sintered at 750 °C, (**b**) initial powder sintered at 850 °C, (**c**) milled powder sintered at 750 °C and (**d**) milled powder sintered at 850 °C.

The fraction of α-phase was determined by image analysis from SEM observations. Ten micrographs were analyzed for each sample. The results of this analysis are summarized in Figure 6. The conclusions from SEM observations were confirmed by image analysis. All sintered conditions contained a significant fraction of the α-phase, except for the initial powder sintered at 850 °C. As both sintering temperatures of 800 °C and 850 °C were above the reported temperature of β-transus (774 ± 14 °C) [1], α-phase should not precipitate during sintering. However, especially the milled powder was contaminated by oxygen, which is an α-stabilizing element and as such it increases the temperature of β-transus. This effect was studied in [31,33] both in Ti-Mo-Cr alloys and in pure Ti. It was found, that 0.4 wt. % of oxygen may increase the temperature of β-transus by 50 °C. The temperature of β-transus of milled powder can be therefore well above 800 °C. Therefore, in the initial powder sintered at 750 °C and in milled powder sintered at 750 °C and 800 °C, α-phase could precipitate during sintering. In initial powder sintered at 800 °C and 850 °C and in milled powder sintered at 850 °C, α-phase precipitated during cooling. High imposed strain in the milled powder provides preferential nucleation sites (dislocations and grain boundaries) for α-phase nucleation and enhances the diffusion promoting the growth of α-phase precipitates [34]. Both these effects—the higher contamination by oxygen and enhanced precipitation of α-phase due to milling—explain the higher amount of α-phase in the sintered milled powder compared to the initial one.

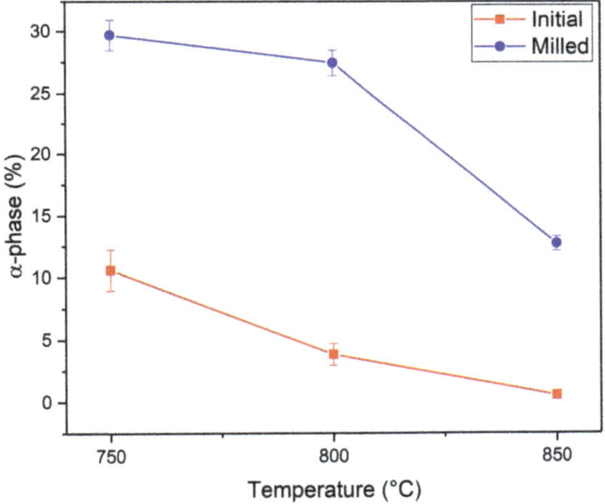

Figure 6. The evolution of α-phase content (vol. %) in sintered samples with sintering temperature.

The fraction of the α-phase decreased with increasing sintering temperatures for both sintered powders. The driving force for α-phase precipitation during sintering decreased as the temperature approached the temperature of β-transus. The amount of α-phase precipitated during cooling was clearly lower at higher temperatures.

The porosity of sintered both initial and milled powders depending on the sintering temperature is shown in Figure 7. Porosity of the initial powder decreased with increasing sintering temperature, while porosity of milled powder remained almost constant at all sintering temperature and was very low even at low sintering temperature. While high sintering temperature of 850 °C was necessary for sintering of the initial powder, the milled powder was well-sintered already at the lowest sintering temperature of 750 °C. It was caused mainly by different shapes of powder particles—the disk-shaped particles after milling are well stacked on each other and have also the higher surface area enhancing the efficiency of sintering as compared to ball particles in the initial powder. As a consequence, lower sintering temperatures were therefore sufficient to obtain fully compacted material. Moreover, the milled powder was better sintered due to enhanced diffusivity in the severely deformed and refined material [34].

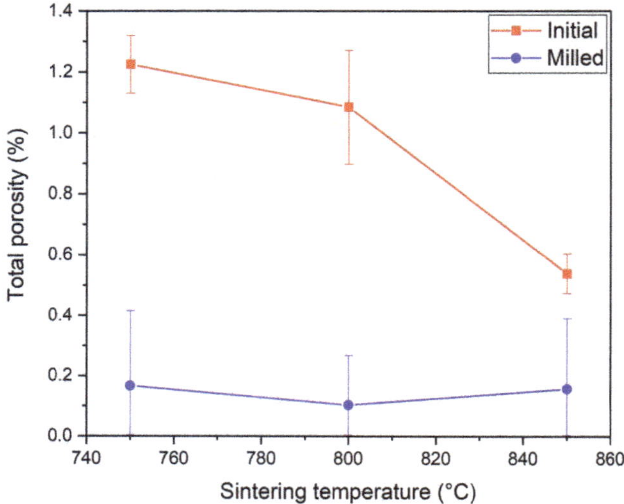

Figure 7. Porosity of sintered initial and milled powder as a function of the sintering temperature.

The dependence of the microhardness on the sintering temperature for both initial and milled powder is shown in Figure 8. Microhardness was affected by residual porosity, by the contamination by oxygen and also by the content of both α-phase and ω-phase. While a decrease of porosity causes an increase in microhardness [19], a decrease in the oxygen content and in the amount of α-phase generally causes a decrease in microhardness [28,35]. The microhardness of sintered initial powder significantly increased with increasing sintering temperature, which was consistent with the decrease of the residual porosity. On the other hand, the amount of α-phase decreased with increasing sintering temperature. However, one might argue that the decrease in the amount of α-phase was compensated by an increase in amount of ω-phase. During cooling after sintering (cf. Figure 1), the Ti-15Mo alloy passes firstly through a region of α-phase precipitation in the temperature region of 750–500 °C [8,36]; and then through a region of ω-phase formation in the range of 450–250 °C [37]. One may therefore assume that ω-phase forms during cooling in the material sintered from the initial powder. On the other hand, the milled powder contained a considerable volume fraction of α-phase. In this case, the remaining β-matrix contained a higher amount of molybdenum due to element partitioning and was therefore comparatively more β-stabilized. Hence, the driving force for ω-formation was lower and the material with the higher amount of α-phase contained the lower amount of the ω-phase and vice versa.

This hypothesis was confirmed by XRD measurement of two conditions: samples sintered at 850 °C from the initial powder and the milled powder. Measured patterns are shown in Figure 9 along with LeBail fits, which allow only qualitative comparison of phase composition. Sintered initial powder contained considerable amount of ω-phase (highlighted by white arrows), while no ω phase was observed in the sintered milled powder. Milled condition contained substantial amount of α phase, which is consistent with SEM observations.

Sintered milled powder had a finer microstructure, higher contamination by oxygen, lower residual porosity and contained higher fraction of α-phase, in comparison with the sintered initial powder. All these effects contributed to the increased microhardness. However, as confirmed by XRD measurements, sintered milled powder did not contain the ω-phase. The microhardness of the sintered milled powder was therefore not enhanced in comparison with the initial powder (sintered at 850 °C) despite its refined microstructure. Coarsening of the microstructure (cf. Figure 5) and the decrease of the amount of α-phase with increasing sintering temperature (cf. Figure 6) did not affect microhardness of milled conditions. The dominant factor affecting microhardness might be increased oxygen content [38].

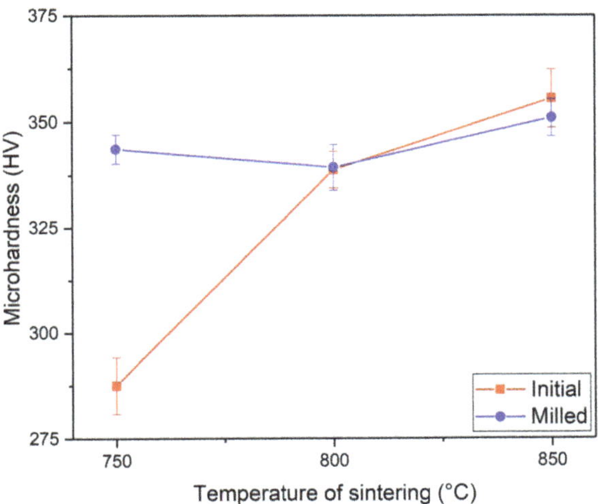

Figure 8. The temperature dependence of microhardness of sintered samples from initial and milled powder.

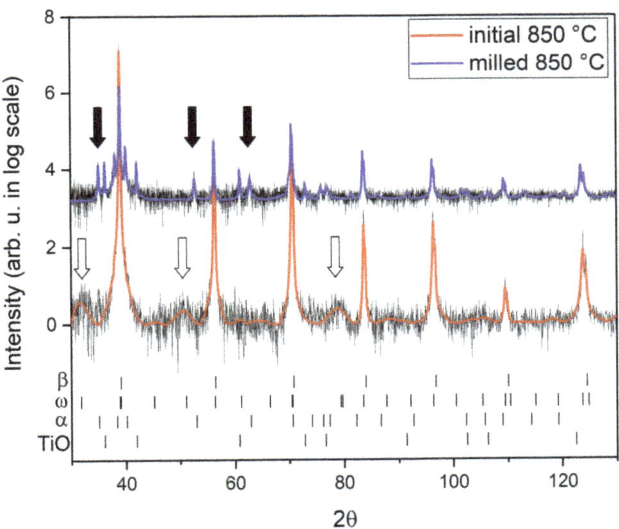

Figure 9. The x-ray diffraction (XRD) patterns of initial and milled powder sintered at 850 °C. Measured data are shown by black lines. LeBail fits are shown in color. White arrows point to the most pronounced peaks of ω-phase, black arrows highlight the most intensive α-phase peaks.

The maximum microhardness of both sintered powders was about 350 HV. It corresponds to the microhardness of α + β-structured Ti-15Mo prepared by conventional processing routes [35]. The microhardness of refined Ti-15Mo prepared by bulk SPD methods can be as high as 500 HV [28].

The results of tensile tests for selected samples are shown in Figure 10. For each sample, two specimens were deformed (noted as spec. 1 and spec. 2 in the Figure 10) and all successfully measured curves are shown. The flow curve for the milled powder sintered at 750 °C could not be shown, because the sample failed already in the elastic region. Milled powder sintered at 850 °C had comparable yield stress to the initial powder sintered at the same temperature. However, the initial powder sintered at 750 °C had lower yield strength. This was consistent with the microhardness measurement and could be explained the same way, namely by the concurrent effect of porosity, oxygen contamination, microstructure and phase composition. The yield strength of both powders sintered at 850 °C was around 1200 MPa. In comparison, coarse grained Ti-12Mo after two step ageing containing both β and α-phase possess yield strength of 1000 MPa [39]. Moreover, severely deformed metastable β-alloy TNTZO containing no α-phase had comparable yield strength over 1100 MPa [40].

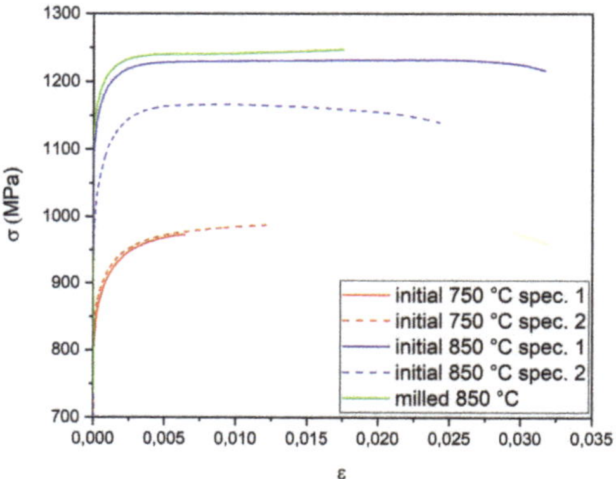

Figure 10. Tensile tests: true stress–true strain curves.

Figure 11 shows a photo of tensile samples after testing. Sintered initial samples broke properly in the active part as marked by red (a) and green (b) arrows in Figure 11. On the other hand, none of the sintered milled powder samples broke in the middle of the active part, as marked by blue (c), yellow (d) and violet (e) arrows in Figure 11. This fact can be explained using the sketch in Figure 2. The cross section of the gauge length was only slightly smaller than the distance between the side of the sample and the holding pin hole as noted by the green line in Figure 2. This would not cause any problem in a standard isotropic material sintered from the initial powder. However, due to stacking of milled powder particles in the sintering die, the material was clearly stronger in the direction parallel to the tensile sample (i.e., perpendicular to the load during sintering). As the result, these samples failed near the pin hole because the flat powder particles were delaminated.

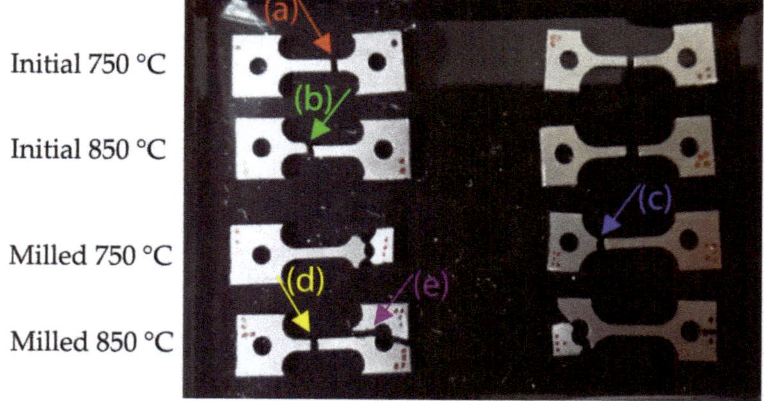

Figure 11. A photo of deformed tensile samples. Letters (**a–e**) correspond to the SEM images in Figure 12.

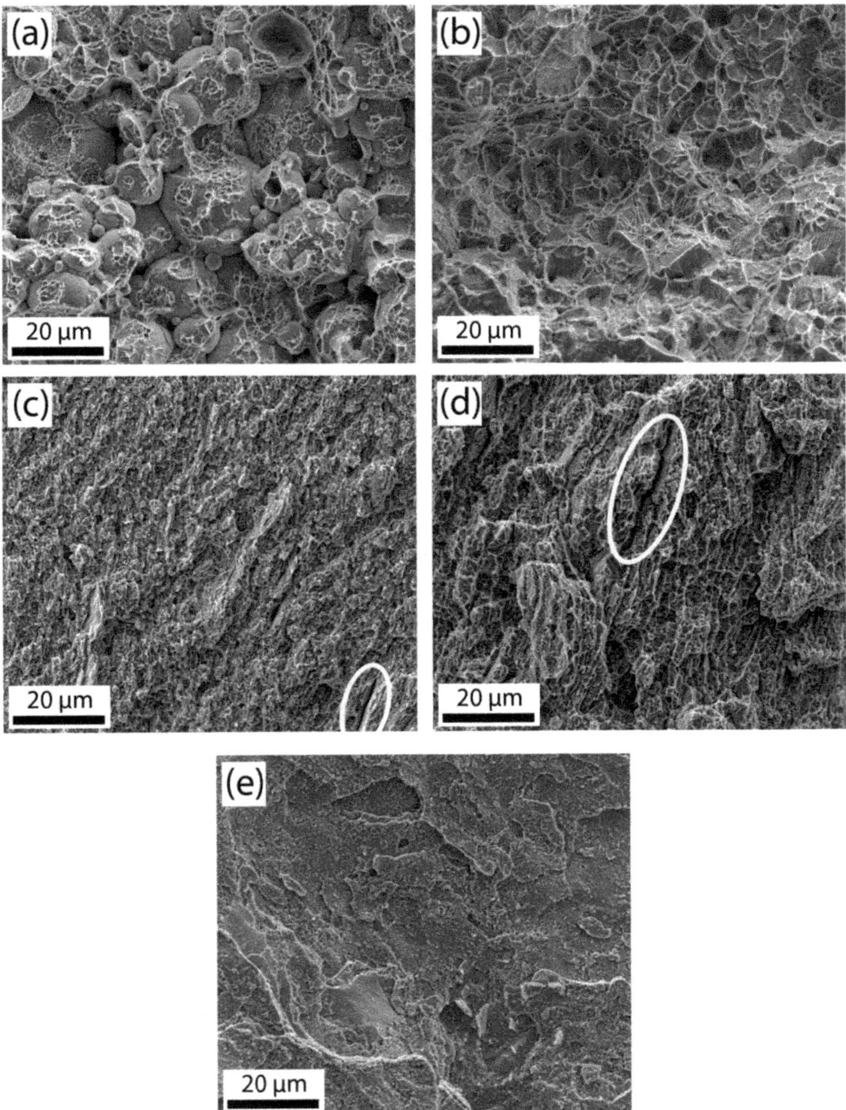

Figure 12. Fracture surfaces of selected specimens after tensile tests: (**a**) initial powder sintered at 750 °C; (**b**) initial powder sintered at 850 °C; (**c**) milled powder sintered at 750 °C; (**d**) milled powder sintered at 850 °C and (**e**) milled powder sintered at 850 °C-parallel to the powder particles.

This hypothesis was confirmed by SEM micrographs of fracture surfaces shown in Figure 12. The corresponding specimen and the position of SEM micrograph on the specimen is shown in Figure 11 by colored arrows. The fracture surface of initial powder sintered at 750 °C is shown in Figure 12a and corresponds to the region marked by the red arrow (a) in Figure 11. The initial round powder particles are clearly visible in Figure 12a. The sample was not fully sintered and the powder particles were joined only at the point of their contact.

A fracture surface of initial powder sintered at 850 °C corresponding to the zone marked by the green arrow (b) in Figure 11 is shown in Figure 12b. It shows a well sintered sample. No pores or initial powder particles were visible and the fracture was typical ductile. Figure 12c,d shows fracture surfaces of milled powder sintered at 750 °C and at 850 °C, respectively. They correspond to zones marked by blue and yellow arrows (c) and (d) in Figure 11, respectively. The fracture surfaces were completely different from those of sintered initial powder. Observed features were significantly finer in the Figure 12c (sintered at 750 °C) when compared to the Figure 12d (sintered at 850 °C), which corresponds well to the observed microstructures in Figure 5c,d. Highlighted regions by ellipses in Figure 12c,d corresponded to zones where a disk-shape powder particle was pulled out. Finally, Figure 12e shows a fracture surface of milled powder sintered at 850 °C oriented parallel to the powder particles as marked by a violet arrow (e) in Figure 11. In this SEM micrograph, planes corresponding to the surfaces of disk-shaped powder particles were visible. Milled powder particles were detached by the flat side of milled disk-shaped powder particles. These observations support our previous assumption of the possible delamination of powder particles.

Although well-sintered powders were ductile, the elongation of all sintered material was very poor, reaching maximum plastic strain of 3%. Commercial bulk Ti-15Mo alloy has a minimum elongation of 10% in aged $\alpha + \beta$ condition [1]. In similar Ti-12Mo alloy, the elongation of the $\alpha + \beta$ condition and the β condition was 9% and 50%, respectively [39]. Low ductility of sintered initial powder could be caused by the presence of pores and ω-phase, which results in the embrittlement of the material. Moreover, oxygen causes the embrittlement of α-phase. This phenomenon can affect the ductility especially in sintered milled powder, which is severely contaminated by oxygen and simultaneously contains high amount of α-phase.

4. Conclusions

In this work, metastable β-Ti alloy Ti-15Mo was prepared by cryogenic milling and subsequent spark plasma sintering. Microhardness measurements and tensile tests were performed to determine mechanical properties of achieved materials.

- Cryogenic milling changed initial ball-shaped powder particles into disk-shaped ones. Powder particles were not refined, but they were intensively deformed, similarly to severe plastic deformation methods (SPD).
- The microstructure of the sintered milled powder consisted of β-grains with the size in range of few micrometers and submicrometer α-phase precipitates.
- The microhardness reached 350 HV and was affected by residual porosity, microstructure, contamination by oxygen and phase composition.
- Both sintered powders possessed the high yield tensile strength of 1200 MPa, but very low ductility.

Author Contributions: A.V. conducted most of the experiments and wrote majority of the manuscript. J.K. performed cryogenic milling and developed the procedure of powder visualization by SEM. K.B. conducted some of the SEM observations and contributed to the discussion of results, T.C. conducted spark plasma sintering of the powders. C.A.C. provided X-ray diffraction experiments. J.S. discussed achieved results and wrote part of the manuscript.

Funding: This research was funded by Czech Science Foundation, grant number 17-20700Y and by Grant Agency of Charles University (GAUK), project No. 20119.

Acknowledgments: The authors wish to acknowledge Beate Kutzner (TU Freiberg) for providing CGHE measurements.

Conflicts of Interest: The authors declare no conflict of interest.

References

1. ATI 15Mo^{TM}. Available online: https://www.atimetals.com/Products/Pages/ati-ti-15mo.aspx (accessed on 10 May 2019).
2. Lütjering, G.; Williams, J.C. *Titanium*; Springer: Berlin, Germany, 2003.

3. Niinomi, M.; Nakai, M.; Hieda, J. Development of new metallic alloys for biomedical applications. *Acta Biomater.* **2012**, *8*, 3888–3903. [CrossRef] [PubMed]

4. Steinemann, S.G. Titanium—The material of choice? *Periodontol. 2000* **1998**, *17*, 7–21. [CrossRef] [PubMed]

5. Kopova, I.; Stráský, J.; Harcuba, P.; Landa, M.; Janeček, M.; Bačáková, L. Newly developed Ti–Nb–Zr–Ta–Si–Fe biomedical beta titanium alloys with increased strength and enhanced biocompatibility. *Mater. Sci. Eng. C* **2016**, *60*, 230–238. [CrossRef] [PubMed]

6. Welsch, G.; Welsch, R.; Collings, E.W. *Materials Properties Handbook: Titanium Alloys*; ASM Internacional: Novelty, OH, USA, 1994.

7. Semiatin, S.L.; Knisley, S.L.; Fagin, P.N.; Barker, D.R.; Zhang, F. Microstructure evolution during alpha-beta heat treatment of Ti-6Al-4V. *Metall. Mater. Trans. A* **2003**, *34*, 2377–2386. [CrossRef]

8. Zháňal, P.; Harcuba, P.; Hájek, M.; Smola, B.; Stráský, J.; Šmilauerová, J.; Veselý, J.; Janeček, M. Evolution of ω phase during heating of metastable β titanium alloy Ti–15Mo. *J. Mater. Sci.* **2018**, *53*, 837–845. [CrossRef]

9. Fang, Z.Z.; Paramore, J.D.; Sun, P.; Chandran, K.S.R.; Zhang, Y.; Xia, Y.; Cao, F.; Koopman, M.; Free, M. Powder metallurgy of titanium—Past, present, and future. *Int. Mater. Rev.* **2018**, *63*, 407–459. [CrossRef]

10. Qian, M.; Froes, F.H. *Titanium Powder Metallurgy*; Elsevier: Amsterdam, The Netherlands, 2015.

11. Guillon, O.; Gonzalez-Julian, J.; Dargatz, B.; Kessel, T.; Schierning, G.; Räthel, J.; Herrmann, M. Field-assisted sintering technology/spark plasma sintering: Mechanisms, materials, and technology developments. *Adv. Eng. Mater.* **2014**, *16*, 830–849. [CrossRef]

12. Yin, W.H.; Xu, F.; Ertorer, O.; Pan, Z.; Zhang, X.Y.; Kecskes, L.J.; Lavernia, E.J.; Wei, Q. Mechanical behavior of microstructure engineered multi-length-scale titanium over a wide range of strain rates. *Acta Mater.* **2013**, *61*, 3781–3798. [CrossRef]

13. Chen, W. Interdiffusion and atomic mobility in bcc Ti–rich Ti–Nb–Zr system. *Calphad* **2018**, *60*, 98–105. [CrossRef]

14. Rechtin, J.; Torresani, E.; Ivanov, E.; Olevky, E. Fabrication of Titanium-Niobium-Zirconium-Tantalium Alloy (TNZT) bioimplant components with controllable porosity by spark plasma sintering. *Materials* **2018**, *11*, 181. [CrossRef]

15. Suryanarayana, C. Mechanical alloying and milling. *Prog. Mater. Sci.* **2001**, *46*, 1–184. [CrossRef]

16. Wilczek, M.; Bertling, J.; Hintemann, D. Optimised technologies for cryogenic grinding. *Int. J. Miner. Process.* **2004**, *74*, S425–S434. [CrossRef]

17. Zhou, F.; Nutt, S.R.; Bampton, C.C.; Lavernia, E.J. Nanostructure in an Al-Mg-Sc alloy processed by low-energy ball milling at cryogenic temperature. *Metall. Mater. Trans. A* **2003**, *34*, 1985–1992. [CrossRef]

18. Ertorer, O.; Topping, T.D.; Li, Y.; Moss, W.; Lavernia, E.J. Nanostructured Ti Consolidated via Spark Plasma Sintering. *Metall. Mater. Trans. A* **2011**, *42*, 964–973. [CrossRef]

19. Kozlík, J.; Becker, H.; Harcuba, P.; Stráský, J.; Janeček, M. Cryomilled and spark plasma sintered titanium: The evolution of microstructure. In *IOP Conference Series: Materials Science and Engineering*; IOP Publishing: Bristol, UK, 2017; Volume 194, p. 012023.

20. Sun, F.; Rojas, P.; Zúñiga, A.; Lavernia, E.J. Nanostructure in a Ti alloy processed using a cryomilling technique. *Mater. Sci. Eng. A* **2006**, *430*, 90–97. [CrossRef]

21. Ertorer, O.; Zúñiga, A.; Topping, T.; Moss, W.; Lavernia, E.J. Mechanical behavior of cryomilled CP-Ti consolidated via quasi-isostatic forging. *Metall. Mater. Trans. A* **2008**, *40*, 91. [CrossRef]

22. Kozlík, J.; Stráský, J.; Harcuba, P.; Ibragimov, I.; Chráska, T.; Janeček, M. Cryogenic milling of Titanium powder. *Metals* **2018**, *8*, 31. [CrossRef]

23. Wen-Bin, F.; Wa, F.; Hong-Fel, S. Preparation of high-strength Mg–3Al–Zn alloy with ultrafine-grained microstructure by powder metallurgy. *Powder Technol.* **2011**, *212*, 161–165. [CrossRef]

24. Kozlík, J.; Harcuba, P.; Stráský, J.; Becker, H.; Šmilauerová, J.; Janeček, M. Microstructure and texture formation in commercially pure titanium prepared by cryogenic milling and spark plasma sintering. *Mater. Charact.* **2019**, *151*, 1–5. [CrossRef]

25. Kent, D.; Wang, G.; Yu, Z.; Ma, X.; Dargusch, M. Strength enhancement of a biomedical titanium alloy through a modified accumulative roll bonding technique. *J. Mech. Behav. Biomed. Mater.* **2011**, *4*, 405–416. [CrossRef]

26. Yilmazer, H.; Niinomi, M.; Nakai, M.; Cho, K.; Hieda, J.; Todaka, Y.; Miyazaki, T. Mechanical properties of a medical β-type titanium alloy with specific microstructural evolution through high-pressure torsion. *Mater. Sci. Eng. C* **2013**, *33*, 2499–2507. [CrossRef] [PubMed]

27. Janeček, M.; Čížek, J.; Stráský, J.; Václavová, K.; Hruška, P.; Polyakova, V.; Gatina, S.; Semenova, I. Microstructure evolution in solution treated Ti15Mo alloy processed by high pressure torsion. *Mater. Charact.* **2014**, *98*, 233–240. [CrossRef]

28. Václavová, K.; Stráský, J.; Polyakova, V.; Stráská, J.; Nejezchlebová, J.; Seiner, H.; Semenova, I.; Janeček, M. Microhardness and microstructure evolution of ultra-fine grained Ti-15Mo and TIMETAL LCB alloys prepared by high pressure torsion. *Mater. Sci. Eng. A* **2017**, *682*, 220–228. [CrossRef]

29. Zháňal, P.; Václavová, K.; Hadzima, B.; Harcuba, P.; Stráský, J.; Janeček, M.; Polyakova, V.; Semenova, I.; Hájek, M.; Hajizadeh, K. Thermal stability of ultrafine-grained commercial purity Ti and Ti-6Al-7Nb alloy investigated by electrical resistance, microhardness and scanning electron microscopy. *Mater. Sci. Eng. A* **2016**, *651*, 886–892. [CrossRef]

30. Bartha, K.; Zháňal, P.; Stráský, J.; Čížek, J.; Dopita, M.; Lukáč, F.; Harcuba, P.; Hájek, M.; Polyakova, V.; Semenova, I.; et al. Lattice defects in severely deformed biomedical Ti-6Al-7Nb alloy and thermal stability of its ultra-fine grained microstructure. *J. Alloys Compd.* **2019**, *788*, 881–890. [CrossRef]

31. Guo, Z.; Malinov, S.; Sha, W. Modelling beta transus temperature of titanium alloys using artificial neural network. *Comput. Mater. Sci.* **2005**, *32*, 1–12. [CrossRef]

32. Miklaszewski, A.; Garbiec, D.; Niespodziana, K. Sintering behavior and microstructure evolution in cp-titanium processed by spark plasma sintering. *Adv. Powder Technol.* **2018**, *29*, 50–57. [CrossRef]

33. Syarif, J.; Rohmannudin, T.N.; Omar, M.Z.; Sajuri, Z.; Harjanto, S. Stability of the beta phase in Ti-Mo-Cr alloy fabricated by powder metallurgy. *J. Min. Metall. Sect. B Metall.* **2013**, *49*, 285–292. [CrossRef]

34. Bartha, K.; Stráský, J.; Harcuba, P.; Semenova, I.; Polyakova, V.; Janeček, M. Heterogeneous precipitation of the α-phase in Ti15Mo alloy subjected to high pressure torsion. *Acta Phys. Pol. A* **2018**, *134*, 790–793. [CrossRef]

35. Václavová, K.; Stráský, J.; Veselý, J.; Gatina, S.; Polyakova, V.; Semenova, I.; Janeček, M. Evolution of microstructure and microhardness in Ti-15Mo β-Ti alloy prepared by high pressure torsion. *Mater. Sci. Forum* **2017**, *879*, 2555–2560. [CrossRef]

36. Zháňal, P.; Harcuba, P.; Šmilauerová, J.; Stráský, J.; Janeček, M.; Smola, B.; Hájek, M. Phase transformations in Ti-15Mo investigated by in situ electrical resistance. *Acta Phys. Pol. A* **2015**, *128*, 779–782. [CrossRef]

37. Šmilauerová, J.; Harcuba, P.; Kriegner, D.; Janeček, M.; Holý, V. Growth kinetics of ω particles in β-Ti matrix studied by in situ small-angle X-ray scattering. *Acta Mater.* **2015**, *100*, 126–134. [CrossRef] [PubMed]

38. Stráský, J.; Zháňal, P.; Václavová, K.; Horváth, K.; Landa, M.; Srba, O.; Janeček, M. Increasing strength of a biomedical Ti-Nb-Ta-Zr alloy by alloying with Fe, Si and O. *J. Mech. Behav. Biomed. Mater.* **2017**, *71*, 329–336. [CrossRef] [PubMed]

39. Sun, F.; Prima, F.; Cloriant, T. High-strength nanostructured Ti–12Mo alloy from ductile metastable beta state precursor. *Mater. Sci. Eng. A* **2010**, *527*, 4262–4269. [CrossRef]

40. Preisler, D.; Václavová, K.; Stráský, J.; Janeček, M.; Harcuba, P. *Microstructure and Mechanical Properties of Ti-Nb-Zr-Ta-O Biomedical Alloy*; Tanger Ltd.: Slezska, Czech Republic, 2016.

Article

Formation and Thermal Stability of ω-Ti(Fe) in α-Phase-Based Ti(Fe) Alloys

Mario J. Kriegel [1,*], Martin Rudolph [1], Askar Kilmametov [2,3], Boris B. Straumal [2,4], Julia Ivanisenko [2], Olga Fabrichnaya [1], Horst Hahn [2] and David Rafaja [1]

[1]   TU Bergakademie Freiberg, Institute of Materials Science, 09599 Freiberg, Germany; m.rudolph@iww.tu-freiberg.de (M.R.); fabrich@ww.tu-freiberg.de (O.F.); rafaja@ww.tu-freiberg.de (D.R.)

[2]   Karlsruhe Institute of Technology (KIT), Institute of Nanotechnology, 76344 Eggenstein-Leopoldshafen, Germany; askar.kilmametov@kit.edu (A.K.); boris.straumal@kit.edu (B.B.S.); julia.ivanisenko@kit.edu (J.I.); horst.hahn@kit.edu (H.H.)

[3]   Institute of Solid State Physics and Chernogolovka Scientific Center, Russian Academy of Sciences, 142432 Chernogolovka, Russia

[4]   Laboratory of Hybrid Nanomaterials, National University of Science and Technology "MISIS", 119049 Moscow, Russia

*   Correspondence: mario.kriegel@iww.tu-freiberg.de; Tel.: +49-3731-39-3106

Received: 4 March 2020; Accepted: 19 March 2020; Published: 21 March 2020

Abstract: In this work, the formation and thermal stability of the ω-Ti(Fe) phase that were produced by the high-pressure torsion (HPT) were studied in two-phase α-Ti + TiFe alloys containing 2 wt.%, 4 wt.% and 10 wt.% iron. The two-phase microstructure was achieved by annealing the alloys at 470 °C for 4000 h and then quenching them in water. Scanning electron microscopy (SEM) and X-ray diffraction (XRD) were utilized to characterize the samples. The thermal stability of the ω-Ti(Fe) phase was investigated using differential scanning calorimetry (DSC) and in situ high-temperature XRD. In the HPT process, the high-pressure ω-Ti(Fe) phase mainly formed from α-Ti. It started to decompose by a cascade of exothermic reactions already at temperatures of 130 °C. The decomposition was finished above ~320 °C. Upon further heating, the phase transformation proceeded via the formation of a supersaturated α-Ti(Fe) phase. Finally, the equilibrium phase assemblage was established at high temperatures. The eutectoid temperature and the phase transition temperatures measured in deformed and heat-treated samples are compared for the samples with different iron concentrations and for samples with different phase compositions prior to the HPT process. Thermodynamic calculations were carried out to predict stable and metastable phase assemblages after heat-treatments at low (α-Ti + TiFe) and high temperatures (α-Ti + β-(Ti,Fe), β-(Ti,Fe)).

Keywords: Ti–Fe; high-pressure torsion; microstructure; high-temperature XRD; differential scanning calorimetry; phase diagram; CalPhaD

---

1. Introduction

Titanium and titanium-base alloys are promising materials for numerous engineering applications, because they have several outstanding properties [1]. In particular, binary Ti–Fe alloys are in the focus of ongoing research as materials for biomedical applications, because they possess excellent corrosion resistance, high wear resistance, biocompatibility, and appropriate mechanical properties, for instance, a low elastic modulus in comparison with other biocompatible metallic materials [2–4]. Furthermore, the binary intermetallic phase TiFe has been presented as a possible material for solid-state hydrogen storage applications [5,6]. Some methods of severe plastic deformation (SPD), such as ball milling and high-pressure torsion (HPT), were utilized to reduce the surface oxidation and to activate the material in order to improve the hydrogen storage properties of TiFe [7,8]. Originally, the SPD methods were utilized to generate bulk nanocrystalline materials that show, in many cases, better physical and mechanical properties than their microcrystalline counterparts [9,10]. Later, it turned out that the unique properties are also facilitated by diffusive and displacive (martensitic) phase transformations, which occur in the material during the HPT process [11–13]. However, the mechanism of the phase transformations and the influence of the initial microstructure on the phases formed after HPT are not fully understood yet.

In the unary Ti system, titanium exists in three modifications: as hexagonal α-Ti (space group (SG) $P6_3/mmc$) that is stable at low temperatures, as cubic β-Ti (SG: $Im\bar{3}m$), which is stable at high temperatures, and as hexagonal ω-Ti (SG: $P6/mmm$), which is stable at high pressures. In the binary Ti–Fe system, two intermediate intermetallic phases TiFe (SG: $Pm\bar{3}m$) and TiFe$_2$ (SG: $P6_3/mmc$) are formed in addition to the phases that are summarized above and in addition to α-Fe/δ-Fe (SG: $Im\bar{3}m$) and γ-Fe (SG: $Fm\bar{3}m$). Furthermore, at least two metastable phases, which are formed during a martensitic transformation upon quenching from the bcc-type β-(Ti,Fe) solid solution, were reported in the Ti-rich part of the Ti–Fe system. The first one is a close-packed hexagonal α'-Ti phase (SG: $P6_3/mmc$) [14–17], which is observed in Ti-rich alloys (>95 wt.% Ti), the other one is the so-called athermal ω-Ti(Fe) phase that exists and in a limited compositional range between 97 wt.% and 95 wt.% Ti [18–22]. For higher Fe concentrations (and lower Ti contents), β-(Ti,Fe) is retained as a metastable phase in the alloys.

Phase transitions and respective transformation pathways that were induced by high-pressure torsion were reported for Ti-rich Ti–Fe alloys containing 1–10 wt.% Fe in different initial states (as-cast [23] and heat-treated [24–28]), and, thus, for different phase compositions prior to the HPT process. The as-cast alloys contained a mixture of α-Ti and β-(Ti,Fe) that were partially transformed into ω-Ti(Fe) during the HPT process. The heat treatments were performed almost exclusively above the eutectoid temperature of ~595 °C (β-(Ti,Fe) $\rightleftharpoons$ α-Ti + TiFe) [14], i.e., in the single-phase β-(Ti,Fe) or in the two-phase α-Ti + β-(Ti,Fe) regions. After annealing, the samples were quenched in water. The retained amount of the high-pressure ω-Ti(Fe) phase after HPT and the transformation pathway varied, depending on the initial phase fractions and the chemical compositions of the quenched phases [26]. Heat-treated alloys with Fe contents below 4 wt.% contained α'-Ti martensite and/or metastable β-(Ti,Fe) after quenching [24–28]. During HPT, α'-Ti and β-(Ti,Fe) transformed partially to ω-Ti(Fe).

The Fe content in β-(Ti,Fe) depends on the overall Fe concentration in the Ti–Fe alloys and it can vary in a relatively broad range. However, if β-(Ti,Fe) contains ~4 wt.% Fe, athermal ω-Ti(Fe) can be formed within the β-(Ti,Fe) grains after quenching [28,29]. The phase transformation β-(Ti,Fe) $\rightarrow$ ω-Ti(Fe) is promoted for an iron content about 4 wt.% Fe, because both crystal structures possess a strong orientation relationship (OR) $\{111\}_\beta\|(0001)_\omega$ and $\langle1\bar{1}0\rangle_\beta\|\langle11\bar{2}0\rangle_\omega$ [18,22], and because the atomic distances within the habitus planes match perfectly together at this composition. Thus, this phase transformation proceeds diffusionless by shearing the crystal structure of β-(Ti,Fe) in the HPT process [26]. For lower Fe contents ($\leq$2 wt.% Fe), the HPT process induces an incomplete ω-Ti(Fe) phase transition [25,28], because the transition of α'-Ti to ω-Ti(Fe) dominates the phase transformation process, which involves the redistribution of iron atoms between α'-Ti and ω-Ti(Fe). The mass transfer impedes finally the phase transformation [26,28].

So far, the thermal stability of HPT-induced ω-Ti(Fe) was predominantly investigated in the samples, which were annealed above the eutectoid temperature (~595 °C) prior to the HPT process and that contained α-Ti, β-(Ti,Fe) and ω-Ti(Fe) in different ratios after the HPT treatment [27,28]. In these samples, the HPT-induced ω-Ti(Fe) phase transformed upon heating at 380 °C into a supersaturated hexagonal α-Ti(Fe) phase. Above 600 °C, α-Ti(Fe) decomposed into the equilibrium phases α-Ti and β-(Ti,Fe), whose fraction varied with the overall composition of the alloy [27].

The thermal stability of ω-Ti(Fe) that was generated by HPT in samples annealed below the eutectoid temperature, i.e., in the α-Ti + TiFe two-phase region, was not investigated in detail yet. Thus, only very little is known regarding the phase transformations in the Ti–Fe alloys with this phase composition. Still, in Reference [24], the binary alloy Ti-4Fe (4 wt.% Fe) was heat-treated below the eutectoid temperature at 470 °C for 750 h and subsequently subjected to HPT. After initial annealing, this alloy contained a mixture of the α-Ti and TiFe phases, which partially transformed into ω-Ti(Fe). Thus, the HPT-deformed microstructure consisted of α-Ti, TiFe and ω-Ti(Fe). In the present work, the HPT-induced formation of ω-Ti(Fe) in initially two-phase alloys (α-Ti + TiFe) is more systematically studied and the stability of the metastable ω-Ti(Fe) phase is investigated upon heating. The transformation pathway was derived from the results of in situ high-temperature X-ray diffraction (HTXRD) and thermal analysis (TA). The experiments were complemented by pressure-dependent thermodynamic calculations that were based on the CalPhaD (calculation of Phase Diagrams) approach, which helped to understand the phase transformations in this system during deformation and heating.

2. Materials and Methods

In the frame of the present work, three alloy compositions (Ti-2Fe, Ti-4Fe, and Ti-10Fe) were prepared and investigated. The numerical values give the iron concentrations in wt.%. The alloys were produced by induction melting of pure materials (Ti: 99.9% and Fe: 99.97%) in vacuum. The cast rod-shaped samples (10 mm in diameter) were cut into the disks of 0.7 mm thickness, which were polished, etched, and annealed for 4000 h at 470 °C in vacuum and subsequently quenched in water. The annealing was carried out in sealed fused silica ampoules, which were evacuated up to the pressure of 4×10^{-4} Pa. The heat-treated samples were deformed by high-pressure torsion (HPT, five rotations) in a Bridgeman anvil-type press (Klement, Lang, Austria). The HPT process was carried out at an ambient temperature, at a pressure of 7 GPa and with the deformation speed of 1 rpm while using a computer-controlled HPT device.

The oxygen content in the samples was determined using carrier gas hot extraction (CGHE) after the HPT process and after the HTXRD measurements. For the CGHE analyses, the GALILEO G8 device (Bruker AXS, Karlsruhe, Germany) was used. The measurements were performed in a graphite crucible while using He as the carrier gas. Before each measurement, the samples were dipped in diluted HCl acid to remove oxide layers from the sample surface. The spatial distribution of the phases was obtained from the scanning electron micrographs for both heat-treated and deformed states. The SEM images were recorded while using back-scattered electrons (SEM/BSE) with the JSM-7800 F (JEOL, Tokyo, Japan) microscope operating at the accelerating voltage of 20 kV. The grain-size distributions of the phases were determined using electron backscatter diffraction (EBSD) that was performed with an EDAX/EBSD system. The analysis of the data was done while utilizing the OIM Analysis software (version 8, AMETEK/EDAX TSL, Mahwah, NJ, USA). The phase identification in the heat-treated and HPT deformed samples was performed by means of X-ray diffraction (XRD) in a D8 Advanced (Bruker AXS, Germany) diffractometer. The diffractometer worked in the Bragg–Brentano geometry, and it was equipped with a sealed X-ray tube with Co anode (wavelength of $CoK\alpha_1 = 0.178897$ nm), with a Johannson-type monochromator in the primary beam that suppressed the spectral line $CoK\alpha_2$ and with a LynxEye one-dimensional detector. The ex situ XRD patterns were analyzed using the Rietveld method [30,31] (whole pattern refinements), as implemented in the TOPAS [32] software package (version 5, Bruker AXS, Karlsruhe, Germany).

The HTXRD experiments were performed at a constant heating rate of 10 K/min in the temperature range between room temperature and 750 °C using a θ-θ Bruker D8 Advance diffractometer (Bruker AXS, Germany) that utilized the Bragg–Brentano geometry, $CuK\alpha$ radiation (wavelengths: 0.154056 nm and 0.154437 nm), a LynxEye XE one-dimensional detector (Bruker AXS, Germany) and the high-temperature chamber MTC HighTemp+ (Bruker AXS, Germany). During the HTXRD measurements, the samples were placed on a resistively heated tantalum strip that was surrounded by an additional Ta radiation heater to reduce the temperature gradients in the sample. A type D thermocouple (W-Re3/W-Re25), which was contacted on the heating strip below the sample position, was used to measure the sample temperature. The sample was placed on a (100)-oriented sapphire plate in order to avoid reactions between the sample and the Ta heater. The chamber was evacuated to a pressure of ~7×10^{-3} Pa to reduce the oxidation of the sample. The HTXRD patterns were recorded in a 2θ range between 33°

and 45° with a step size of 0.02° in order to speed up the measurement and reduce the difference in the annealing time and in the temperature during the acquisition of individual diffraction patterns. The temperature difference between the beginning and the end of each measurement was 20 K at the acquisition time of 120 s per diffraction pattern.

Differential scanning calorimetry (DSC) was utilized to determine phase transformation temperatures in the heat-treated samples before and after the HPT process. For DSC, the samples were placed in a platinum crucible with a thin Al_2O_3 inlet to avoid reactions of the samples with the Pt crucible. The measurements were performed in inert Ar atmosphere (99.999% + Varian cleaning system) while using the differential scanning calorimeter DSC Pegasus 404C (Netzsch, Selb, Germany). Before each experiment, the calorimeter was evacuated and back-filled with Ar several times to remove the remaining oxygen from the DSC chamber. The measurements were done in the temperature range between 35 °C and 1100 °C using the same heating rate of 10 K/min as used for the HTXRD measurements. The temperatures that were recorded in the DSC device were calibrated using melting points of pure metals (In, Sn, Al and Au).

3. Results

3.1. Characterization of the Initial State of the Samples

After annealing at 470 °C, i.e., below the temperature of the eutectoid reaction $\beta \rightleftharpoons \alpha$-Ti + TiFe, the alloys Ti-2Fe, Ti-4Fe, and Ti-10Fe show a two-phase α-Ti + TiFe microstructure (see Figure 1a–c). Some TiFe particles were ordered in chains along the boundaries of the α-Ti grains (see, for example, Figure 1a). This phenomenon is called incomplete or complete grain boundary wetting by a second solid phase and it has been previously observed in various Ti-based alloys [23,33–37]. In the corresponding SEM/BSE micrographs, the TiFe grains appear to be bright due to a higher Fe content and, thus, a higher mean atomic number of TiFe as compared to α-Ti. Moreover, no coarse-grain microstructure of TiFe was observed, even though the samples were annealed for 4000 h at 470 °C. The averaged grain sizes of α-Ti, determined using EBSD were (3.1 ± 2.0) μm, (3.0 ± 2.0) μm, and (1.4 ± 1.0) μm in the alloys Ti-2Fe, Ti-4Fe, and Ti-10Fe, respectively. The sizes of the TiFe grains were (0.6 ± 0.2) μm, (0.7 ± 0.3) μm, and (2.1 ± 1.0) μm in the same samples. The limited growth of the α-Ti and TiFe grains can be explained by either very slow recrystallization and grain-growth kinetics or by very slow diffusion velocities at the heat-treatment temperature of 470 °C. The phase fractions of α-Ti and TiFe were concurrently determined from the whole XRD patterns using the Rietveld method and from the chemical composition of the respective alloy using the lever rule for comparison (see Table 1) [14].

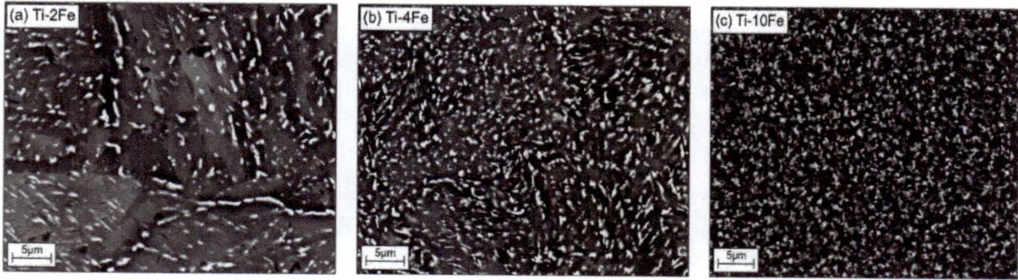

Figure 1. Initial microstructures of samples Ti-2Fe (**a**), Ti-4Fe (**b**), and Ti-10Fe (**c**) annealed at 470 °C for 4000 h. The white grains belong to TiFe, the gray areas to α-Ti.

Upon the Rietveld refinement, the lattice parameters of α-Ti and TiFe were calculated in addition to the phase composition that was fit together with the degree of the preferred orientation {0001} of the α-Ti crystallites. The March–Dollase model described this texture, which was probably caused by the foregoing sample preparation and annealing [38]. Applying this model, the phase fractions obtained from XRD for Ti-2Fe and Ti-4Fe agree very well with the phase fractions calculated from the chemical composition (Table 1). In alloy Ti-10Fe; however, the apparent texture was much more complex, because bad grain statistics caused the differences in diffracted intensities. Consequently, it was not possible to describe the texture satisfactorily using the March–Dollase model, which led to a larger uncertainty in the calculated phase fractions of α-Ti and TiFe (see Table 1). Assuming that the equilibrium state was achieved in all samples, the chemical compositions of α-Ti and TiFe are the same for all of investigated alloys (independent of the iron content). The lattice parameters of α-Ti and TiFe were $a_{\alpha\text{-Ti}} = 0.2951(2)$ nm, $c_{\alpha\text{-Ti}} = 0.4691(2)$ nm and $a_{\text{TiFe}} = 0.2978(2)$ nm, respectively. The obtained lattice parameters $a_{\alpha\text{-Ti}}$ and $c_{\alpha\text{-Ti}}$ correspond to the maximum Fe content in α-Ti at 470 °C. The lattice parameter a_{TiFe} agrees with the reference value for TiFe [39].

Table 1. Comparison of the phase fractions in the samples annealed at 470 °C for 4000 h, which were determined from the X-ray diffraction (XRD) measurements and calculated using the thermodynamic description of the binary Ti–Fe system by application of the lever rule. All of the phase amounts are given in wt.%. The uncertainty of the XRD phase analysis utilizing Rietveld refinement falls generally within a range of 1% to 3%.

Samples	Phase Fractions			
	Measured by XRD		Calculated	
	α-Ti	TiFe	α-Ti	TiFe
Ti-2Fe	95	5	95.9	4.1
Ti-4Fe	92	8	92.6	7.4
Ti-10Fe	77	23	81.3	18.7

The phase transition temperatures of the annealed alloys were determined by means of DSC upon heating at the heating rate of 10 K/min and they are shown in Figure 2. The endothermic heat effect registered at ~584 °C corresponds to the eutectoid reaction $\beta \rightleftharpoons \alpha$-Ti + TiFe that was detected in all of the samples, but with different extents. The increase of the heat amount with increasing Fe content is related to a higher amount of TiFe taking part in the eutectoid reaction. The β-transus temperatures (marked by crosses in Figure 2) decrease with an increasing Fe content, because the eutectoid point is located at higher iron contents (between 11 wt.% and 14 wt.% Fe) [14,40–43].

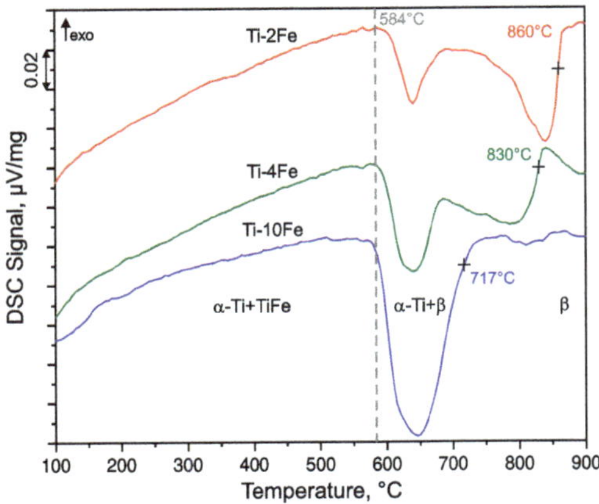

Figure 2. Differential scanning calorimetry (DSC) heating curves of alloys Ti-2Fe, Ti-4Fe, and Ti-10Fe measured with the heating rate of 10 K/min. The DSC curves were shifted vertically for better visibility. The dashed line indicates the temperature of the eutectoid reaction $\beta \rightleftharpoons \alpha$-Ti + TiFe. The crosses indicate the β-transus temperatures (solvus temperatures of the β phase), which were determined as inflection points of the respective DSC curve.

3.2. Characterization of the Deformed Samples after HPT Process

Microstructural changes in the HPT-deformed alloys were characterized while using SEM (Figure 3a–c). Apparently, the TiFe phase (white grains) is only slightly affected by the HPT process, because the α-Ti matrix mainly absorbed the deformation energy. After HPT, the grains of the α-Ti matrix are strongly refined. This behavior is expected, because the intermetallic compound TiFe is much harder (481 ± 44 HV0.025 [44]) than the soft α-Ti matrix (~180 HV0.5 [45]). Still, the chains of the TiFe precipitates, which were initially ordered in the annealed samples, were destroyed (compare the micrographs in Figures 1 and 3). The oxygen content within the deformed samples measured while using CGHE was very low, namely 0.0049(3) wt.%, 0.0055(3) wt.%, and 0.0024(3) wt.% for the alloys Ti-2Fe, Ti-4Fe, and Ti-10Fe, respectively.

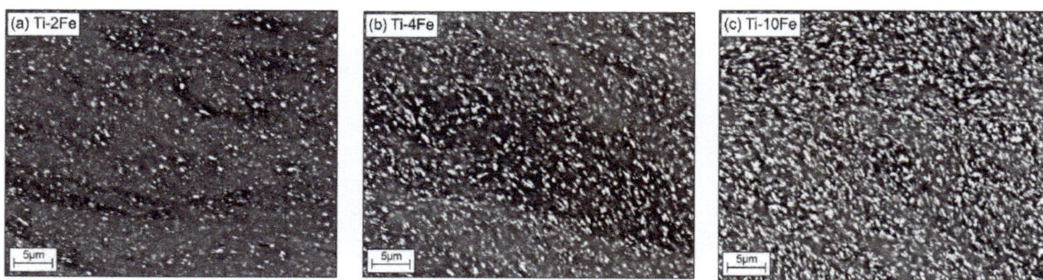

Figure 3. Microstructures of the high-pressure torsion (HPT) samples Ti-2Fe (**a**), Ti-4Fe (**b**), and Ti-10Fe (**c**) as seen by SEM/BSE. Prior to the HPT process, the samples were annealed at 470 °C for 4000 h. The white grains are TiFe, the gray areas correspond to the α-Ti and ω-Ti(Fe) phases.

The phase analysis using XRD confirmed that the HPT process produced ω-Ti(Fe). The comparison of the phase fractions in the annealed (Table 1) and HPT-treated samples (Table 2) shows that the amount of TiFe only decreased slightly, while the amount of α-Ti was drastically reduced during the HPT process. Thus, the ω-Ti(Fe) phase predominantly developed from α-Ti. In the alloys Ti-2Fe and Ti-4Fe, approximately 50 wt.% of α-Ti, was transformed into ω-Ti(Fe). Higher Fe contents and, consequently, a higher amount of the TiFe phase present in the respective alloys impede the HPT-induced phase transition, as can be seen on the lower relative amount of ω-Ti(Fe) in alloy Ti-10Fe (Table 2). As the lattice parameters of α-Ti in the annealed samples does not depend on the chemical composition of the alloy (see Section 3.1) the HPT-induced phase transformation α-Ti + TiFe $\rightarrow$ ω-Ti(Fe) + TiFe must be impeded by a higher amount of TiFe because the concentration of Fe in α-Ti is the same.

Table 2. Phase fractions (wt.%) in HPT-deformed samples, as determined using XRD. $n^{\alpha \rightarrow \omega} = \omega/(\alpha + \omega)$ is the fraction of α-Ti, which transformed to ω-Ti(Fe). The errors of the XRD phase analysis (1% to 3%) were estimated based on the goodness of fit.

Samples	Phase Fractions XRD			
	α-Ti(Fe)	TiFe	ω-Ti(Fe)	$n^{\alpha \rightarrow \omega}$
Ti-2Fe	45	4	51	0.53
Ti-4Fe	45	5	50	0.53
Ti-10Fe	52	17	31	0.37

After the HPT treatment, the line positions of the α-Ti(Fe) and TiFe phases were found to be shifted towards lower diffraction angles in all alloys. Such an increase of the unit cell volume of the phases after the deformation by HPT was already detected in earlier observations [24–26]. A large amount of defects and lattice distortions are generated during the severe plastic deformation in the HPT process [9,46], which lead to an increase of the lattice parameter of α-Ti and TiFe were $a_{\alpha\text{-Ti}} =$

0.2956(1) nm, $c_{\alpha\text{-Ti}}$ = 0.4694(1) nm, and a_{TiFe} = 0.2982(1) nm, respectively. The refined lattice parameters of ω-Ti(Fe), $a_{\omega\text{-Ti(Fe)}}$ = 0.4620(1) nm and $c_{\omega\text{-Ti(Fe)}}$ = 0.2829(1) nm, obey the relationship

$$c_{\omega}/a_{\omega} = \sqrt{3}/\left(2\sqrt{2}\right) \tag{1}$$

which indicates that the crystal structure of ω-Ti(Fe) is pseudo-cubic. Consequently, the XRD lines $10\bar{1}1$ and $11\bar{2}0$ from ω-Ti(Fe) are located at the same position. Because of the orientation relationship $\{111\}_{bcc}\|(0001)_{\omega}$ and $1\bar{1}0_{bcc}\|11\bar{2}0_{\omega}$ between the hexagonal ω phase and the cubic bcc lattice [18,22,47,48], the lattice parameters of ω-Ti(Fe) can be expressed in terms of a cubic bcc lattice parameter

$$a_{\omega} = \sqrt{2}a_{bcc} \text{ and } c_{\omega} = \left(\sqrt{3}/2\right)a_{bcc} \tag{2}$$

with a_{bcc} = 0.3267(1) nm. After HPT deformation, a satisfying agreement between the measured and refined XRD data was achieved, even for Ti-10Fe, when the preferred orientation was implemented into the Rietveld refinement using TOPAS [32], as described above. It was found that the α-Ti phase possesses a {0001} texture, which is, however, almost negligible for Ti-2Fe.

3.3. Thermal Stability of ω-Ti(Fe) Produced by the HPT Process

Complementary DSC and high-temperature XRD measurements were performed for the description of the thermal stability of the deformation-induced ω-Ti(Fe) phase. The DSC curves of the HPT-deformed samples were recorded upon heating at the heating rate of 10 K/min, and they are shown in Figure 4. The HTXRD patterns (Figure 5) were measured at the same heating rate (10 K/min). For HTXRD, the sample temperatures were calibrated with the aid of the initial temperature at the beginning of the measurements (25 °C) and while using the temperature of the endothermic effect at 562 °C that corresponds to the eutectoid reaction $\beta \rightleftharpoons \alpha$-Ti + TiFe and the transformation of the intermetallic phase TiFe. The transformation of TiFe can be detected by both techniques, DSC and XRD. The reaction temperature should be the same for all of the investigated alloys due to the invariance of the eutectoid reaction. Therefore, the same temperature calibration procedure can be applied to all investigated alloys.

Prior to the DSC and HTXRD measurements, all of the samples contained a mixture of α-Ti(Fe), TiFe and ω-Ti(Fe). The denotation α-Ti(Fe) emphasizes an increased iron solubility in α-Ti, due to (i) the HPT process and (ii) the reconversion of ω-Ti(Fe) $\rightarrow$ α-Ti(Fe) upon heating. These phenomena will be discussed in detail below. The first DSC effect observed upon heating was an exothermic peak occurring at approx. 130 °C (Figure 4), which was accompanied by the sharpening of the originally extremely broad XRD lines from α-Ti(Fe) (Figure 5). Concurrently, the XRD lines from α-Ti(Fe) became more intense at the expense of the XRD lines from ω-Ti(Fe), which indicates the onset of the reconversion of ω-Ti(Fe) to α-Ti(Fe). At 320 °C, the phase transition ω-Ti(Fe) $\rightarrow$ α-Ti(Fe) proceeds tremendously. The transition is completed at temperatures that were slightly above ~350 °C. In contrast to HTXRD, DSC did not recognize the end of the decomposition of ω-Ti(Fe), because it is not accompanied with a noticeable thermal effect. The ω-Ti(Fe) $\rightarrow$ α-Ti(Fe) transition is a continuous process, thus the heat release is spread over a broad temperature range.

After the decomposition of ω-Ti(Fe), all of the samples exhibited a two-phase microstructure containing hexagonal α-Ti(Fe) and cubic TiFe. The appearance of an additional diffraction line (0002 of α-Ti, c.f., Figure 5) at lower diffraction angles and the presence of anisotropic (*hkl*-dependent) line broadening upon further heating indicate changes in the Fe concentration in the α-Ti(Fe) phase. The incorporation of Fe into the hexagonal α-Ti lattice mainly leads to a reduction of the lattice parameter $c_{\alpha\text{-Ti}}$, while the lattice parameter $a_{\alpha\text{-Ti}}$ remains nearly unaffected [25,27]. Accordingly, the diffraction line 0002, which appears at temperatures above 400 °C, and that is located at a lower diffraction angle in Figure 5b, corresponds to the equilibrium α-Ti phase that only exhibits a negligible solubility for Fe [14]. On the other side, the hexagonal α-Ti(Fe) phase, which is present in all alloys after the HPT process and that is formed by the back-transformation of ω-Ti(Fe), should possess increased iron solubility.

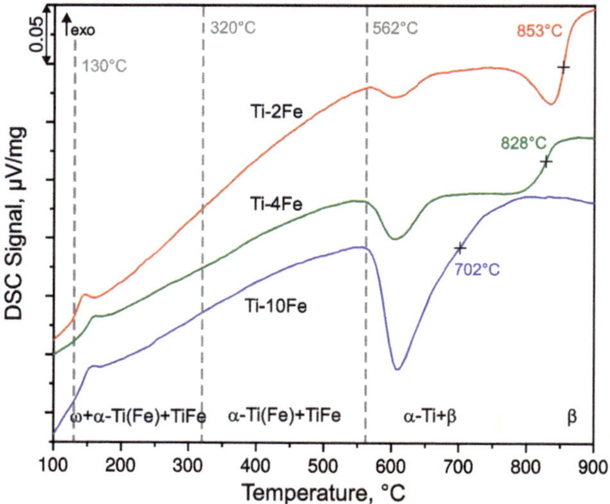

Figure 4. DSC heating curves of alloys Ti-2Fe, Ti-4Fe, and Ti-10Fe measured with the heating rate of 10 K/min. The dashed lines at 130 °C and 320 °C indicate the beginning and the end of the ω back-transformation as concluded from high-temperature X-ray diffraction (HTXRD). The dashed line at 562 °C marks the eutectoid reaction β-(Ti,Fe) ⇌ α-Ti(Fe) + TiFe. The temperatures marked by crosses indicate the β-transus temperatures (solvus temperatures of the β phase), which were determined as inflection points of the respective DSC curves.

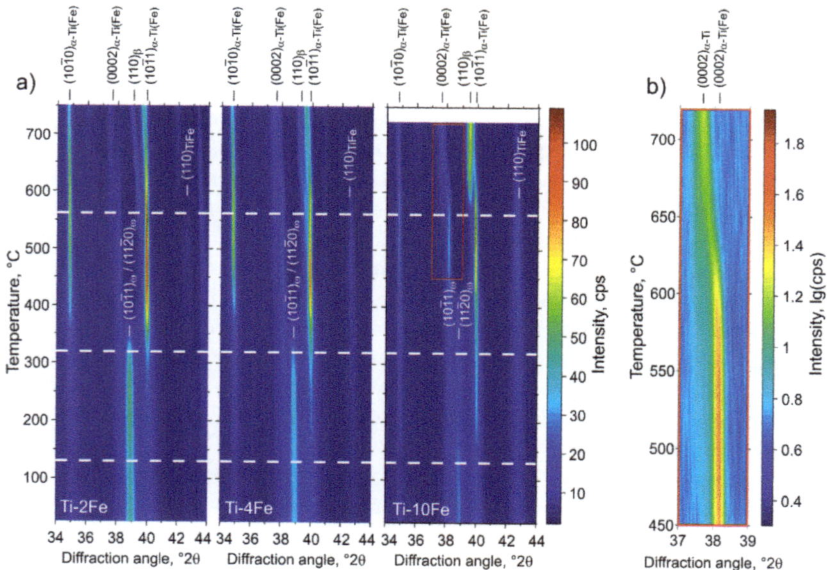

Figure 5. Low-angle part of the HTXRD patterns of Ti-2Fe, Ti-4Fe, and Ti-10Fe that were originally annealed for 4000 h at 470 °C and subjected to HPT. The positions of diffraction lines originating from the phases α-Ti(Fe), β, ω and TiFe are indicated by markers at the top or inside the figures. The temperature axes of the HTXRD measurements were calibrated according to the DSC measurements as described in the text. The dashed lines in (**a**) indicate the transformation temperatures upon heating from Figure 4. (**b**) illustrates the change of the 0002 line position in Ti-10Fe (see inset in (**a**)) at high temperatures.

At the temperature of 562 °C, the mixture of α-Ti, α-Ti(Fe) and TiFe transforms via an eutectoid reaction into a two-phase mixture of β-(Ti,Fe) and α-Ti (Figure 5). In the DSC measurements (Figure 4), the temperature of the eutectoid reaction was determined from its onset point. Above the eutectoid temperature, the amount of β-(Ti,Fe) continuously increases with further heating. The transus temperatures of β-(Ti,Fe) determined using DSC depends on the iron content in the respective alloy (Figure 4). A comparison of the eutectoid temperature and the β-(Ti,Fe)-transus temperatures that were measured for the heat-treated (Figure 2) and severe plastically deformed samples (Figure 4) reveals that both temperatures are slightly shifted towards lower values after the HPT process. Figure 6 shows a comparison of the transition temperatures measured by DSC before and after HPT.

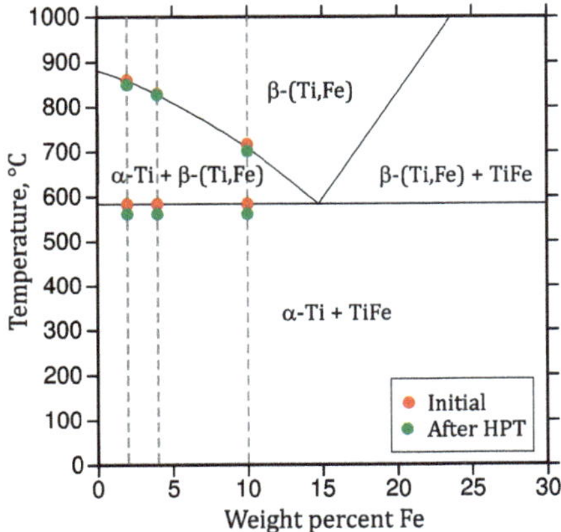

Figure 6. Partial phase diagram of the Ti-rich corner of the binary Ti–Fe system. The circles indicate the measured phase transition temperatures (DSC) of the samples in the initial state (red) and after deformation by HPT (green).

The sequence of the phase transitions is illustrated on the temperature dependence of the integral intensities of the XRD lines measured for individual phases (Figure 7), i.e., $10\bar{1}1_\omega/11\bar{2}0_\omega$, 110_{TiFe}, 110_β, and $10\bar{1}0_\alpha/0002_\alpha/10\bar{1}1_\alpha$. For ω-Ti(Fe) and α-Ti(Fe), the sums of the integral intensities of the measured lines were considered. For convenience, the integral intensities were converted into the phase compositions by normalizing the phase composition of the respective alloy to the phase composition from Table 2. This 'external standard' method neglects the effect of the possible changes in the preferred orientation of crystallites due to the sample recrystallization and the effect of the Debye–Waller factor on the phase composition, as the corresponding factors influencing the diffracted intensities are assumed to remain constant. However, it gives a good overview of the phase transitions, as can be seen from the almost monotonous and definitely reasonable trend of the TiFe phase fraction.

The beginning of the transformation ω-Ti(Fe) → α-Ti(Fe), which was observed during the DSC measurement as an exothermic effect at 130 °C, results in a rapid decrease of the ω-Ti(Fe) phase fraction and in a concurrent increase of the α-Ti(Fe) phase fraction. In alloy Ti-4Fe, the DSC peak from the exothermal effect is slightly shifted towards higher temperatures. Upon further heating, ω-Ti(Fe) continuously transforms into α-Ti(Fe). Above a temperature of 400 °C, all of the alloys exhibit a two-phase microstructure containing the hexagonal α-Ti(Fe) and the cubic TiFe phase. The phase amount of TiFe increases with increasing Fe content in the alloy. At temperatures above 500 °C and below the eutectoid temperature, the supersaturated α-Ti(Fe) releases iron, which is subsequently solved in the cubic β-(Ti,Fe) phase. Thus, β-(Ti,Fe) was formed by this process already below the eutectoid temperature. At 562 °C, the eutectoid reaction occurs and TiFe transforms into the equilibrium phases α-Ti + β-(Ti,Fe). A further temperature increase should lead to a continuously increasing phase amount of β-(Ti,Fe) and a decreasing amount of α-Ti. This behavior was not observed during the

HTXRD measurements, which is caused by the proceeding oxidation of the samples, which stabilizes α-Ti. Therefore, the results of HTXRD measurements can be compared with the results of the DSC measurements only up to temperatures of ~650 °C. In the alloys Ti-2Fe, Ti-4Fe, and Ti-10Fe that were subjected to HTXRD, the CGHE analysis revealed the oxygen concentrations of 0.0374(3) wt.%, 0.0262(3) wt.%, and 0.0032(3) wt.%, respectively, which are much higher than prior to HTXRD.

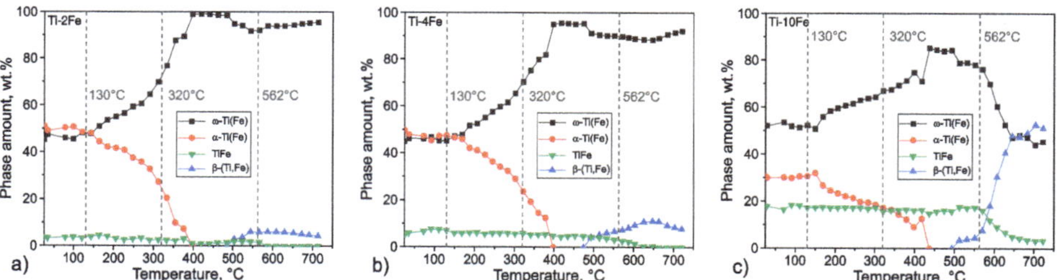

Figure 7. Temperature dependences of the phase composition in alloys Ti-2Fe (**a**), Ti-4Fe (**b**), and Ti-10Fe (**c**) obtained from the HTXRD measurements carried out upon heating. The integral intensities were normalized to the phase compositions at the initial state of samples after HPT (Table 2). The intensities of the diffraction lines 10$\bar{1}$0, 0002, and 10$\bar{1}$1 of α-Ti and α-Ti(Fe), and 10$\bar{1}$1 and 11$\bar{2}$0 of ω-Ti(Fe) were summed up. The temperature axes of the HTXRD measurements were calibrated according to the DSC measurements.

3.4. Thermodynamic Calculations

A pressure-dependent thermodynamic description of the binary Ti–Fe system was generated with the aid of the CalPhaD (Calculation of Phase Diagrams) approach for a better understanding of the experimental observations and especially for validation of the phase stabilities and reverse transformation of ω-Ti(Fe) to α-Ti [49]. The details of the thermodynamic calculations will be described elsewhere [50]. Exemplarily, the temperature-pressure (*t–p*) phase diagram is shown for alloy Ti-4Fe (see Figure 8a). Figure 8b illustrates the effect of the chemical composition for the hydrostatic pressure of 10 GPa. In contrast to unary *t–p* phase diagrams, which solely contain single-phase regions, two-phase regions can be present in those binary *t–p* phase diagrams (calculated for a given binary composition). The black lines represent either solvus lines or three-phase equilibria, with varying pressure and temperature. The presence of two-phase regions become obvious if a binary *t–p* phase diagram (Figure 8a) is compared with a *t–w(Fe)* phase diagram (Figure 8b). In Figure 8a, the vertical dashed line marks the pressure value of 10 GPa in Figure 8b the alloy composition of Ti-4Fe.

For an ambient temperature, the thermodynamic calculations revealed that the α-Ti + TiFe two-phase mixture, which is stable at ambient pressure, transforms at the pressure of ~0.8 GPa into a two-phase mixture of ω-Ti(Fe) and TiFe. Thus, the applied pressure during the HPT (7 GPa) should be sufficient for initiating the phase transformation of α-Ti + TiFe to ω-Ti(Fe) + TiFe. The high-pressure phase persists, even after the HPT process, being stabilized by the interaction with other phases that are present in the HPT samples.

In alloys that were annealed at high temperatures (800 °C) and subsequently quenched [26], the high-temperature phase assemblage was retained for the iron contents ≥4 wt.%. In that case, the transformation pathway was found to proceed from β-(Ti,Fe) or from an α-Ti + β-(Ti,Fe) mixture to a β-(Ti,Fe) + ω-Ti(Fe) mixture [26,28,29]. However, it was also reported that minor amounts of α-Ti are preserved after the HPT process. The initiation of the ω-Ti(Fe) transformation was found to be very easy, which means that the phase transformation should already occur at low pressures [26]. The metastable extension of the β-(Ti,Fe) + ω-Ti(Fe) region was thermodynamically calculated in order to predict this behavior (see black dashed line in Figure 8a) by suspending the formation of the TiFe phase. Thermodynamic calculations revealed that the phase transformation into the high-pressure phase assemblage should occur at room temperature already at the atmospheric pressure. This indicates that the transformation to the high-pressure ω-Ti(Fe) phase should already occur at very low pressures using HPT, which was also experimentally observed [26]. It is worth noting that the CalPhaD calculations reflect the equilibrium state in the samples under hydrostatic pressures, which is far away from the sample state that is generated in the HPT process. The main features of the HPT process are (i) the large portion of torsional stain, which mainly induces shear stresses and (ii) the short process times at ambient temperatures. However, the predicted transformations pathways are comparable with those experimentally observed—the trends are correct.

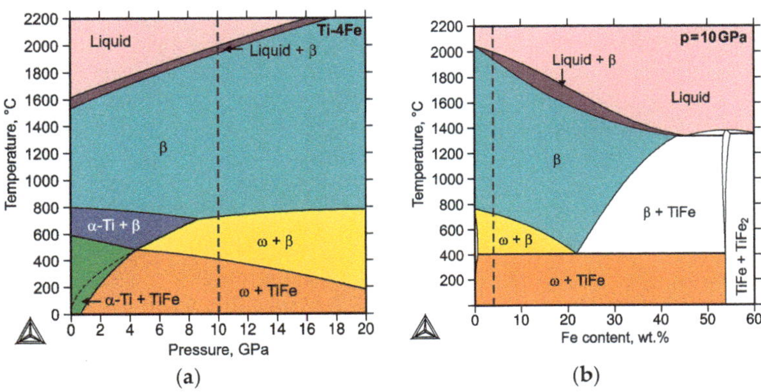

Figure 8. (**a**) Temperature-pressure phase diagram calculated for the composition Ti-4Fe (wt.%) and (**b**) temperature–composition phase diagram calculated for the pressure of 10 GPa. The vertical dashed lines mark the pressure value of 10 GPa (in (**a**)) and the alloy composition of Ti-4Fe (in (**b**)). The black dashed line in the bottom left corner of panel (**a**) indicates the metastable extension of the β-(Ti,Fe) + ω-Ti(Fe) region.

4. Discussion

4.1. HPT-Induced Formation of ω-Ti(Fe) in Samples Containing α-Ti As a Dominant Phase

The ω-Ti(Fe) phase can be produced in different ways in a HPT process. In previous studies, in which Ti-rich Ti–Fe samples were heat treated at temperatures above the eutectoid reaction [25–28], ω-Ti(Fe) was formed either from a single β-(Ti,Fe) phase or from a mixture of α′-Ti(Fe) martensite and β-(Ti,Fe). In Reference [26], it was shown that the iron concentration and the phase composition prior to the HPT process influence strongly the amount of the HPT-induced ω-Ti(Fe). For 4 wt.% of Fe, the transformation β-(Ti,Fe) → ω-Ti(Fe) already occurred at low pressures and it was almost completed after few HPT rotations, because the coincidence of the lattice parameters of pseudo-cubic ω-Ti(Fe) with the lattice parameter of β-(Ti,Fe) facilitates the diffusionless phase transition at this particular Fe concentration. For 2 wt.% Fe, still approximately 80% of the sample was transformed to ω-Ti(Fe) [26]. In the current study, the fractions of the ω-Ti(Fe) phase in samples Ti-2Fe and Ti-4Fe (~50%, c.f., Table 2) were significantly lower than the amount of ω-Ti(Fe) in samples with the same Fe content that were annealed at high temperatures [26]. The phase fractions of ω-Ti(Fe) produced in samples with 10 wt.% Fe are very similar for different annealing temperatures and, therefore, for different phase compositions prior to the HPT process.

The microstructure characterization using XRD and SEM confirmed that all of the samples under study (Ti-2Fe, Ti-4Fe, and Ti-10Fe) contained α-Ti and TiFe after annealing at 470 °C for 4000 h. The HPT induced the formation of the high-pressure ω-Ti(Fe) phase, which also remained stable at ambient conditions. The ω-Ti(Fe) phase was mainly produced from severely plastically deformed α-Ti, which absorbed the majority of the deformation energy. The TiFe grains were only slightly affected by the HPT process. Still, the TiFe precipitates, which were ordered in chains along the grain boundaries of the α-Ti phase in the annealed samples, were more randomly distributed after the HPT treatment, and the amount of crystalline TiFe was slightly reduced (c.f., Tables 1 and 2). After the long-term annealing, i.e., under thermodynamically equilibrium conditions, α-Ti can accommodate approximately 9.4×10^{-3} wt.% Fe. This means that α-Ti is practically free of iron in the samples that were annealed at 470 °C and that the ω-Ti(Fe) formation in these samples should be inhibited by the lack of iron in the parent phase. On the other hand, the shift of the diffraction line $0002_{\alpha\text{-Ti(Fe)}}$ towards lower diffraction angles, which is visible in Figure 5, indicates that the iron content in back-transformed α-Ti(Fe) varies upon heating and, consequently, ω-Ti(Fe) did not originate from pure α-Ti, but from supersaturated α-Ti(Fe) like in the samples from Reference [27].

Two approaches were used in order to estimate the iron content in ω-Ti(Fe). In the first one, the iron content in ω-Ti(Fe) was determined from the phase fractions before and after the HPT process. This approach is based on the assumptions that the application of pressure does not significantly change the homogeneity range of TiFe and that the residual hexagonal α-Ti(Fe), which is still present after HPT, contains the same amount of iron, like the high-pressure ω-Ti(Fe) phase. In the second approach, the iron content was estimated from the measured lattice parameter of ω-Ti(Fe) and from the Vegard dependence of $a_{\omega\text{-Ti(Fe)}}$ known from literature.

The phase amount of α-Ti(Fe) is significantly reduced due to the formation of ω-Ti(Fe) (compare Tables 1 and 2). However, the HPT process also reduced the amount of TiFe. Even though the decrease of the phase fraction of TiFe after HPT is small, their variation falls outside the estimated error limit. Using the law of the mass conservation, the initiation of the phase transitions due to HPT can be expressed, as follows:

$$m^{\alpha}_{Ini} \times w^{\alpha}_{Ini} + m^{TiFe}_{Ini} \times w^{TiFe}_{Ini} = m^{\alpha}_{HPT} \times w^{\alpha}_{HPT} + m^{TiFe}_{HPT} \times w^{TiFe}_{HPT} + m^{\omega}_{HPT} \times w^{\omega}_{HPT} \tag{3}$$

In Equation (3), m^{φ}_i (in wt.%) and w^{φ}_i (in wt.% of Fe) are the fractions of the involved phases (α, TiFe and ω) and their chemical compositions in the sample state (initial, after HPT), respectively. While assuming a constant chemical composition of TiFe before and after HPT ($w^{TiFe}_{Ini/HPT} = w^{TiFe}_{Ini} = w^{TiFe}_{HPT}$), and the same chemical composition of α-Ti(Fe) and ω-Ti(Fe) in HPT processed samples ($w^{\alpha}_{HPT} = w^{\omega}_{HPT}$), Equation (3) can be written as:

$$w^{\omega}_{HPT} = \frac{m^{\alpha}_{Ini} \cdot w^{\alpha}_{Ini} + \left(m^{TiFe}_{Ini} - m^{TiFe}_{HPT}\right) \cdot w^{TiFe}_{Ini/HPT}}{m^{\alpha}_{HPT} + m^{\omega}_{HPT}} \tag{4}$$

For $w^{\alpha}_{Ini} = 9.4 \times 10^{-3}$ wt.% Fe, $w^{TiFe}_{Ini/HPT} = 53.6$ wt.% Fe, and for the phase compositions, according to Tables 1 and 2, the Fe content in ω-Ti(Fe) was determined to be $w^{\omega}_{HPT} \cong 1$ wt.%.

The comparison of the lattice parameters was used as a complementary approach for the estimation of the iron content in ω-Ti(Fe). In commercially pure HPT-deformed Ti samples, the lattice parameters of ω-Ti were $a_{\omega\text{-Ti}} = 0.4627$ nm and $c_{\omega\text{-Ti}} = 0.2830$ nm [25]. The lattice parameters of ω-Ti(Fe) measured for the iron-containing alloys in this study were $a_{\omega\text{-Ti(Fe)}} = 0.4620(1)$ nm and $c_{\omega\text{-Ti(Fe)}} = 0.2829(1)$ nm. This confirms that iron atoms that are dissolved in the ω-Ti(Fe) phase lead to a reduction of the elementary cell volume, as already stated in References [25,26]. Moreover, the change of $a_{\omega\text{-Ti(Fe)}}$ is larger than the change of $c_{\omega\text{-Ti(Fe)}}$. At the iron contents of 4 wt.%, the pseudo-cubic lattice parameter of ω-Ti(Fe) coincides with the lattice parameter of β-(Ti,Fe) [26]. The whole dependence of the lattice parameter of β-(Ti,Fe) on the iron content was described in References [14,26,29]. The lattice parameters $a_{\omega\text{-Ti(Fe)}} = 0.4603(6)$ nm and $c_{\omega\text{-Ti(Fe)}} = 0.2819(1)$ nm containing 4 wt.% Fe were calculated using Equation (2) from the lattice parameter of β-(Ti,Fe) with the same amount of Fe ($a_{bcc} = 0.3255(2)$ nm) [29]. Assuming a linear dependence of the lattice parameter $a_{\omega\text{-Ti(Fe)}}$ on the Fe concentration between Fe-free ω-Ti and ω-Ti(Fe) containing 4 wt.% Fe, the iron content in the sample under study was estimated to be approximately 1.2 wt.%. This value is in good agreement with the iron concentration of ~1 wt.% that was concluded from the difference in the phase fractions before and after the HPT deformation. The lattice parameter $a_{\omega\text{-Ti(Fe)}}$ was used for the estimation of the Fe concentration in ω-Ti(Fe), because it is more sensitive to the iron concentration than $c_{\omega\text{-Ti(Fe)}}$.

4.2. Thermal Stability of the HPT-Deformed Microstructure

At ambient conditions, all of the HPT-deformed samples contained a mixture of three phases, i.e., α-Ti(Fe), TiFe and ω-Ti(Fe). The phase amounts were different, depending on the alloy composition. Upon heating, a relatively large exothermal DSC effect was registered at ~130 °C (onset temperature). However, in alloy Ti-4Fe, the beginning of the exothermal DSC effect was less abrupt than for the other alloys (compare Figure 4). The exothermal effect is related with the initiation of the decomposition process of the pressure-induced ω-Ti(Fe) phase and with the release of the deformation energy. Similar observations have already been made for pure Ti [51] and for Ti-1Fe [23]. At a temperature of ~320 °C, the phase amount of ω-Ti(Fe) decreases abruptly. Between 400 °C and 450 °C, this metastable phase completely disappeared (Figure 7). In a previous study [27] that was devoted to the investigation of the thermal stability of ω-Ti(Fe) in HPT deformed metastable β-(Ti,Fe) and α-Ti + β-(Ti,Fe) alloys, a cascade of exothermal DSC effects was observed between 150 °C and 450 °C. These effects originated from the gradual transformations of ω-Ti(Fe) to α-Ti(Fe) and from the defect annihilation and recrystallization processes [27]. Moreover, the strongest exothermal effect was observed at around 380 °C, whereas the released heat increases with increasing iron content within the investigated alloys. These results demonstrate that ω-Ti(Fe) possesses a lower thermal stability, if it is formed from α-Ti (present work) than if it is formed from β-(Ti,Fe). The differences in the phase composition of the HPT samples might cause this difference in the stability of the pressure-induced ω-Ti(Fe) phase, but also by the differences in the iron concentration in ω-Ti(Fe), because ω-Ti(Fe) originating from β-(Ti,Fe) possesses a higher Fe concentration than ω-Ti(Fe) stemming from α-Ti(Fe). The ω-Ti(Fe) phase, which was formed from the α-Ti + TiFe two-phase alloys contained approximately 1 wt.% Fe after HPT, whereas the ω-Ti(Fe) generated from metastable β-(Ti,Fe) alloys was found to contain ~4 wt.% Fe [26–28].

In the intermediate range between 400 °C and 562 °C, all of the alloys exhibit a two-phase α-Ti(Fe) + TiFe microstructure. At ~500 °C, which is still below the eutectoid reaction temperature, a shift of the line 0002 from α-Ti(Fe) towards lower diffraction angles was detected by HTXRD, together with the appearance of diffraction lines of β-(Ti,Fe). This phenomenon was already observed in former studies [27,29], where it was assigned to a variation of the iron solubility in the α-Ti(Fe) phase. The decomposition of the supersaturated α-Ti(Fe) phase to the equilibrium α-Ti phase is related to the shift of the diffraction line $0002_{\alpha\text{-Ti(Fe)}}$ towards smaller diffraction angles. In the present work, this effect is much less pronounced, because the phase amount of ω-Ti(Fe) only approaches approx. 50% after the HPT process. A pronounced endothermic heat effect corresponding to the eutectoid reaction β-(Ti,Fe) $\rightleftharpoons$ α-Ti + TiFe was registered at ~584 °C in the annealed, but not deformed, samples. The eutectoid reaction temperature is in good agreement with References [14,40–43], where this reaction was reported between 583 °C and 590 °C. After the HPT deformation, the measured temperature of the eutectoid reaction was found to be lowered to ~562 °C. The same behavior was observed for the β-(Ti,Fe)-transus temperature. Thus, strong deformations and/or small grain sizes of the α-Ti(Fe) and TiFe phases, which are characteristic microstructural features after the HPT process, lead to a shift of the equilibrium reaction temperatures towards lower values. Moreover, the development of the phase fractions upon heating (Figure 7) shows, in contrast to the phase diagram (Figure 6), that the phase amount of β-(Ti,Fe) does not continuously increase. At temperatures above 600 °C in the HTXRD measurements, the phase amount of β-(Ti,Fe) stagnated, and even decrease in Ti-2Fe and Ti-4Fe. The reason for that behavior is a slight oxidation of the samples during the HTXRD measurement. The CGHE measurements revealed increased oxygen content inside the samples after the HTXRD measurements (see Section 3.2), which confirms that the oxidation of the samples at high-temperatures, even under high vacuum, could not be prevented. Oxygen is an element stabilizing the α-Ti phase [52]. Therefore, α-Ti is stabilized at temperature above 600 °C and, thus, no single-phase β-(Ti,Fe) state was generated inside the samples upon further heating in the HTXRD device.

5. Conclusions

The high-pressure torsion (HPT) treatment of two-phase Ti–Fe alloys consisting of a mixture of α-Ti and TiFe led to the formation of the high-pressure ω-Ti(Fe) phase mainly at the expense of α-Ti. The concentration of iron in ω-Ti(Fe), which was ~1 wt.%, was alternatively estimated from the difference in the phase fractions before and after the HPT process and from the measured lattice parameters of ω-Ti(Fe). The overall iron concentration in the samples (2 wt.%, 4 wt.%, and 10 wt.% Fe) primarily influenced the amount of TiFe, which inhibited the phase transition α-Ti $\rightarrow$ ω-Ti(Fe) to some extent, in particular in sample containing 10 wt.% Fe. The comparison with previous investigations, which were carried out on samples that were annealed above the eutectoid reaction, i.e., on samples mainly containing β-(Ti,Fe), has shown that the high-pressure ω-Ti(Fe) phase can be much more easy produced from β-(Ti,Fe) than from α-Ti. Thermodynamic calculations confirmed this experimental finding. The combination of *in situ* high-temperature X-ray diffraction and differential scanning calorimetry revealed that ω-Ti(Fe) starts to decompose exothermically already at 130 °C. Still, it survives up to ~320 °C, where its amount tremendously decreases. The thermal stability of ω-Ti(Fe) in the samples under study was lower than in the alloys, which were annealed above the eutectoid reaction. Nevertheless, the decomposition pathway via a supersaturated α-Ti(Fe) phase was also observed in the present work.

Author Contributions: Conceptualization, M.J.K., D.R. and B.B.S.; methodology, A.K., M.J.K. and B.B.S.; software, M.J.K. and M.R..; validation, M.J.K. and M.R.; formal analysis, M.J.K. and M.R.; investigation, M.J.K. and M.R.; resources, D.R. and H.H.; writing—original draft preparation, M.J.K. and M.R.; writing—review and editing, O.F., D.R., B.B.S., J.I.; supervision, O.F. and D.R.; funding acquisition, J.I., O.F., H.H., B.B.S. and D.R. All authors have read and agreed to the published version of the manuscript.

Funding: This research was funded by the German Research Foundation (grant numbers RA 1050/20-1, IV 98/5-1, HA 1344/32-1, FA 999/1-1), Russian Foundation for Basic Research (grant number 18-03-00067) and State tasks of the Russian federal ministry of education and science (grant number n/a), and by the Karlsruhe Nano Micro Facility.

Acknowledgments: The authors would like to thank Alena Gornakova from the Institute of Solid State Physics (Russian Academy of Sciences) for her assistance in the preparation of the alloys, Th. Kreschel from the Institute of Iron and Steel Technology (TU Bergakademie Freiberg) for the determination of the oxygen content in the samples by means of carrier hot gas extraction method, and Ch. Schimpf and A. Walnsch from the Institute of Materials Science (TU Bergakademie Freiberg) for conventional X-ray measurements and SEM/EBSD investigations.

Conflicts of Interest: The authors declare no conflict of interest.

References

1. Lütjering, G.; Williams, J.C. *Titanium*, 2nd ed.; Springer: Berlin/Heidelberg, Germany, 2007; ISBN 978-3-540-71397-5.

2. Ehtemam-Haghighi, S.; Prashanth, K.G.; Attar, H.; Chaubey, A.K.; Cao, G.H.; Zhang, L.C. Evaluation of mechanical and wear properties of TixNb7Fe alloys designed for biomedical applications. *Mater. Des.* **2016**, *111*, 592–599. [CrossRef]

3. Abdelrhman, Y.; Gepreel, M.A.-H.; Kobayashi, S.; Okano, S.; Okamoto, T. Biocompatibility of new low-cost ($\alpha+\beta$)-type Ti-Mo-Fe alloys for long-term implantation. *Mater. Sci. Eng. C* **2019**, *99*, 552–562. [CrossRef] [PubMed]

4. Zhang, L.-C.; Chen, L.-Y. A review on biomedical titanium alloys: Recent progress and prospect. *Adv. Eng. Mater.* **2019**, *21*, 1801215. [CrossRef]

5. Sandrock, G. A panoramic overview of hydrogen storage alloys from a gas reaction point of view. *J. Alloy. Compd.* **1999**, *293*, 877–888. [CrossRef]

6. Sujan, G.K.; Pan, Z.; Li, H.; Liang, D.; Alam, N. An overview on TiFe intermetallic for solid-state hydrogen storage: Microstructure, hydrogenation and fabrication processes. *Crit. Rev. Solid State Mater. Sci.* **2019**, 1–18. [CrossRef]

7. Edalati, K.; Akiba, E.; Horita, Z. High-pressure torsion for new hydrogen storage materials. *Sci. Technol. Adv. Mater.* **2018**, *19*, 185–193. [CrossRef]

8. Emami, H.; Edalati, K.; Matsuda, J.; Akiba, E.; Horita, Z. Hydrogen storage performance of TiFe after processing by ball milling. *Acta Mater.* **2015**, *88*, 190–195. [CrossRef]

9. Valiev, R.Z.; Islamgaliev, R.K.; Alexandrov, I.V. Bulk nanostructured materials from severe plastic deformation. *Prog. Mater. Sci.* **2000**, *45*, 103–189. [CrossRef]

10. Zhilyaev, A.P.; Langdon, T.G. Using high-pressure torsion for metal processing: Fundamentals and applications. *Prog. Mater. Sci.* **2008**, *53*, 893–979. [CrossRef]

11. Bachmaier, A.; Pippan, R. High-Pressure Torsion Deformation Induced Phase Transformations and Formations: New Material Combinations and Advanced Properties. *Mater. Trans.* **2019**, *60*, 1256–1269. [CrossRef]

12. Mazilkin, A.; Straumal, B.; Kilmametov, A.; Straumal, P.; Baretzky, B. Phase transformations induced by severe plastic deformation. *Mater. Trans.* **2019**, *60*, 1489–1499. [CrossRef]

13. Sauvage, X.; Chbihi, A.; Quelennec, X. Severe plastic deformation and phase transformations. *J. Phys. Conf. Ser.* **2010**, *240*, 12003. [CrossRef]

14. Murray, J.L. The Fe-Ti (iron-titanium) system. *Bull. Alloy Phase Diagr.* **1981**, *2*, 320–334. [CrossRef]

15. Dobromyslov, A.V.; Elkin, V.A. Martensitic transformation and metastable β-phase in binary titanium alloys with d-metals of 4–6 periods. *Scr. Mater.* **2001**, *44*, 905–910. [CrossRef]

16. Levi, I.; Shechtman, D. The microstructure of rapidly solidified Ti-Fe melt-spun ribbons. *Metall. Trans. A* **1989**, *20*, 2841–2845. [CrossRef]

17. Stupel, M.M.; Ron, M.; Weiss, B.Z. Phase identification in titanium-rich Ti-Fe system by Mössbauer spectroscopy. *J. Appl. Phys.* **1976**, *47*, 6–12. [CrossRef]

18. Banerjee, S.; Mukhopadhyay, P. *Phase Transformations*; Elsevier Science: Amsterdam, The Netherlands, 2007.

19. Heo, T.W.; Shih, D.S.; Chen, L.Q. Kinetic pathways of phase transformations in two-phase Ti alloys. *Metall. Mater. Trans. A* **2014**, *45*, 3438–3445. [CrossRef]

20. Matyka, J.; Faudot, F.; Bigot, J. Study of iron solubility in α titanium. *Scr. Metall.* **1979**, *13*, 645–648. [CrossRef]

21. Nosova, G.I.; D'yakonova, N.B.; Lyasotskii, I.V. Metastable phases of electron type in titanium alloys with 3d-metals. *Met. Sci. Heat Treat.* **2006**, *48*, 427–432. [CrossRef]

22. Sikka, S.K.; Vohra, Y.K.; Chidambaram, R. Omega phase in materials. *Prog. Mater. Sci.* **1982**, *27*, 245–310. [CrossRef]

23. Straumal, B.B.; Kilmametov, A.R.; Ivanisenko, Y.; Gornakova, A.S.; Mazilkin, A.A.; Kriegel, M.J.; Fabrichnaya, O.B.; Baretzky, B.; Hahn, H. Phase transformations in Ti-Fe alloys Induced by high-pressure torsion. *Adv. Eng. Mater.* **2015**, *17*, 1835–1841. [CrossRef]

24. Straumal, B.B.; Kilmametov, A.R.; Ivanisenko, Y.; Mazilkin, A.A.; Valiev, R.Z.; Afonikova, N.S.; Gornakova, A.S.; Hahn, H. Diffusive and displacive phase transitions in Ti-Fe and Ti-Co alloys under high pressure torsion. *J. Alloy. Compd.* **2018**, *735*, 2281–2286. [CrossRef]

25. Kilmametov, A.; Ivanisenko, Y.; Straumal, B.; Mazilkin, A.A.; Gornakova, A.S.; Kriegel, M.J.; Fabrichnaya, O.B.; Rafaja, D.; Hahn, H. Transformations of α′ martensite in Ti-Fe alloys under high pressure torsion. *Scr. Mater.* **2017**, *136*, 46–49. [CrossRef]

26. Kilmametov, A.R.; Ivanisenko, Y.; Mazilkin, A.A.; Straumal, B.B.; Gornakova, A.S.; Fabrichnaya, O.B.; Kriegel, M.J.; Rafaja, D.; Hahn, H. The α → ω and β → ω phase transformations in Ti-Fe alloys under high-pressure torsion. *Acta Mater.* **2018**, *144*, 337–351. [CrossRef]

27. Kriegel, M.J.; Kilmametov, A.; Rudolph, M.; Straumal, B.B.; Gornakova, A.S.; Stöcker, H.; Ivanisenko, Y.; Fabrichnaya, O.; Hahn, H.; Rafaja, D. Transformation pathway upon heating of Ti-Fe alloys deformed by high-pressure torsion. *Adv. Eng. Mater.* **2018**, *20*, 1700933. [CrossRef]

28. Chong, Y.; Deng, G.; Shibata, A.; Tsuji, N. Microstructure Evolution and Phase Transformation of Ti-1.0 wt%Fe Alloy with an Equiaxed α + β Initial Microstructure during High-Pressure Torsion and Subsequent Annealing. *Adv. Eng. Mater.* **2019**, *21*. [CrossRef]

29. Kriegel, M.J.; Kilmametov, A.; Klemm, V.; Schimpf, C.; Straumal, B.B.; Gornakova, A.S.; Ivanisenko, Y.; Fabrichnaya, O.; Hahn, H.; Rafaja, D. Thermal stability of athermal ω-Ti(Fe) produced upon quenching of β-Ti(Fe). *Adv. Eng. Mater.* **2019**, *21*, 1800158. [CrossRef]

30. Rietveld, H.M. Line profiles of neutron powder-diffraction peaks for structure refinement. *Acta Crystallogr.* **1967**, *22*, 151–152. [CrossRef]

31. Rietveld, H.M. A profile refinement method for nuclear and magnetic structures. *J. Appl. Crystallogr.* **1969**, *2*, 65–71. [CrossRef]

32. Coelho, A.A. *TOPAS, Version 5*; Bruker AXS GmbH: Karlsruhe, Germany, 2014.

33. Gornakova, A.S.; Prokofiev, S.I.; Kolesnikova, K.I.; Straumal, B.B. Formation regularities of grain-boundary interlayers of the α-Ti phase in binary titanium alloys. *Russ. J. Non-Ferr. Met.* **2016**, *57*, 229–235. [CrossRef]
34. Gornakova, A.S.; Prokofiev, S.I.; Straumal, B.B.; Kolesnikova, K.I. Growth of (αTi) grain-boundary layers in Ti-Co alloys. *Russ. J. Non-Ferr. Met.* **2016**, *57*, 703–709. [CrossRef]
35. Gornakova, A.S.; Straumal, B.B.; Nekrasov, A.N.; Kilmametov, A.; Afonikova, N.S. Grain Boundary Wetting by a Second Solid Phase in Ti-Fe Alloys. *J. Mater. Eng. Perform.* **2018**, *27*, 4989–4992. [CrossRef]
36. Gornakova, A.S.; Straumal, B.B.; Prokofiev, S.I. Coarsening of (αTi) + (βTi) Microstructure in the Ti-Al-V Alloy at Constant Temperature. *Adv. Eng. Mater.* **2018**, *20*, 1800510. [CrossRef]
37. Gornakova, A.S.; Straumal, A.B.; Khodos, I.I.; Gnesin, I.B.; Mazilkin, A.A.; Afonikova, N.S.; Straumal, B.B. Effect of composition, annealing temperature, and high pressure torsion on structure and hardness of Ti–V and Ti-V-Al alloys. *J. Appl. Phys.* **2018**, *125*, 82522. [CrossRef]
38. March, A. Mathematische Theorie der Regelung nach der Korngestalt bei affiner Deformation. *Z. Krist.* **1932**, *81*, 285–297.
39. Thompson, P.; Reilly, J.J.; Hastings, J.M. The application of the Rietveld method to a highly strained material with microtwins TiFeD1.9. *J. Appl. Crystallogr.* **1989**, *22*, 256–260. [CrossRef]
40. Kubaschewski, O. *IRON—Binary Phase Diagrams*; Springer: Berlin/Heidelberg, Germany, 1982; ISBN 978-3-662-08026-9.
41. Bo, H.; Wang, J.; Duarte, L.; Leinenbach, C.; Liu, L.B.; Liu, H.S.; Jin, Z.P. Thermodynamic re-assessment of Fe-Ti binary system. *Trans. Nonferrous Met. Soc. China* **2012**, *22*, 2204–2211. [CrossRef]
42. Jonsson, S. Assessment of the Fe-Ti system. *Metall. Trans.* **1998**, *29*, 361–370. [CrossRef]
43. Hari Kumar, K.C.; Wollaiits, P.; Delaey, L. Thermodynamic reassessment and calculation of Fe-Ti phase diagram. *Calphad* **1994**, *18*, 223–234. [CrossRef]
44. Schlieter, A.; Kühn, U.; Eckert, J.; Löser, W.; Gemming, T.; Friák, M.; Neugebauer, J. Anisotropic mechanical behavior of ultrafine eutectic TiFe cast under non-equilibrium conditions. *Intermetallics* **2011**, *19*, 327–335. [CrossRef]
45. Kao, Y.L.; Tu, G.C.; Huang, C.A.; Liu, T.T. A study on the hardness variation of α-and β-pure titanium with different grain sizes. *Mater. Sci. Eng. A* **2005**, *398*, 93–98. [CrossRef]
46. Ungár, T.; Schafler, E.; Hanák, P.; Bernstorff, S.; Zehetbauer, M. Vacancy production during plastic deformation in copper determined by in situ X-ray diffraction. *Mater. Sci. Eng. A* **2007**, *462*, 398–401. [CrossRef]
47. Devaraj, A.; Nag, S.; Srinivasan, R.; Williams, R.E.A.; Banerjee, S.; Banerjee, R.; Fraser, H.L. Experimental evidence of concurrent compositional and structural instabilities leading to ω precipitation in titanium-molybdenum alloys. *Acta Mater.* **2012**, *60*, 596–609. [CrossRef]
48. Hickman, B.S. The formation of omega phase in titanium and zirconium alloys: A review. *J. Mater. Sci.* **1969**, *4*, 554–563. [CrossRef]
49. Lukas, H.; Fries, S.G.; Sundman, B. *Computational Thermodynamics: The Calphad Method*; Cambridge University Press: Cambridge, UK, 2007; ISBN 9780521868112.
50. Kriegel, M.J.; Fabrichnaya, O. Pressure dependent modelling of the phase stabilities in the binary Ti-Fe system. *Calphad* **2020**. in preparation.
51. Ivanisenko, Y.; Kilmametov, A.; Rösner, H.; Valiev, R.Z. Evidence of α → ω phase transition in titanium after high pressure torsion. *Int. J. Mater. Res.* **2008**, *99*, 36–41. [CrossRef]
52. Polmear, I.; StJohn, D.; Nie, J.-F.; Qian, M. (Eds.) *Light Alloys*, 5th ed.; Butterworth-Heinemann: Boston, MA, USA, 2017; ISBN 978-0-08-099431-4.

Article

Microstructural and Mechanical Stability of a Ti-50.8 at.% Ni Shape Memory Alloy Achieved by Thermal Cycling with a Large Number of Cycles

Anna Churakova [1,2,*] and Dmitry V. Gunderov [1,2]

[1] Institute of Molecule and Crystal Physics—Subdivision of the Ufa Federal Research Centre of the Russian Academy of Sciences (IMCP UFRC RAS), 151 October Av., 450075 Ufa, Russia; dimagun@mail.ru
[2] Institute of physics of advanced materials, Ufa State Aviation Technical University, 12 K. Marx St., 450008 Ufa, Russia
* Correspondence: churakovaa_a@mail.ru; Tel.: +7-927-339-1086

Received: 30 December 2019; Accepted: 4 February 2020; Published: 6 February 2020

Abstract: The influence of thermal cycling (TC) with a large number of cycles on the microstructure, the parameters of martensitic transformations (MTs), and the mechanical properties of a Ti-50.8 at.% Ni shape-memory alloy in coarse-grained (CG) and ultrafine-grained (UFG) states was investigated. The effect of microstructural and mechanical stability was found in both coarse-grained and ultrafine-grained states starting from the 100th cycle of martensitic transformations. In addition, an unusual temperature change was observed in martensitic transformations occurring with the formation of an intermediate R phase.

Keywords: TiNi alloy; thermal cycling; ultrafine-grained structure; microstructural and mechanical stability

---

1. Introduction

The phenomenon of shape memory is characteristic of some alloys based on titanium, iron, copper, or manganese. Particular attention to materials with the shape memory effect (SME) is associated with their ability of recovering significant inelastic deformations when heated. The most famous representative of such materials is titanium nickelide. Its unique properties are widely used in world industries and in medicine [1,2]. The SMEs in titanium nickelide are caused by the B2–B19′ thermoelastic martensitic transformations (MTs) occurring in the temperature range close to room temperature [1–4]. The cycle of martensitic transformations upon cooling and heating leads to the generation of dislocations in the crystal lattice. Understanding the nature of the influence of multiple cycles of "cooling and heating" below and above the points of martensitic transformation—thermal cycling (TC)—on the structure and properties of materials is of great importance, particularly for TiNi alloys and their products. The phenomenon of phase hardening (PH)—the accumulation of dislocations during martensitic transformations—does not seem trivial in the case of martensitic transformation with a reversible motion of martensitic boundaries. The term "thermoelastic transformation" in the strict sense does not imply irreversible changes in the structure. At the same time, in real metallic materials, including TiNi alloys, a certain increase in the dislocation density occurs during multiple MT cycles, which, in turn, is accompanied by a change in the martensitic transformation temperature and an increase in the dislocation yield strength of the alloys under mechanical loading [5–7].

The design of products with SME makes certain demands on the physico-mechanical and functional properties and their stability. The properties in alloys with shape memory can be further improved by forming an ultrafine-grained (UFG) state using severe plastic deformation (SPD) methods, in particular, equal channel angular pressing (ECAP) [8–13]. Because TiNi system alloys are the most common in technological applications and have the best set of properties among alloys with a shape memory effect, the effect of thermal cycles (TCs) on their structure and properties has been studied for many years. In TiNi alloys, the transformation of B2 into B19′ is characterized by the incompatibility of lattice deformation, which contributes to the emergence of local stresses at the phase boundary, and stress relaxation leads to the accumulation of plastic deformation and, as a consequence, irreversible changes in the kinetics of martensitic transformations with each thermal cycling cycle [14]. The first works [5,6] were devoted to the influence of TCs on the structure and characteristic temperatures of martensitic transformations, and the mechanical characteristics in the TiNi alloy. It was previously shown that thermal cycling through the interval of martensitic transformations leads to a change in the staging of the transformation [15–19]. Alloy $Ti_{50.0}Ni_{50.0}$ undergoes B19′ → B2 transformation upon cooling. However, after several thermal cycles during cooling, the alloy begins to experience a multi-stage B2→R→B19′ transformation. At the same time, other studies report slightly different dependences of the transformation temperatures upon thermal cycling under an applied load [20,21]. There are additional studies of the "thermocyclic training" of TiNi alloys to enhance memory effects [20,22–24]. However, the studies were carried out mainly on alloys in a coarse-grained (CG) state, or in states with a small degree of deformation, and there is a limited number of studies on the processes of accumulation of dislocations and change in properties during thermal cycling of TiNi alloys in UFG and nanocrystalline (NC) states [25]. In addition, the conducted studies did not determine how many cycles were required to obtain optimal characteristics of the properties in these alloys. The studies in this work were aimed at determining the optimal number of thermal cycles necessary to obtain stability of the structure and the physico-mechanical properties of the TiNi alloy in coarse-grained and ultrafine-grained states.

2. Materials and Methods

As a research material, a two-component alloy of the TiNi system was chosen—the stoichiometric alloy Ti-50.8 at.% Ni enriched with nickel, manufactured by MATEK-SMA Ltd. (Moscow, Russia). It has a bcc lattice ordered by type B2 (CsCl) and a phase enriched with nickel Ti_2Ni_3 [1,4], and the chemical composition of the alloy is presented in Table 1. To obtain a solid solution and eliminate processing history, quenching was carried out from the homogeneity region (heating at a temperature of 800 °C in a Nabertherm furnace for 1 h) into the water. The average grain size of the hardened alloy was about 20 ± 2 μm.

Table 1. Chemical composition of the alloy of titanium nickelide, % (by atomic).

Main Elements		Impurity							
Ni	Ti	Fe	Si	C	N	O	H	Co	Remainder
50.8	Remainder	0.27	0.13	0.09	0.04	0.18	0.012	-	0.27

To carry out the deformation by the ECAP method, the equipment of the Ufa State Aviation Technical University (USATU) design in isothermal mode was used. To form the UFG structure, quenching samples of cylindrical TiNi alloys (Ø20 mm, length 100 mm) were subjected to 8 passes along the B_c route at 450 °C, and the channel intersection angle (φ) was 120° [9]. Thermal cycling of the samples in different initial states was carried out as follows: the samples were successively immersed in liquid nitrogen (-196 °C), then they were heated to a temperature of 150 °C, which is actually lower and higher than the temperatures M_f of direct and A_f reverse martensitic transformation. The number of heating–cooling thermal cycles ranged from 0 to 250. The thickness of the samples subjected to TC in the section was less than 1 mm to ensure their rapid heating and cooling. The exposure time at the heating and cooling temperatures was 5 min [25]. Quantitative and qualitative analyses of the initial structure were carried out using an OLYMPUS GX51 metallographic microscope. To detect the microstructure, an etchant with a composition of 60% H_2O + 35% HNO_3 + 5% HF was used. Using the random secant method, the sizes of structural elements were calculated.

X-ray diffraction studies of the samples were carried out on a Rigaku Ultima IV diffractometer (U = 40 kV and I = 35 mA) at room temperature in the angle range 2θ = 30°–120°. The main structure parameters were determined by the Rietveld method using the Materials Analysis Using Diffraction (MAUD) program. The dislocation density was calculated by processing X-ray diffraction data using MATLAB software. The formula was used to calculate the density of dislocations [8]: $\rho = 2\sqrt{3} < \varepsilon^2 >^{1/2}/Db$, where $<\varepsilon^2>^{1/2}$ represents the microdistortions, D is the average grain size, and b is the Burgers vector.

The fine structure of the material was studied at room temperature using a JEOL JEM-2100 transmission microscope ("JEOL Ltd.", Tokyo, Japan) with an accelerating voltage of 200 kV. Samples for thin foils cut by the electro-erosion method were made by double-sided jet electrolytic polishing using a Tenupol-5 device ("Struers", Copenhagen, Denmark) in a solution of 10% perchloric acid ($HClO_4$) and 90% butanol ($CH_3(CH_2)_3OH$).

The average size of structural elements (grains, subgrains, martensitic twins) was estimated using the "GrainSize" software package by measuring chord lengths, the relative measurement error of which did not exceed 5%. In this work, the calorimetric testing of the material was carried out on a Mettler Toledo high-sensitivity differential scanning calorimeter ("Mettler Toledo", Columbus, OH, USA) on samples weighing up to 50 mg (diameter 3.5 mm, thickness 0.5–0.7 mm), and the change in heat flux was studied during cooling and heating in the temperature range from −196 °C to 150 °C at a rate of 10 °C/min. The temperatures of the beginning (M_s and A_s) and the end (M_f and A_f) of the direct and reverse transformations were determined by standard tangent methods (ASTM 2004-05).

The microhardness Hv in this work was determined by the Vickers method on a Micromet 5101 instrument with a diamond indenter ("Buehler", Lake Bluff, OH, USA). Mechanical tensile tests of small flat specimens in compliance with all dimensional ratios with a working part of 1 × 0.25 × 4 mm were carried out with a strain rate of 1×10^{-3} s^{-1} on a special installation of the USATU design at room temperature. According to the test results, the strength characteristics (phase yield stress σ_m, dislocation yield strength σ_{YS}, ultimate tensile strength σ_{UTS}) and ductility (elongation, δ) were determined. The difference between the dislocation and phase yield limits allows one to estimate the stress $\sigma_{reac} = \sigma_{YS} - \sigma_m$ (estimated reactive stress) [4,25]. The length of the plateau at the phase yield stress stage was adopted as an estimate of the recovery strain ε_{rec} (Figure 1). The values of the conditional stresses of phase yield σ_m, dislocation yield stress σ_{YS}, and ultimate tensile strength σ_{UTS} are calculated as average statistical values for three samples.

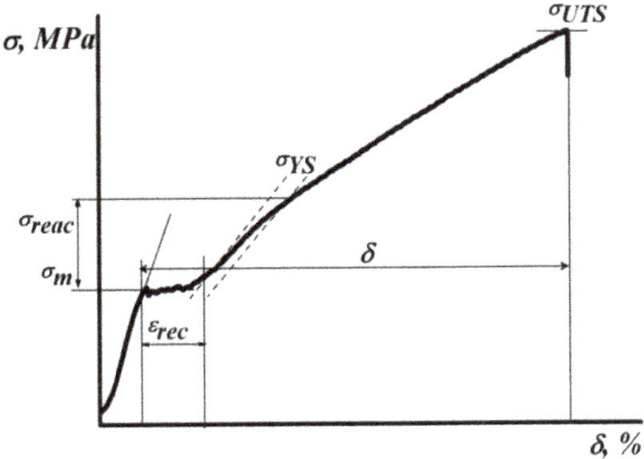

Figure 1. Diagram of a tensile stress indicating mechanical characteristics.

3. Results

The phase composition of TiNi alloys significantly affects the microstructure, functional and mechanical properties, as well as the processes that can occur during thermal cycling. It is known that with an increase in the Ni content, the tendency to phase hardening and, accordingly, the sensitivity of the structure to the accumulation of defects decrease. In this paper, the features of the influence of TCs on the characteristics of the alloy are considered on the example of an understated Ti-50.8 at.% Ni alloy up to 250 cycles.

3.1. Study of the Microstructure of the Ti-50.8 at.% Ni Alloy under Various Conditions

3.1.1. Structure in the Coarse-Grained State

In the initial state, after quenching, the alloy at room temperature has a predominantly equiaxial structure of the B2 phase (austenite) with an average grain size of about 20 ± 5 µm (Figure 2a). Moreover, globular inclusions 0.5–1 µm in size are observed in the structure inside and at the grain boundaries. Optical microscopy (OM) and scanning electron microscopy (SEM) (Figure 3) fail to accurately assess changes in the structure after repeated thermal cycling (Figure 2b); therefore, the TEM method was used (Figure 4).

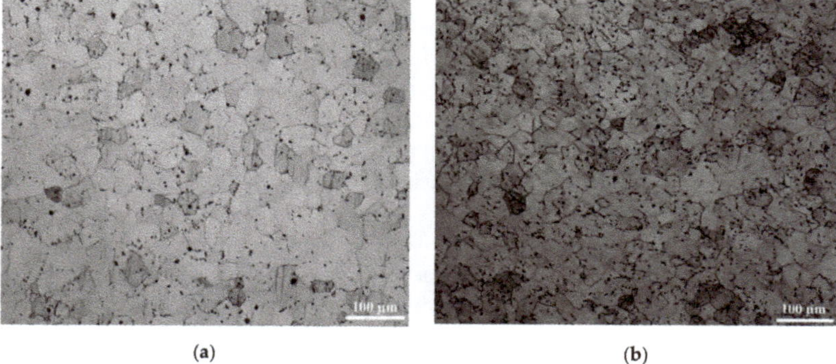

(a) (b)

Figure 2. Microstructure of the Ti-50.8 at.% Ni alloy obtained by optical microscopy (OM): (**a**) in the initial coarse-grained (CG) state, (**b**) after thermal cycling with a maximum number of cycles ($n = 250$).

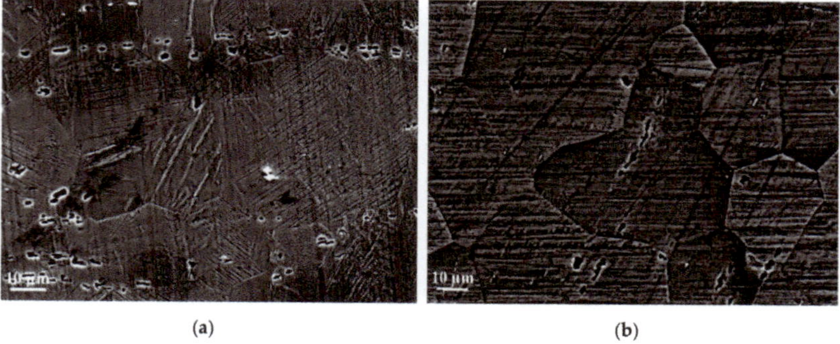

(a) (b)

Figure 3. SEM image of the microstructure of the Ti-50.8 at.% Ni alloy in the coarse-grained state: initial state (**a**), 250 thermal cycles (**b**).

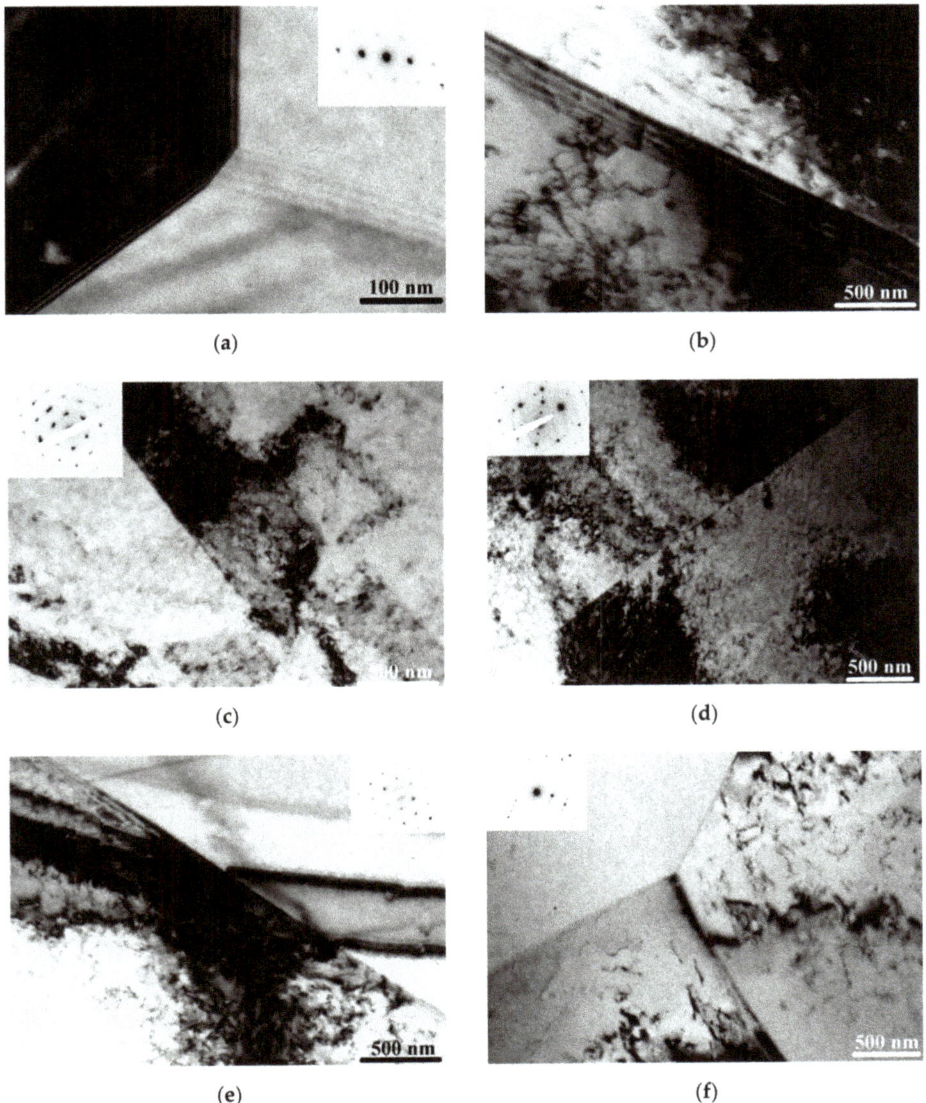

Figure 4. TEM images of the microstructure of the Ti-50.8 at.% Ni alloy in a coarse-grained state with different numbers of thermal cycles: (**a**) $n = 0$, (**b**) $n = 50$, (**c**) $n = 100$, (**d**) $n = 150$, (**e**) $n = 200$, (**f**) $n = 250$.

According to the obtained TEM data in the CG state without thermal cycling in the microstructure of the alloy, grain boundaries and triple-grain junctions free of dislocations are observed (Figure 4a). After thermal cycling in the temperature range MT B2→B19′ with a number of cycles equal to 50, developed dislocation clusters are present that form the so-called "dislocation forest". The average grain size decreased insignificantly and became about 18 ± 3 microns in size. With an increase in the number of thermal cycles in the structure, the formation of large clusters of dislocations and dislocation walls, which are formed during phase hardening, was observed. This was first noticeable near the grain boundaries (Figure 4b). In this case, there is a broadening of the extinction contours, which is also associated with an increase in the level of internal stresses and distortions of the crystal lattice.

The average grain size at the maximum number of cycles compared with the initial value decreased by about 45% (the assessment was carried out according to OM and SEM). Thermal cycling with the maximum number of thermal cycles preserves the dislocation structure in the form of clusters and irregular walls and dislocation tangles (Figure 4f).

Figure 5 presents structures after $n = 200$ and $n = 250$ cycles with fields with extinction contours, the width of which reaches 150 ± 20 nm, which may indicate a high density of defects accumulated during multiple martensitic transformations.

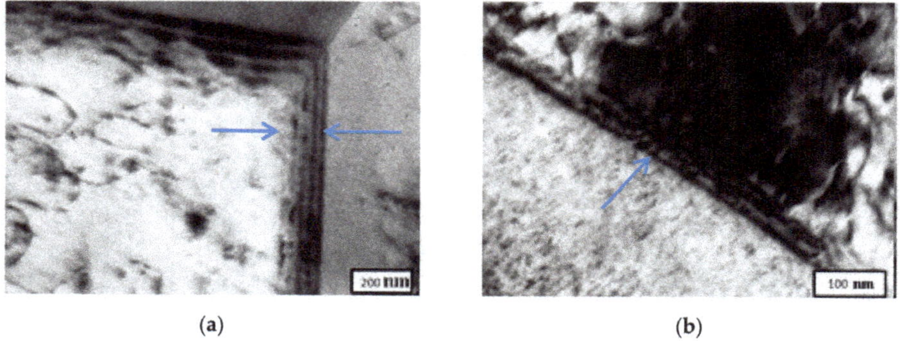

(a) (b)

Figure 5. Microstructures of the Ti-50.8 at.% Ni alloy in a coarse-grained state after thermal cycling with $n = 200$ (**a**) and $n = 250$ (**b**) with fields of the extinction contours.

Figure 6 shows a graph of the average grain size versus the number of cycles.

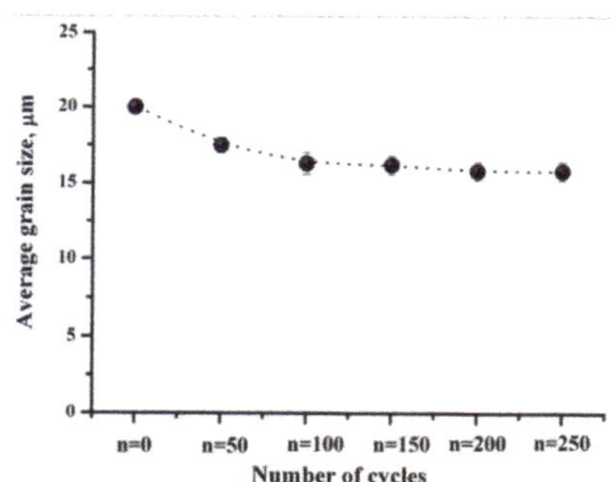

Figure 6. Graph of changes in the average grain size with an increase in thermal cycles in the CG state.

In addition, starting from 100 cycles of martensitic transformation, sections of the structure are observed in which packets of martensitic plates are visible (plate thickness in the range of 50–300 nm). Present in individual plates at high magnifications were composite (001) B19′ nanotwins of type I with a width of up to 5 nm. The formation of nanotwins is probably associated with saturation of the structure after a certain number of cycles.

3.1.2. Microstructure in the Ultrafine-Grained State

After applying the ECAP method, the transformation of the initial coarse-grained structure into an inhomogeneous grain–subgrain ultrafine-grained structure with an increased dislocation density was observed (Figure 7). The microdiffraction pattern corresponds to an ultrafine-grained structure with the presence of diffuse cords.

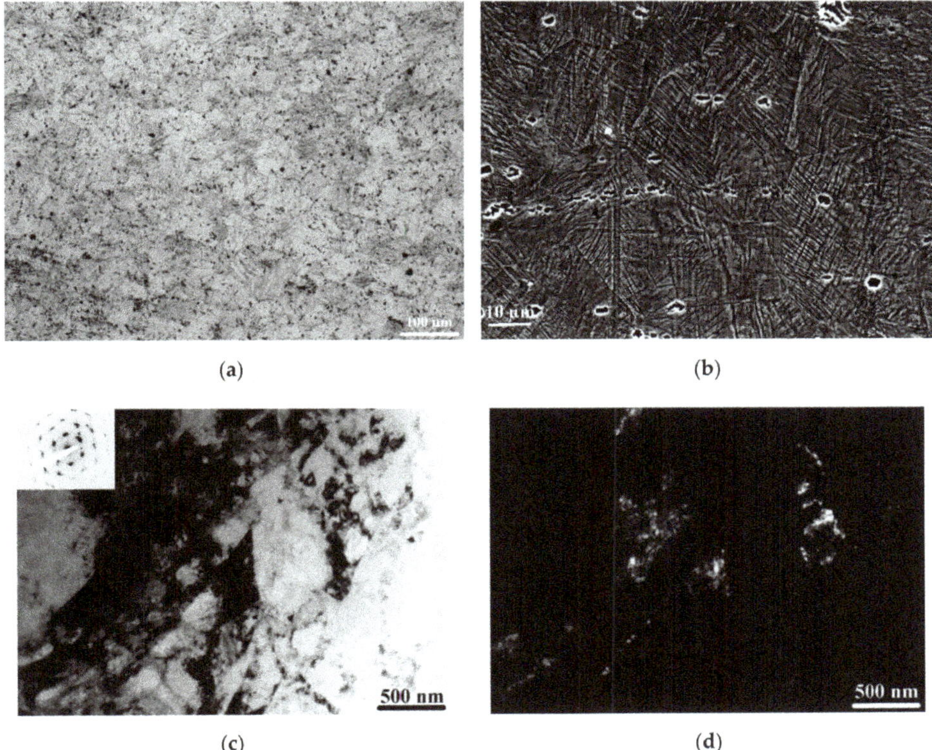

(a)   (b)

(c)   (d)

Figure 7. Microstructure of the Ti-50.8 at.% Ni alloy in the ultrafine-grained (UFG) state: (**a**) optical microscopy (OM), (**b**) scanning electron microscopy (SEM), (**c**) bright-field image, (**d**) dark-field image.

As the number of martensitic transformation cycles increases, grains are observed in the structure with predominantly nonequilibrium boundaries, which may indicate a highly defective structure (Figure 8). Nevertheless, complex dislocation structures formed in the grains—various accumulations and tangles—and the density of dislocations increased. The average size of structural elements in the UFG state without thermal cycling was 320 ± 15 nm, which decreased to 260 ± 20 nm after maximum thermal cycling (Figure 9).

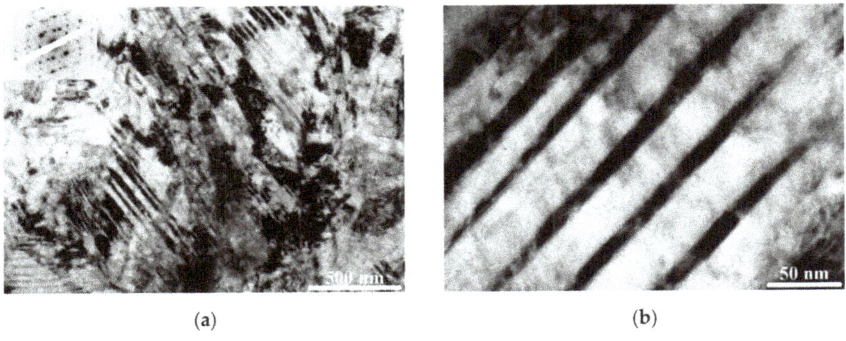

(a)   (b)

Figure 8. *Cont.*

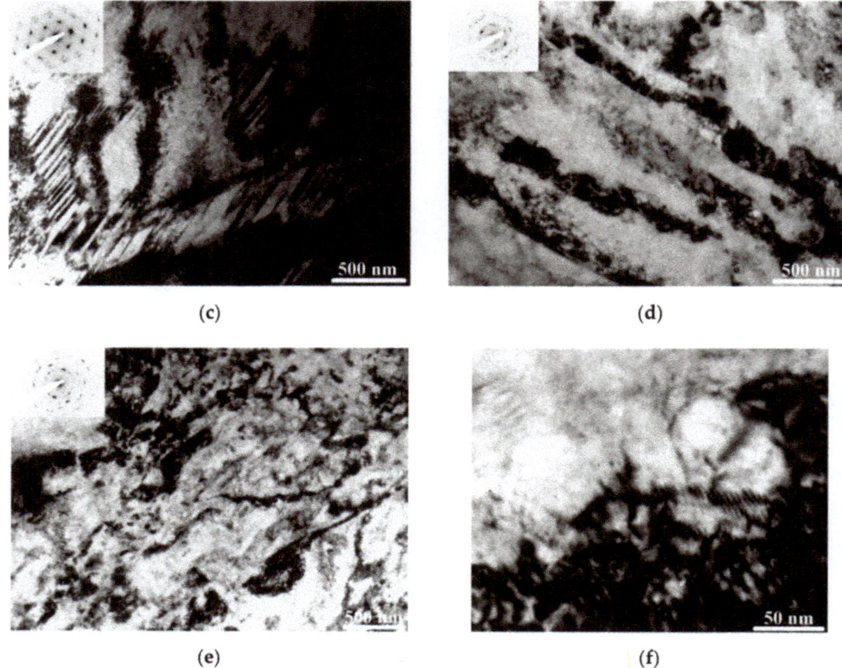

(c)   (d)

(e)   (f)

Figure 8. TEM images of the microstructure of the Ti-50.8 at.% Ni alloy in the UFG state: (**a**) $n = 100$, (**b**) $n = 100$ with nanotwins, (**c**) $n = 150$, (**d**) $n = 200$, (**e**) $n = 250$, (**f**) $n = 250$ with large magnification.

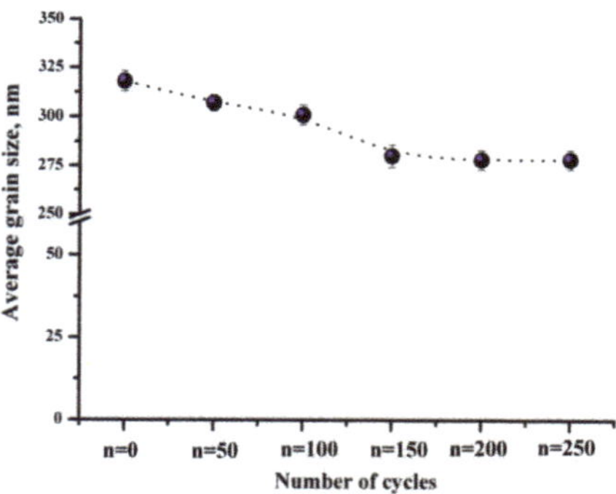

Figure 9. Changes in the average grain size with an increase in thermal cycles in the UFG state.

3.2. X-ray Analysis of the TiNi Alloy

To determine the effect of multiple martensitic transformations on the structural characteristics of the Ti-50.8 at.% Ni alloy, an X-ray diffraction analysis was performed at room temperature. X-ray diffraction patterns of the alloy indicate that the main phase in the CG state is B2-austenite (Figure 10). After TC with the maximum number of cycles, the X-ray diffraction pattern indicates a change in the phase composition; instead of peaks on the B2 phase, a doublet peak of phase B19′ is observed. A slight broadening of all peaks and a decrease in their intensity were observed (Figure 10). This can be explained by structural changes in the alloy—an increase in the density of dislocations accumulating during multiple cycles and an increase in internal microdistortions.

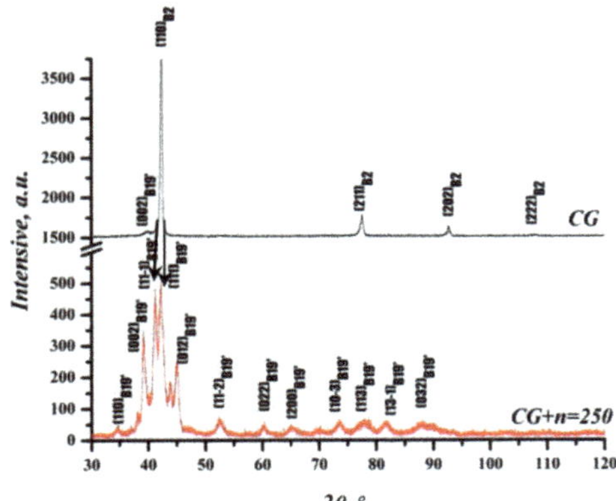

Figure 10. XRD patterns of the Ti-50.8 at.% Ni alloy in the coarse-grained state before thermal cycling and after maximum thermal cycling.

The X-ray diffraction pattern of the alloy in the UFG state also corresponds to the B2 phase of austenite (Figure 11). Moreover, there were more lines in this state than in the coarse-grained state. After TC, there was a broadening of the main peaks (Figure 11), which is caused by distortions of the crystal lattice and large values of microdistortions. However, there was no change in the phase composition of the alloy. The presence of only the B2 phase in the X-ray diffraction pattern in the ultrafine-grained state after the maximum number of cycles can be explained by the fact that, in this state, the structure is more homogeneous and the reverse transformation proceeds with the formation of an intermediate R phase with lower transformation energy. This all contributes to the reverse martensitic transformation, and in the UFG state it completely ends. In the case of the coarse-grained state, the structure is more heterogeneous, which affects the heterogeneity of the martensitic transformation; therefore, part of the material may contain a martensitic phase, the presence of which is recorded by X-ray diffraction analysis.

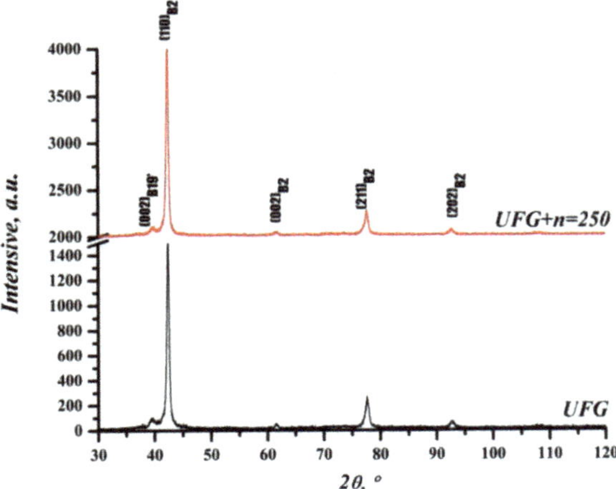

Figure 11. XRD patterns of the Ti-50.8 at.% Ni alloy in the ultrafine-grained state before thermal cycling and after maximum thermal cycling.

Based on the obtained X-ray diffraction data, the following structural parameters were calculated: coherent scattering regions (CSRs), lattice parameter (a), magnitude of the root-mean-square microdistortions of the crystal lattice ($<\varepsilon^2>^{1/2}$), and dislocation density (ρ) [26]. The results are shown in Table 2. An analysis of the structural parameters showed that in both states there is a decrease in the CSR values, an increase in internal microdistortions, and an increase in the dislocation density related to them. The dislocation density increased more in the ultrafine-grained state than in the coarse-grained state, which suggests that a higher density of grain boundaries and a smaller grain size contribute to the intensity of defect accumulation.

Table 2. Structural parameters of the Ti-50.8 at.% Ni alloy.

State	Parameters of Structure			
	Parameter Lattice a, Å	CSR, nm	$<\varepsilon^2>^{1/2} \times 10^{-4}$	$\rho \times 10^{15}$, m^{-2}
CG	3.013 ± 0.001	97 ± 2	0.8 ± 0.1	0.5 ± 0.1
CG + TC	2.895 ± 0.001 (monoclinic)	37 ± 2	2.2 ± 0.1	1.6 ± 0.1
Δ	0.118	60	1.4	1.1
UFG	3.011 ± 0.003	35 ± 3	2.7 ± 0.1	5.3 ± 0.15
UFG + TC	3.013 ± 0.001	19 ± 2	3.4 ± 0.1	7.1 ± 0.1
Δ	0.002	16	0.7	1.8

Δ = Parameter difference between the initial state of the alloy and the state after TC.

3.3. Differential Scanning Calorimetry (DSC)

According to the data obtained in the coarse-grained state, during direct martensitic transformation (DMT), one distinct exothermal peak appears on the DSC curves. During the reverse martensitic transformation (RMT), an endothermic peak is observed associated with the appearance of the high-temperature austenitic phase B2 from the martensitic phase B19′. After thermal cycling with $n =$ 100 cycles, a peak was observed from the intermediate R phase during direct martensitic transformation and a decrease in the temperatures of martensitic transformations (M_s, A_f). After the maximum number of cycles, a multidirectional change in temperatures was observed, including a slight decrease in the temperatures of the beginning of the direct conversion (M_s) and the end of the reverse (A_f), and an increase in the temperatures of the end of the direct conversion (M_f) and the beginning of the reverse (A_s) (Table 3). In the ultrafine-grained state, a decrease in martensitic temperatures was observed with an increase in the number of cycles from 0 to 250; at the same time, in this state, a transformation with the formation of an intermediate R phase proceeded without thermal cycling, and thermal cycling reduces temperatures—R_s (the temperature of the start of R transformation with direct MT) and R_f (the temperature of the end of the R transformation with direct MT). In this case, the decrease in the temperatures of direct martensitic transformation in the ultrafine-grained state was lower than that in the coarse-grained state. The graphs of the DSC curves of Ti-50.8 at.%Ni alloy are plotted according to Table 3 (Figure 12).

Table 3. Results of differential scanning calorimetry (DSC) of the Ti-50.8 at.% Ni alloy.

States	Number of Cycles	M_s, °C	M_f, °C	R_s, °C	R_f, °C	A_s, °C	A_f, °C	A_f^R, °C
CG	$n = 0$	3.07	−60.11	-	-	−25.07	26.83	-
	$n = 100$	−38.60	−98.32	8.46	-	−29.55	8.47	-
	$n = 250$	−49.63	−70.23	11.62	−29.5	−18.33	−8.14	-
UFG	$n = 0$	−13.52	−80.96	39.44	2.18	6.11	33.16	-
	$n = 100$	−16.98	−74.27	38.60	2.24	7.83	36.65	-
	$n = 250$	−59.72	−83.09	15.10	−32.07	−19.87	−9.87	7.97

A_f^R = Temperature of the end of the reverse transformation with an intermediate R phase.

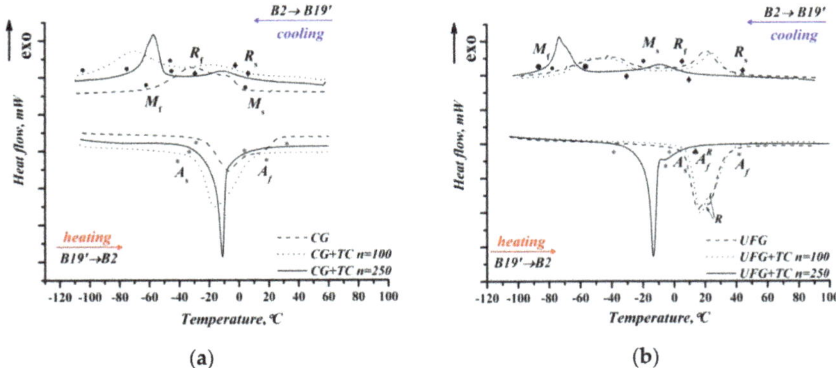

(a)   (b)

Figure 12. DSC curves of the Ti-50.8 at.% Ni alloy in CG (**a**) and UFG (**b**) states.

3.4. Mechanical and Functional Characteristics of Ti-50.8 at.% Ni Alloy during Thermal Cycling

In both CG and UFG states of the Ti-50.8 at.%Ni alloy, the values of microhardness as a result of multiple martensitic transformations increased slightly compared to the state before thermal cycling (Figure 13). However, in the first 50–100 cycles, a more intense increase in microhardness is characteristic, then the values stabilize and saturation occurs after 100 cycles of transformations.

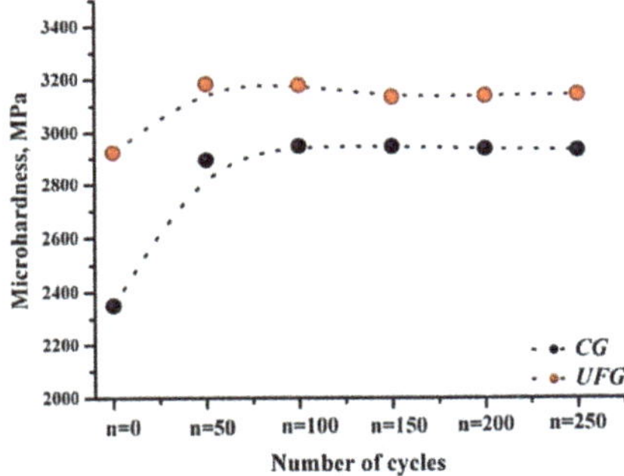

Figure 13. The dependence of microhardness on the number of thermal cycles in various states.

The results of mechanical tensile tests are presented in Figures 14 and 15 for CG and UFG states. The functional characteristics determined from the analysis of mechanical tensile tests are presented in the form of graphs of the dependence of the estimated reactive stress (σ_{reac}) and plateau length at the phase yield stage, used as an estimate of the recovery deformation (ε_{rec}) on the number of thermal cycles (Figure 16).

The most sensitive characteristic to thermal cycling is the yield strength. In both states of the alloy, it increases with an increase in the number of cycles. UFG states up to TC are characterized by higher values of strength and yield strength due to the contribution of grain boundary hardening. The tensile strength in the CG state increased to 150 cycles, then the values decreased and did not change. This is probably due to the fact that after a certain number of cycles (in CG state, $n = 150$), the material is saturated and no further increase in characteristics is observed. In the UFG state, with an increase in the number of cycles of multiple transformations, the values of the tensile strength increased until reaching 100 thermal cycles; then, a decrease in the parameter was observed, which can be explained by the heterogeneity of the formed structure.

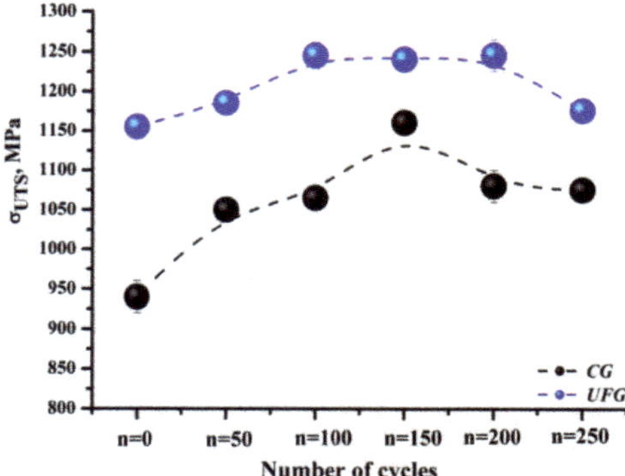

Figure 14. The dependence of the ultimate tensile stress of the alloy on the number of thermal cycles.

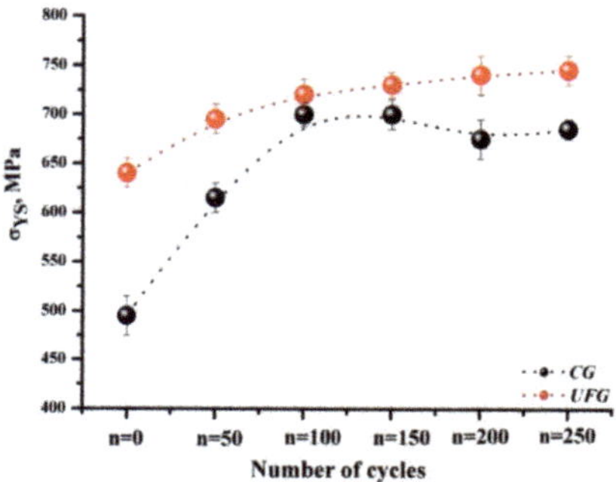

Figure 15. The dependence of the yield stress of the alloy on the number of thermal cycles.

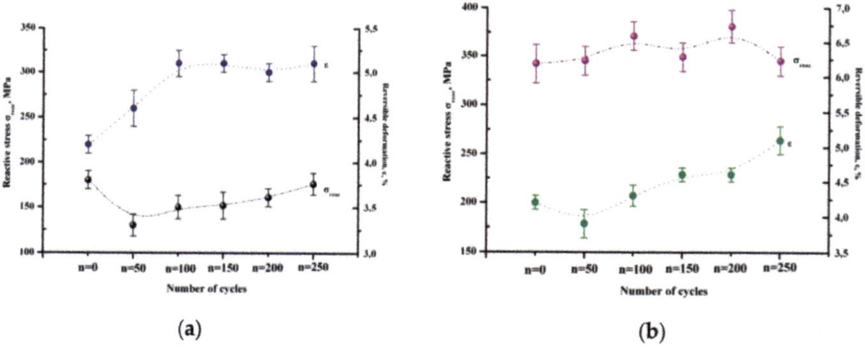

(a)   (b)

Figure 16. The dependence of the functional characteristics of the alloy on the number of thermal cycles: (**a**) coarse-grained state, (**b**) ultrafine-grained state.

According to the constructed graphs (Figure 16), the functional characteristics (the magnitude of the reversible deformation and the estimated reactive stress) increased slightly and then remained stable both in the CG and in the UFG states with an increase in thermal cycles.

4. Discussion

The results of this study showed that during multiple martensitic transformations, dislocations were generated and accumulated in the Ti-50.8 at.% Ni alloy. This process occurs intensively in the first

100 cycles, after which a saturation effect is observed and the structure and mechanical characteristics practically do not change. The saturation effect also leads to the stabilization of martensite, which is reflected in changes in the temperatures of martensitic transformations. Thus, the obtained results allow us to state that, as a result of thermal cycling to 100–150 cycles, a change in structural parameters and mechanical and functional characteristics is observed, but a subsequent increase in the number of cycles, in general, results in the stabilization of the alloy in both coarse and ultrafine grains. Thus, it can be said that, for this alloy, 100–150 cycles are sufficient to enhance the mechanical and functional characteristics and form a stable structure.

5. Conclusions

1. As a result of thermal cycling in the Ti-50.8 at.% Ni alloy, an increase in the dislocation density occurs, internal stresses in the CG and UFG states increase, the size of the structural components decreases slightly, which is associated with the formation of dislocation walls and sub-boundaries.
2. When studying the microhardness of titanium nickelide in the CG and UFG states as a result of multiple cycles, the values slightly increase compared to the state before thermal cycling. The first 100 cycles are characterized by a more intense increase in microhardness, then the values stabilize.
3. As a result of mechanical tensile tests for the alloy in both CG and UFG states, the mechanical properties increase slightly, especially the yield strengths and phase yield stresses. Saturation occurs after 150 cycles. Furthermore, in the states under study as a result of repeated martensitic transformations, the functional characteristics—the estimated reactive stress and the length of the phase yield area, which determines the magnitude of the reversible deformation—remain stable.
4. The Ti-50.8 at.% Ni alloy in the UFG state is more attractive for applications, since in this state a higher level of properties is obtained compared to the coarse-grained state. In addition, the UFG state shows greater stability during thermal cycling with a large number of cycles.

Author Contributions: Investigation, writing—original draft preparation, A.C.; writing—review and editing, D.V.G. All authors have read and agreed to the published version of the manuscript.

Funding: This research was funded by a grant of the Republic of Bashkortostan, the Russian Federation, for young scientists, grant number 28 of 07.03.2019. The reported study was funded partially by the Russian Foundation for Basic Research (RFBR) according to the research project No. 19-58-80018 BRICS_t.

Acknowledgments: Anna Churakova acknowledges Natalya Nikolaevna Resnina (St. Petersburg State University) for conducting the DSC research.

Conflicts of Interest: The authors declare no conflict of interest. The funders had no role in the design of the study; in the collection, analyses, or interpretation of data; in the writing of the manuscript; or in the decision to publish the results.

References

1. Khachin, V.N.; Pushin, V.G.; Kondratyev, V.V. *Titanium Nickelide: Structure and Properties*; Nauka: Moscow, Russia, 1992; p. 161.
2. Brailovski, V.; Prokoshkin, S.; Terriaultet, P.; Trochu, F. *Shape Memory Alloys: Fundamentals, Modeling, Applications*; Ecolede technology esuperieure (ETS): Montréal, QC, Canada, 2003; p. 851.
3. Otsuka, K.; Wayman, C.M. *Shape Memory Materials*; Cambridge University Press: Cambridge, UK, 1999; p. 284.
4. Gunther, V.E.; Dambaev, G.C.; Sysolyatin, P.G.; Ilyushenov, V.N. *Medical Materials and Implants with Shape Memory*; TSU: Tomsk, Russia, 1998; p. 487.
5. Miyazaki, S.; Igo, Y.; Otsuka, K. Effect of thermal cycling on the transformation temperatures of Ti-Ni alloys. *Acta Metall.* **1986**, *34*, 2045–2051. [CrossRef]
6. Erofeev, V.; Ya, E.V.; La, M.; Yu, P. Phase hardening during martensitic transformation of titanium nickelide. *FMM* **1982**, *53*, 963–965.
7. Mironov, Y.P.; Erokhin, P.G.; Kul'kov, S.N. Evolution of the crystal structure during phase hardening of titanium nickelide. *Univ. News Phys.* **1997**, *2*, 100–104.

8. Valiev, R.Z.; Alexandrov, I.V. *Bulk Nanostructured Metallic Materials: Preparation, Structure and Properties*; Akademkniga: Moscow, Russia, 2007; p. 398.

9. Valiev, R.Z.; Islamgaliev, R.K.; Alexandrov, I.V. Bulk nanostructured materials from severe plastic deformation. *Prog. Mater. Sci.* **2000**, *45*, 103–189. [CrossRef]

10. Tatyanin, E.V.; Kurdyumov, V.G.; Fedorov, V.B. Obtaining an amorphous TiNi alloy during shear deformation under pressure. *FMM* **1986**, *62*, 133–137.

11. Valiev, R.Z.; Gunderov, D.V.; Pushin, V.G. Metastable nanostructured SPD TiNi alloys with unique properties. *J. Metastable Nanocryst. Mater.* **2005**, *24*, 7–12.

12. Prokoshkin, S.D.; Khmelevskaya, I.Y.; Trubitsyna, I.B.; Dobatkin, S.V.; Tata'yanin, E.V.; Stolyarov, V.V.; Prokof'ev, E.A. Structure evolution during intense plastic deformation of shape memory alloys based on TiNi. *FMM* **2004**, *97*, 84–90.

13. Valiev, R.Z.; Gunderov, D.V.; Lukyanov, A.V.; Pushin, V.G. Mechanical behavior of nanocrystalline TiNi alloy produced by SPD. *J. Mater. Sci.* **2012**, *47*, 7848–7853. [CrossRef]

14. Otsuka, K.; Ren, X. Physical metallurgy of Ti-Ni-based shape memory alloys. *Prog. Mater. Sci.* **2005**, *50*, 511–678. [CrossRef]

15. Wagner, M.F.-X.; Dey, S.R.; Gugel, H.; Frenzel, J.; Somsen, C.; Eggeler, G. Effect of low-temperature precipitation on the transformation characteristics of Ni-rich NiTi shape memory alloys during thermal cycling. *Intermetallics* **2010**, *8*, 1172–1179. [CrossRef]

16. Tang, W.; Sandström, R. Analysis of the influence of cycling on TiNi shape memory alloy properties. *Mater. Des.* **1993**, *14*, 103–113. [CrossRef]

17. Pelton, A.R.; Huang, G.H.; Moinec, P.; Sinclaird, R. Effects of thermal cycling on microstructure and properties in Nitinol. *Mater. Sci. Eng. A* **2012**, *532*, 130–138. [CrossRef]

18. Wayman, C.M.; Cornelis, I.; Shimizu, K. Transformation behavior and the shape memory in thermally cycled TiNi. *Scr. Metall.* **1972**, *6*, 115–122. [CrossRef]

19. Wasilewski, R.J.; Butler, S.R.; Hanlon, J.E. On the martensitic transformation in TiNi. *Metal Sci.* **1967**, *1*, 104–110. [CrossRef]

20. Jones, N.G.; Dye, D. Martensite evolution in a NiTi shape memory alloy when thermal cycling under an applied load. *Intermetallics* **2011**, *19*, 1348–1358. [CrossRef]

21. McCormick, P.G.; Liu, Y. Thermodynamic analysis of the martensitic transformation in NiTi— II. Effect of transformation cycling. *Acta Metal. Mater.* **1994**, *42*, 2407–2413. [CrossRef]

22. Belyaev, S.; Resnina, N.; Sibirev, A. Peculiarities of residual strain accumulation during thermal cycling of TiNi alloy. *J. Alloys Compd.* **2012**, *542*, 37–42. [CrossRef]

23. Belyaev, S.; Resnina, N.; Zhuravlev, R. Deformation of Ti-51.5at.% Ni alloy during thermal cycling under different thermal-mechanical conditions. *J. Alloys Compd.* **2013**, *577*, S232–S236. [CrossRef]

24. Belyaev, S.; Resnina, N. Stability of mechanical behavior and work performance in TiNi-based alloys during thermal cycling. *Inter. J. Mater. Res.* **2013**, *104*, 11–17. [CrossRef]

25. Churakova, A.A.; Gunderov, D.V.; Dmitriev, S.V. Microstructure transformation and physical and mechanical properties of ultrafine-grained and nanocrystalline TiNi alloys in multiple martensitic transformations B2-B19'. *Materialwissenschaft Werkstofftechnik* **2018**, *49*, 769–778. [CrossRef]

26. Scardi, P.; Lutterotti, L.; Di Maggio, R. Size-Strain and quantitative analysis by the Rietveld method. *Adv. X-Ray Anal.* **1991**, *35A*, 69–76. [CrossRef]

Article

Effect of the High-Pressure Torsion (HPT) and Subsequent Isothermal Annealing on the Phase Transformation in Biomedical Ti15Mo Alloy

Kristína Bartha [1,*], Josef Stráský [1], Anna Veverková [1], Pere Barriobero-Vila [2], František Lukáč [3], Petr Doležal [4], Petr Sedlák [5], Veronika Polyakova [6], Irina Semenova [6] and Miloš Janeček [1]

[1] Department of Physics of Materials, Charles University, 12000 Prague, Czech Republic; josef.strasky@gmail.com (J.S.); annaterynkova@gmail.com (A.V.); janecek@met.mff.cuni.cz (M.J.)
[2] Institute of Materials Research, German Aerospace Center (DLR), 51147 Cologne, Germany; Pere.BarrioberoVila@dlr.de
[3] Institute of Plasma Physics, Czech Academy of Sciences, 18000 Prague, Czech Republic; lukac@ipp.cas.cz
[4] Department of Physics of Condensed Matter, Charles University, 12000 Prague, Czech Republic; petr.dolezal@mff.cuni.cz
[5] Institute of Thermomechanics, Czech Academy of Sciences, 18000 Prague, Czech Republic; petr.sedlak@fjfi.cvut.cz
[6] Institute of Physics of Advanced Materials, Ufa State Aviation Technical University, 450000 Ufa, Russia; vnurik@gmail.com (V.P.); semenova-ip@mail.ru (I.S.)
* Correspondence: kristina.bartha@met.mff.cuni.cz; Tel.: +420-95155-1361

Received: 17 October 2019; Accepted: 5 November 2019; Published: 7 November 2019

Abstract: Ti15Mo metastable beta Ti alloy was solution treated and subsequently deformed by high-pressure torsion (HPT). HPT-deformed and benchmark non-deformed solution-treated materials were annealed at 400 °C and 500 °C in order to investigate the effect of UFG microstructure on the α-phase precipitation. Phase evolution was examined using laboratory X-ray diffraction (XRD) and by high-energy synchrotron X-ray diffraction (HEXRD), which provided more accurate measurements. Microstructure was observed by scanning electron microscopy (SEM) and microhardness was measured for all conditions. HPT deformation was found to significantly enhance the α phase precipitation due the introduction of lattice defects such as dislocations or grain boundaries, which act as preferential nucleation sites. Moreover, in HPT-deformed material, α precipitates are small and equiaxed, contrary to the α lamellae in the non-deformed material. ω phase formation is suppressed due to massive α precipitation and consequent element partitioning. Despite that, HPT-deformed material after ageing exhibits the high microhardness exceeding 450 HV.

Keywords: β titanium alloys; severe plastic deformation; α phase precipitation; phase composition; high energy synchrotron X-ray diffraction

---

1. Introduction

The interest in metastable β-Ti alloys has gradually increased due to their high specific strength, which make them ideal for applications in the aerospace industry [1]. Manufacturing of medical implants and devices is another high-added-value field and constitutes a prospective application of metastable β titanium alloys [2]. These alloys offer higher strength levels than commonly used $\alpha + \beta$ alloys due to controlled precipitation of tiny particles of α phase [3]. The strength of titanium and Ti-based alloys can be further improved by achieving an ultra-fine grained (UFG) structure via severe plastic deformation methods (SPD) [4–7]. Manufacturing of UFG metastable β-Ti alloys is of significant interest as demonstrated by recent reports [8–11]. This study focuses on the effect of microstructure refinement by SPD on the precipitation of α phase upon subsequent thermal treatment.

The α phase forms in metastable β Ti alloys by a standard mechanism of nucleation and growth. The kinetics of α phase formation in a solution-treated material is generally given by chemical composition of the alloy and the temperature of ageing. However, α phase precipitation also strongly depends on the microstructure since α phase particles nucleate at preferential sites such as grain and subgrain boundaries and dislocations [12,13].

In some β-Ti alloy, nanosized ω phase particles can also act as preferential nucleation sites for precipitation of α phase, although the exact mechanism of ω formation has not been fully resolved yet [14–18]. It is therefore of significant interest to investigate α phase precipitation in the presence of high concentration of lattice defects (UFG material) in an alloy prone to the ω phase formation. Binary metastable Ti15Mo alloy used in this study consists, similarly to other β-Ti alloys with similar degree of β stabilization, of a mixture of β and ω_{ath} (athermal ω) phases in the solution-treated (ST) condition [19,20]. This ω_{ath} phase forms during quenching of the alloy from the temperatures above β transus by a displacive diffusionless mechanism [21]. However, it was also reported that ω phase can form as a result of high deformation, and is referred to as deformation induced ω [8,22]. The $\beta \rightarrow \omega_{ath}$ transformation is reversible up to a temperature of 110 °C [23]. Upon ageing at higher temperatures, the ω_{ath} particles become stabilized by diffusion i.e., in Ti15Mo alloy by expelling Mo. This phase is referred to as ω_{iso} (isothermal ω) [24]. The size of particles of ω_{iso} phase is typically in the range of few nanometers up to 100 nm [19,24].

Solution-treated Ti15Mo alloy was processed by high-pressure torsion (HPT) [5] in order to achieve ultra-fine grained (UFG) microstructure with a high density of lattice defects. The equivalent von Mises strain achieved by HPT is heterogeneous and can be calculated according to Equation (1) [25]:

$$\varepsilon_{vM} = \frac{2\pi N r}{\sqrt{3}h} \tag{1}$$

where r represents the distance from the center of the sample, h is the thickness of the specimen, and N is the number of revolutions. The equivalent inserted von Mises strain after a single HPT rotation ($N = 1$, $r = 10$ mm and thickness h of 1 mm) ranges from 0 (the exact center) up to 35 = 3500% (periphery). Such extreme strain results in a dislocation density exceeding $\rho = 5 \times 10^{14}$ m^{-2} and grain size in the range of hundreds of nanometers [11,26].

The objective of this study was to investigate the effect of SPD on the mechanisms and kinetics of α phase precipitation and to compare it with the precipitation in the non-deformed solution-treated material. Both laboratory X-ray diffraction (XRD) and high-energy XRD using synchrotron radiation (HEXRD) were employed for this experimental study.

2. Materials and Methods

Ti15Mo (wt%) alloy was supplied by Carpenter Co. (Richmond, VA, USA) in the form of a rod with the diameter of 10 mm. The as-received material was solution treated (ST) in an inert Ar atmosphere at the temperature of 810 °C for 4 h and subsequently water quenched. The cylindrical samples of the length of 5 mm were first cut from the rod and pressed with 6 GPa to achieve disk-shaped samples with the diameter of 20 mm and the thickness of approximately 1 mm. Disk samples were subsequently subjected to HPT deformation at room temperature at Ufa State Aviation Technical University (USATU), Russian Federation. Note that this pre-HPT deformation induced deformation of about 75% is significantly lower than the actual HPT deformation. The detailed description of the HPT method can be found elsewhere [5]. For this study, samples after $N = 1$ HPT rotation were prepared.

ST and HPT deformed samples (hereafter referred to as non-deformed and HPT-deformed, respectively) were aged at temperatures of 400 °C and 500 °C for 1, 4, and 16 h. The ageing was performed by immersing the samples to preheated salt-bath (i.e., with very high heating rate) without air access and subsequently water quenched. HPT samples for scanning electron microscopy (SEM) and microhardness study were prepared from the periphery part of the disks (>5 mm from the center) where the imposed strain is maximum. For XRD measurements, a quarter of a disk-shaped sample was used. All samples were mechanically grinded and polished by standard methods followed by a three-step vibratory polishing.

The microstructure of the specimens was observed using SEM Zeiss Auriga Compact Cross Beam (Jena, Germany) equipped with the energy dispersive spectroscopy (EDS) detector operated at 4 kV. K-line and L-line for Ti and Mo, respectively, were used for quantification of EDS data.

XRD measurement of the non-deformed Ti15Mo alloy was performed employing a Bruker D8 Advance powder X-ray diffractometer using Cu K_α radiation (Bruker AXS GmbH, Karlsruhe, Germany), with a variable divergent slit and a Sol X detector. The width of the beam was 6 mm and the sample was rotated, allowing it to probe the whole specimen surface. XRD measurements of the HPT-deformed Ti15Mo alloy were carried out on Bruker D8 Discover powder X-ray diffractometer. Vertical Bragg-Brentano geometry (2.5° Soller slits in both primary and secondary beam and 0.24° divergence slit) with filtered Cu K_α radiation was used. Beam size of 20 mm × 5–15 mm (depending on the θ angle) was used. Note that the beam size is comparable to the size of the HPT disk. Diffraction patterns in both cases were collected at room temperature in the 2θ range from 30° to 130° and were analyzed using LeBail approach in the program Jana2006 (Václav Petříček, Michal Dušek and Lukáš Palatinus, Institute of Physics Academy of Sciences, Prague, Czech Republic).

The HEXRD measurement was carried out at the P07-HEMS beamline of PETRA III (Deutsches Elektronen-Synchrotron, Hamburg, Germany) [27] using the energy of 100 keV (λ = 0.124 Å) in a transmission mode perpendicular to the HPT surface. Patterns of entire Debye–Scherrer rings were acquired ex-situ from the bulk of Ti15Mo samples at room temperature. A PerkinElmer XRD 1621 image plate detector was used. The samples were kept fixed during the acquisition and measured 5 mm from the center of the sample with an incident beam of slit size of 1 × 1 mm^2. The acquired diffraction patterns were processed by Rietveld structural refinement using FullProf software (Juan Rodriguez Carvajal, ILL Grenoble, France). Azimuthal averaging over 360° was performed first. HEXRD diffractograms after azimuthal averaging were treated by the March–Dollase approach in order to obtain at least rough estimates of volume fractions of individual phases. In fact, β phase peaks, which are the most intensive, could be fitted without the March–Dollase 'adjustment'. Nevertheless, diffractions of evolving α and ω phases had to be treated by the March–Dollase approach to achieve reasonable agreement with the measured data [28]. Even with the use of the March–Dollase approach the resulting R-factors of the fit accuracy range between 15–20 for α and ω phases, while R-factor for β phase is around 10.

The microhardness of the samples was measured using the Vickers method with the use of microhardness tester Qness Q10a (Golling, Austria). Note that all samples from HPT disks were cut from the area at least 5 mm distant from the center where microhardness of HPT N = 1 condition is saturated [8]. For each specimen 0.5 kg load and the dwell time of 10 s were applied. At least 20 indents were evaluated for each sample in order to get satisfactory statistical results.

3. Results

3.1. Initial Conditions

Microstructure

The initial microstructure of the ST Ti15Mo alloy is shown in Figure 1. A coarse-grained (CG) structure consisting of grains of the average size of ~50 μm is well visible due to the channeling contrast. In addition, brighter and darker and areas are also visible in the material due to the chemical contrast (marked by yellow arrows in Figure 1). These chemical inhomogeneities were investigated by EDS. Table 1 summarizes the results of the EDS point analysis. Note that results were obtained by standardless EDS in which the measured spectra are compared to the data collected from standards by the EDS manufacturer under different conditions (namely different beam conditions). Such data are therefore subjected to systematic error and as such they are not fully reliable in terms of exact quantitative Mo content determination. However, the relative difference in chemical composition between different areas is accurate and unambiguous.

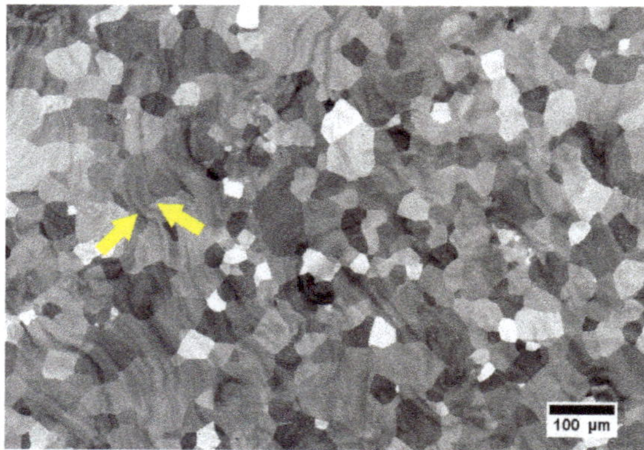

Figure 1. Scanning electron microscopy–back-scattered electrons (SEM–BSE) micrograph of the solution-treated (ST) Ti15Mo alloy (the yellow arrows indicate chemical inhomogeneities in the material).

Table 1. Chemical composition of the darker and brighter bands (marked by yellow arrows in Figure 1) as determined by energy dispersive spectroscopy (EDS) point analysis.

Element	Brighter Part (wt.%)	Darker Part (wt.%)
Ti	83.3 ± 0.5	85.7 ± 0.4
Mo	16.7 ± 0.5	14.3 ± 0.4

Local chemical inhomogeneities in the ST material were also investigated by EDS mapping. In Figure 2a, several β-grains and darker and lighter areas (visible especially in the top left corner of the image) are visible due to channeling contrast and chemical contrast (Z-contrast), respectively. EDS mapping confirms the chemical inhomogeneity—darker areas in Figure 2a contain less Mo as shown in Figure 2b. Variations in the local content of Mo (β stabilizing element) may affect the phase stability of the β phase matrix.

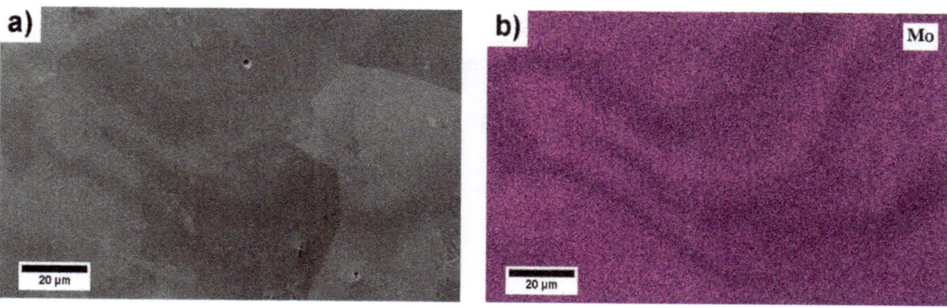

Figure 2. Local chemical inhomogeneities in ST Ti15Mo alloy: (**a**) scanning electron microscopy–secondary electrons (SEM–SE) micrograph of the area of interest, (**b**) corresponding element map of Mo using EDS.

The chemical inhomogeneities were also studied in the HPT-deformed sample. The SEM–BSE micrograph in Figure 3 clearly shows lighter and darker bands corresponding to the chemical composition differences, which were also confirmed by EDS. Darker areas with lower Mo were formed from curly bands in the non-deformed material (Figure 1). In HPT deformed material, they are elongated in the direction of the deformation (Figure 3).

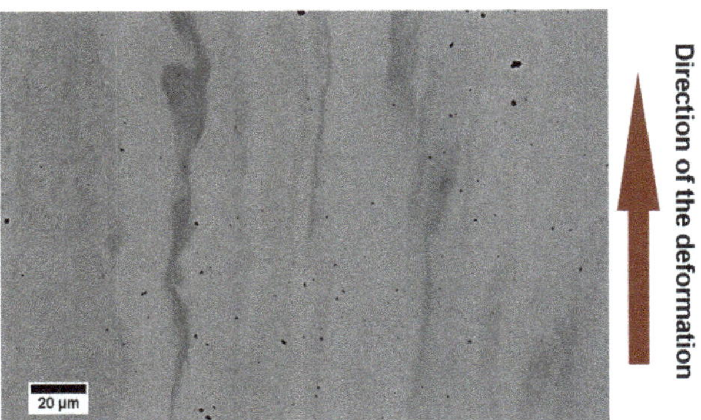

Figure 3. SEM–BSE micrograph of Ti15Mo alloy after high-pressure torsion (HPT) processing. Surface of HPT disk with highlighted direction of deformation (azimuthal direction). Darker and brighter areas are caused by the difference in chemical composition (black dots are polishing artefacts).

3.2. Non-Deformed and HPT-Deformed Material after Ageing

3.2.1. The Evolution of Microstructure during Ageing

SEM–BSE micrographs in Figure 4 show the evolution of the microstructures of the non-deformed material after ageing at 400 °C and 500 °C for 1–16 h. The microstructure consists of coarse-grained β matrix. At least one triple-junction is shown in each image. After ageing at 400 °C for 1 and 4 h, only β matrix is observed. In the specimen aged for a longer time (400 °C/16 h), nanometer-sized precipitates are seen in the SEM–BSE micrograph (note the higher magnification of this micrograph). These small ellipsoidal particles are particles of ω phase, which are visible due to chemical partitioning—ω phase particles are slightly Mo depleted [29]. Ageing at the higher temperature of 500 °C resulted in a precipitation of continuous and coarse α phase along grain boundaries (hereafter referred to as grain boundary α or GB α), which is also Mo depleted and appears as a long dark particle along the former β/β boundary (indicated by a yellow arrow). In the vicinity of the β grain boundaries, α phase particles with a typical lamellar morphology precipitated. Small ellipsoidal particles in the grain interior belong to ω phase. The contrast of these particles increases with the increasing ageing time due to ongoing chemical partitioning. After ageing at 500 °C for 16 h, tiny ellipsoidal ω particles are clearly seen in grain interiors, GB α is visible along the former β/β grain boundaries and α lamellae span from the GB α to the grain interiors. In conclusion, ω particles with ellipsoidal morphology can be observed by SEM in the grain interior after ageing at 500 °C and the fraction of α phase particles with lamellar morphology increases with increasing time of ageing at 500 °C. The coexistence of all three β, α, and ω phases is observed.

Figure 5 shows the microstructures of the HPT-deformed samples after ageing. Already after ageing at 400 °C/1 h, significant differences between the non-deformed and HPT-deformed specimens can be observed. In the non-deformed material, there is no evidence of α phase particles. On the other hand, small and equiaxed α particles already precipitated in HPT sample. In specimens aged at 400 °C for longer times of 4 h and 16 h, the volume fraction of the α phase increased and, simultaneously, α precipitates coarsened. Moreover, the precipitation is not homogeneous—some areas contain clearly more α phase particles. Ageing at 500 °C resulted in the formation of larger α particles, which are generally equiaxed, but not round—rather polygonal and sharp edged.

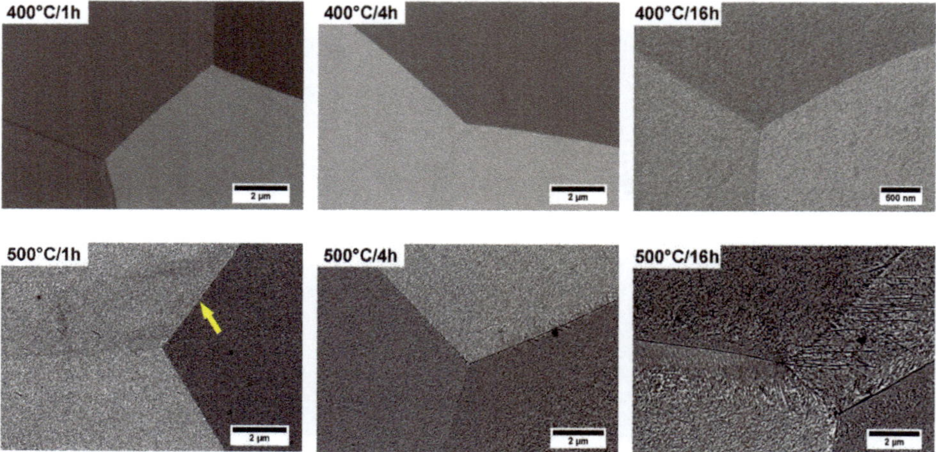

Figure 4. SEM–BSE micrographs of the non-deformed samples after ageing.

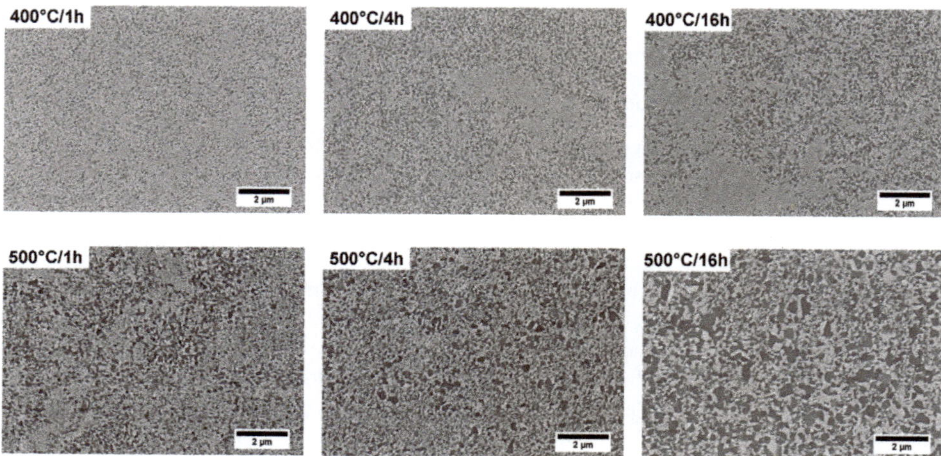

Figure 5. SEM–BSE micrographs of the samples deformed by high-pressure torsion (HPT-deformed samples) after ageing.

3.2.2. Evolution of Phase Composition during Ageing

The phase composition of the non-deformed and HPT-deformed samples before and after ageing is shown in laboratory XRD patterns in Figures 6 and 7, respectively. Both the measured (thin black curves) and fitted (colored curves) XRD patterns are shown. The interplanar distance is displayed on the horizontal axis for the comparison to the HEXRD data, while the y-axis shows the intensity in a logarithmic scale, allowing one to distinguish small peaks. The most important peaks, which best describe the evolution of emerging phases, are marked with arrows—full and open arrows for α and ω phase, respectively. A quantitative determination of phase content is not possible. The non-deformed specimen contains large grains with the size of hundreds of micrometers while HPT-deformed material is severely plastically deformed with high dislocation density and high internal stress resulted in the broadening of XRD peaks. Moreover, in both conditions, the grains have a preferred orientation as can be revealed from the relative intensity of the peaks. Therefore, laboratory XRD patterns could not be successfully fitted by any other method than the simple LeBail approach. However, several qualitative comparisons can be made.

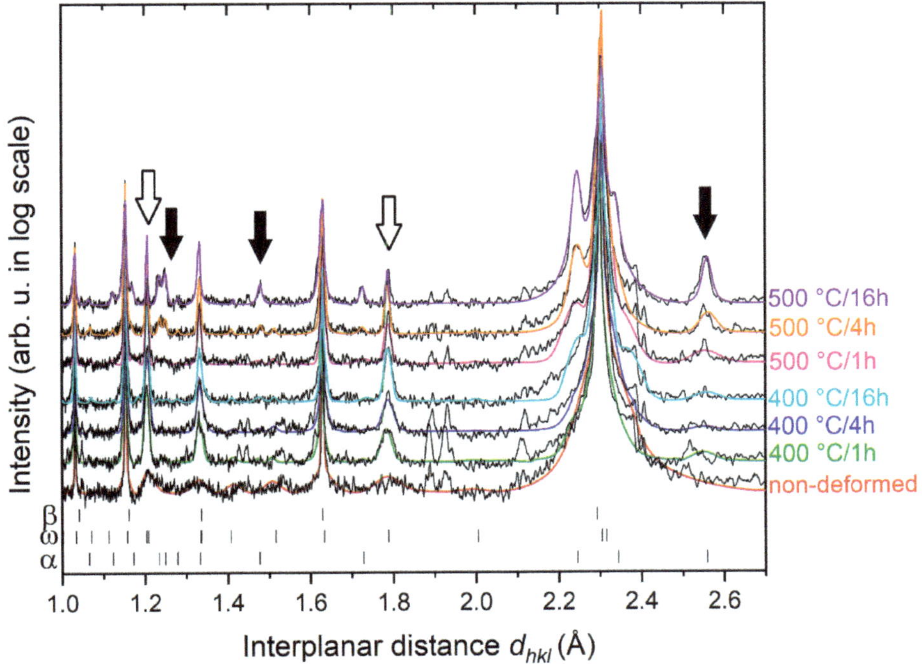

Figure 6. X-ray diffraction (XRD) patterns (in log-scale) of aged conditions of the non-deformed Ti15Mo alloy. Black thin curves correspond to data, colored curves are numerical fits. Non-deformed without ageing (red curve) and aged conditions (other colored curves) are displayed. The patterns are vertically shifted for clarity. The most important peaks are marked by full and open arrows for α and ω phase, respectively. Two unfitted peaks around $d_{hkl} = 1.9$ originated from the sample holder.

The non-deformed Ti15Mo alloy contains a mixture of β and ω phases. However, the identification of the ω phase content is difficult due to overlapping peaks of β and ω peaks. Nevertheless, ω peaks $(11\text{–}22)_{\omega}$ and $(11\text{–}21)_{\omega}$ can be observed at the inter planar distances $d_{hkl} \approx 1.2$ Å and $d_{hkl} \approx 1.8$ Å, respectively, as shown in Figure 6. Ageing of the non-deformed specimen at 400 °C/1 h (red curve in Figure 8) resulted in an increase of the intensity and narrowing of ω peaks (open arrows). In addition, small α phase peaks are also visible (full black arrows). XRD patterns of the non-deformed material aged at 400 °C for 1, 4, and 16 h are very similar. In the specimen aged at 500 °C for 1 h, the α phase is clearly present (full arrows in Figure 6). Moreover, its volume fraction increases with increasing ageing time (4–16 h), as also confirmed by SEM observations (cf. Figure 5); ω phase is still present in the specimen aged 500 °C even for the longest time of 16 h reported in [29].

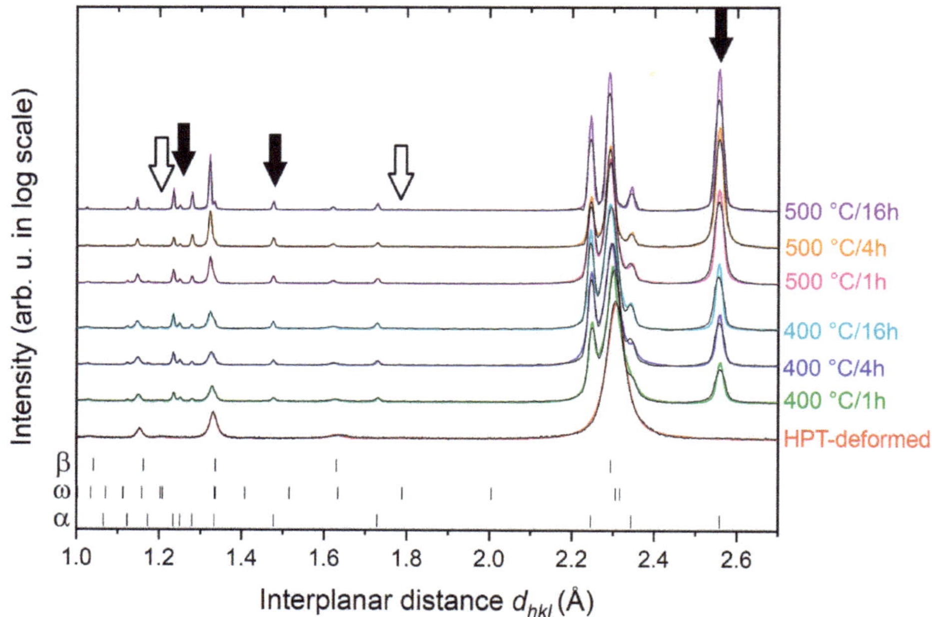

Figure 7. XRD patterns of aged conditions of the HPT-deformed Ti15Mo alloy: HPT-deformed without ageing (black curve) and aged under different conditions (colored curves) are displayed. The patterns are vertically shifted for clarity. The most important peaks are marked by full and open arrows for α and ω phase, respectively.

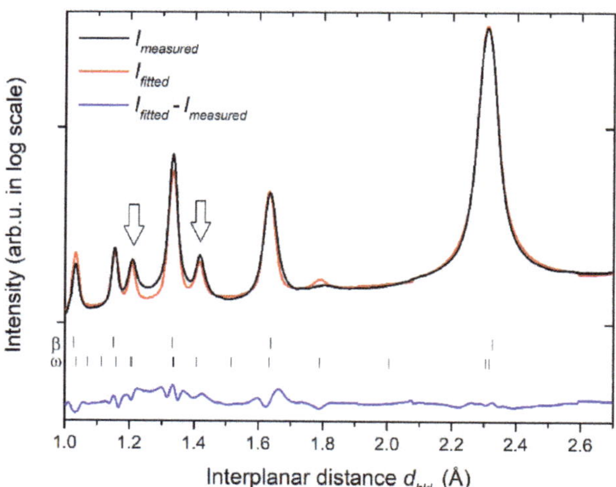

Figure 8. High-energy synchrotron X-ray diffraction (HEXRD) pattern of the HPT-deformed sample.

The XRD pattern of the HPT deformed specimen exhibits significantly broadened peaks due to enhanced dislocation density and reduced crystallite size in this specimen. Moreover, the peaks are slightly shifted to different values of interplanar distances due to residual stresses in the deformed material—the direction of the shift depends on the type of the residual stress [30].

The ω phase content in the HPT sample seems to be inferior to that in the non-deformed sample; only a tiny peak can be resolved at the interplanar distance $d_{hkl} \approx 1.2$ Å. However, the most intensive peaks of the ω phase coincide with the peaks of the β phase.

In order to obtain more precise information about volume fraction of individual phases, HEXRD measurement was carried out on the HPT-deformed sample. In contrast to the laboratory XRD, HEXRD provides a better signal-to-noise ratio and the simultaneous measurement of the scattering signal in various directions due to the use of a 2D detector and subsequent azimuthal averaging. Consequently, a better resolution of small peaks, namely those of the ω phase, is achieved. Figure 8 shows the HEXRD pattern of the HPT specimen. Both the measured and fitted intensities as well as the difference curve between the fitted and measured intensity are shown in Figure 8. The results indicate that the HPT-deformed alloy is a two-phase material with volume fractions of the β and ω phase of 72% and 28%, respectively (the error of the volume fractions estimation is approximately ±5%).

In order to get more accurate results, the selected HPT-deformed specimen after ageing at 400 °C/1 h was examined using HEXRD. Both the measured and fitted intensity and the difference between the fitted and measured data are displayed in Figure 9.

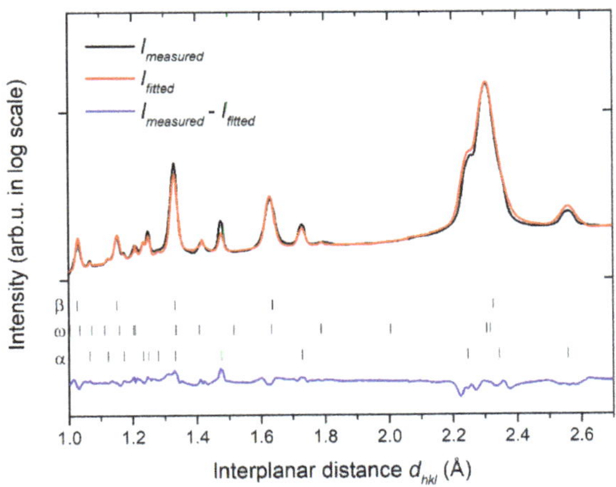

Figure 9. HEXRD pattern of the HPT-deformed sample after aging 400 °C/1 h.

The volume fractions of individual phases in HPT-deformed sample before and after ageing at 400 °C/1 h are summarized in Table 2. The non-aged HPT-deformed material contains a high-volume fraction of the ω phase (28%). After ageing, the volume fraction decreases to approximately 9%. This is caused by enhanced volume fraction of the α phase, which reaches 23% in the aged condition.

Table 2. Volume fraction of individual phases in HPT-deformed Ti15Mo alloy as determined from high energy synchrotron X-ray diffraction (HEXRD). The experimental errors are also shown.

Material	Volume Fraction of the β Phase	Volume Fraction of the ω Phase	Volume Fraction of the α Phase
Ti15Mo HPT	72% ± 5%	28% ± 5%	-
Ti15Mo HPT + ageing 400 °C/1 h	67% ± 5%	9% ± 3%	23% ± 4%

3.2.3. Microhardness Evolution during Ageing

Figure 10 shows the dependence of Vickers microhardness on ageing of the non-deformed and HPT-deformed samples. Ageing of both samples at 400 °C resulted in an abrupt increase of microhardness. In the non-deformed specimen, the microhardness increases with increasing ageing time at 400 °C. On the other hand, in the HPT-deformed sample aged at 400 °C for 1 h the microhardness reaches the maximum (500 HV). With increasing ageing time at 400 °C the microhardness continuously decreases. Specimens aged at 500 °C exhibit lower microhardness than specimens aged at 400 °C. The microhardness of HPT sample aged at 500 °C for 16 h even drops below the microhardness of the non-aged HPT specimen.

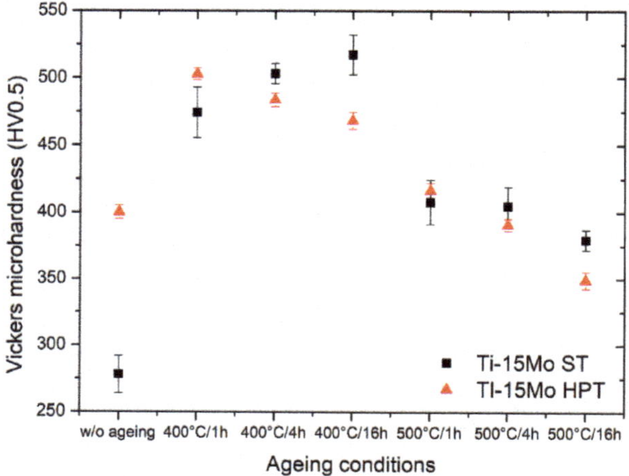

Figure 10. The evolution of the microhardness during ageing in the non-deformed and HPT-deformed samples.

4. Discussion

4.1. Enhanced α Phase Precipitation

A significant difference in the evolution of α phase in the non-deformed and the HPT-deformed conditions was observed. α phase is known to precipitate preferentially along the grain boundaries as so-called grain boundary α (GB α) [31]. Enhanced α phase precipitation was also found in pre-deformed materials due to the high dislocation density [32,33]. High concentration of defects in the HPT-deformed condition reduce the energy barrier for the nucleation of α phase. The growth of an α nuclei and its coarsening is controlled by the diffusion of Mo (β stabilizing element) in the β matrix [34]. It is well-known known that the pipe diffusion along dislocation cores as well as the diffusion rate along grain boundaries are several orders of magnitude higher than the bulk diffusion [35]. As a consequence, the growth of the α precipitates along grain boundaries is also accelerated. The enhanced precipitation of the α phase in severely deformed metastable β Ti alloys was reported in several studies [36,37]. In the coarse-grained material, α phase particles precipitate in the form of lamellae, because certain mutual orientations of neighboring α and β lattices are associated with the significantly lower interfacial energy and therefore, lamellar shape is optimal for the reduction of the total interfacial energy of a precipitate [38]. On the other hand, α particles in the HPT-deformed materials are equiaxed, but not round—detailed inspection of Figure 5 reveals that particles are rather polygonal and sharp edged. It is assumed that α phase particles nucleate at triple junctions and all observed α particles are in fact GB α.

Ageing of HPT-deformed material (particularly at 400 °C/16 h) resulted in an inhomogeneous precipitation of α particles. Such inhomogeneity was reported to be caused by shear bands formed in the HPT deformed material [36,37,39,40]. However, we did not find any shear bands in the HPT material. On the other hand, we observed chemical inhomogeneities both in the non-deformed and in the HPT-deformed sample. In the latter case, the inhomogeneities are extended in the direction of HPT deformation (cf. Figure 3). As a consequence, the nucleation of the α phase particles may be therefore promoted in the areas depleted in Mo, even if the shape and the scale of precipitation inhomogeneities in Figure 5 cannot be directly compared to Mo-depleted regions in Figure 3 due to very different magnification (zone of observation).

In the specimen aged at 500 °C, ω phase was retained in the non-deformed material while it was completely absent in the HPT-deformed specimen. Enhanced precipitation of the α phase results in the rejection of the β stabilizing Mo to the surrounding β matrix causing a thermodynamic stabilization of the β matrix and suppression of the formation of the ω phase HPT deformed material [26]. Similar behavior, i.e., the preferred α phase precipitation over the formation of the ω phase, was observed in Ti-25Nb-2Mo–4Sn alloy deformed by cold-rolling [41].

4.2. Microhardness Evolution

Microhardness is significantly increased by HPT deformation as discussed in detail in [8] due to the microstructural refinement, the introduction of high dislocation density, and increased content of ω phase. Ageing of both non-deformed and HPT-deformed materials at 400 °C resulted in microhardness increase. Moreover, similar microhardness values were observed in both conditions. However, the similar increase of the microhardness can be attributed to different effects in both conditions.

The hardening of the non-deformed material aged at 400 °C is caused by the ω particles—the nano-sized ω particles are stabilized by diffusion, their size increases, and they act as much stronger obstacles for motion of dislocations. One may assume that a moving dislocation can pass through (cut) ω_{ath} particles (known as Friedel effect [42]) as they are small and coherent. It is well known that the shear stress required for a dislocation to pass through a precipitate increases with its increasing size (within the Friedel's limit) and/or with increasing strength of the obstacle to dislocation motion [42,43]. Due to this and also because of the increasing volume fraction of ω phase, the hardness of the non-deformed material increases with increasing ageing time at 400 °C.

In HPT-deformed material, ageing at 400 °C already for 1 h results in the precipitation of tiny α phase particles, which are incoherent and cause significant Orowan strengthening. On the other hand, ω phase content is relatively low. The decreasing microhardness of the HPT material aged at 400 °C for longer times (4 and 16 h) may be related to the coarsening of the $\alpha + \beta$ microstructure. Both β matrix grains and α phase precipitates coarsen with increasing ageing time. The same process is even more pronounced during ageing at 500 °C. The microhardness was found to monotonically decrease with increasing ageing temperature and time. The maximum microhardness is therefore achieved in the HPT specimen aged at 400 °C. In a recent study [44], HPT deformation of Ti15Mo/TiB composite was performed at 400 °C and very high microhardness values (650 HV after $N = 1$ HPT revolution) were achieved.

The microhardness of the non-deformed material aged at 500 °C is inferior to that of the material aged at 400 °C for all ageing times. The relative decrease of microhardness of the non-deformed material with increasing temperature of ageing may be attributed to the decreasing volume fraction and increasing size of ω particles, whose size is well beyond the Friedel's limit.

Severe plastic deformation of the parent β phase and the introduction of high density of defects significantly accelerates the α phase precipitation. An important additional effect of this enhancement is the reduction of the content of ω phase and its disappearance at comparatively low ageing temperatures.

5. Conclusions

Metastable β titanium Ti15Mo alloy was prepared by HPT and subsequently aged at 400 °C and 500 °C. Phase transformations were observed by XRD and SEM. The following conclusions can be drawn from this study:

Precipitation of the α phase in the HPT-deformed material is significantly enhanced by the high density of lattice defects such as dislocations or grain boundaries, which act as preferential nucleation sites for α phase precipitation.

α phase particles in the non-deformed material precipitate in the form of lamellae, while in the HPT-deformed material, α precipitates are small, equiaxed, and polygonal in shape suggesting that they all formed as grain boundary α at the β grains triple junctions.

Two-phase α + β microstructure continuously coarsens with increasing temperature and time of ageing.

Deformation by HPT significantly increases the microhardness due to microstructural refinement, but also due to the formation of ω phase.

During annealing, the microhardness of the non-deformed material is governed mainly by the evolution of ω phase. In the HPT-deformed material, the main strengthening mechanism is the precipitation of fine α phase particles.

Author Contributions: K.B. conducted SEM experiments, interpreted the results, and wrote the majority of the manuscript. J.S. and M.J. discussed achieved results and wrote parts of the manuscript. A.V. cooperated on SEM observations and microhardness measurements. P.S. interpreted achieved mechanical properties. P.B.-V. conducted HEXRD experiment. F.L. and P.D. conducted laboratory XRD experiments. V.P. and I.S. prepared material by HPT.

Funding: This work was financially supported by the Czech Science Foundation under the project 17-04871S and by ERDF, project No. CZ.02.1.01/0.0/0.0/15_003/0000485. The Deutsches Elektronen-Synchrotron (DESY) is acknowledged for the provision of synchrotron radiation facilities in the framework of the proposal I-20150533 EC. Partial financial support by the Czech Ministry of Education, Youth and Sports under the project LTARF18010 is also gratefully acknowledged.

Conflicts of Interest: The authors declare no conflict of interest.

References

1. Boyer, R.R. An overview on the use of titanium in the aerospace industry. *Mater. Sci. Eng. A* **1996**, *213*, 103–114. [CrossRef]
2. Steinemann, S.G. Titanium—the material of choice? *Periodontol 2000* **1998**, *17*, 7–21. [CrossRef]
3. Weiss, I.; Semiatin, S.L. Thermomechanical processing of beta titanium alloys—An overview. *Mater. Sci. Eng. A* **1998**, *243*, 46–65. [CrossRef]
4. Valiev, R.Z.; Islamgaliev, R.K.; Alexandrov, I.V. Bulk nanostructured materials from severe plastic deformation. *Prog. Mater. Sci.* **2000**, *45*, 103–189. [CrossRef]
5. Zhilyaev, A.P.; Langdon, T.G. Using high-pressure torsion for metal processing: Fundamentals and applications. *Prog. Mater. Sci.* **2008**, *53*, 893–979. [CrossRef]
6. Kawabe, Y.; Muneki, S. Strengthening and Toughening of Titanium Alloys. *ISIJ Int.* **1991**, *31*, 785–791. [CrossRef]
7. Valiev, R.Z.; Langdon, T.G. Principles of equal-channel angular pressing as a processing tool for grain refinement. *Prog. Mater. Sci.* **2006**, *51*, 881–981. [CrossRef]
8. Václavová, K.; Stráský, J.; Polyakova, V.; Stráská, J.; Nejezchlebová, J.; Seiner, H.; Semenova, I.; Janeček, M. Microhardness and microstructure evolution of ultra-fine grained Ti-15Mo and TIMETAL LCB alloys prepared by high pressure torsion. *Mater. Sci. Eng. A* **2017**, *682*, 220–228. [CrossRef]
9. Kent, D.; Wang, G.; Yu, Z.; Ma, X.; Dargusch, M. Strength enhancement of a biomedical titanium alloy through a modified accumulative roll bonding technique. *J. Mech. Behav. Biomed.* **2011**, *4*, 405–416. [CrossRef]
10. Yilmazer, H.; Niinomi, M.; Nakai, M.; Cho, K.; Hieda, J.; Todaka, Y.; Miyazaki, T. Mechanical properties of a medical β-type titanium alloy with specific microstructural evolution through high-pressure torsion. *Mater. Sci. Eng. C* **2013**, *33*, 2499–2507. [CrossRef]
11. Janeček, M.; Čížek, J.; Stráský, J.; Václavová, K.; Hruška, P.; Polyakova, V.; Gatina, S.; Semenova, I. Microstructure evolution in solution treated Ti15Mo alloy processed by high pressure torsion. *Mater. Charact.* **2014**, *98*, 233–240. [CrossRef]
12. Ivasishin, O.M.; Markovsky, P.E.; Semiatin, S.L.; Ward, C.H. Aging response of coarse- and fine-grained β titanium alloys. *Mater. Sci. Eng. A* **2005**, *405*, 296–305. [CrossRef]
13. Makino, T.; Chikaizumi, R.; Nagaoka, T.; Furuhara, T.; Makino, T. Microstructure development in a thermomechanically processed Ti15V3Cr3Sn3Al alloy. *Mater. Sci. Eng. A* **1996**, *213*, 51–60. [CrossRef]

14. Barriobero-Vila, P.; Requena, G.; Schwarz, S.; Warchomicka, F.; Buslaps, T. Influence of phase transformation kinetics on the formation of α in a β-quenched Ti–5Al–5Mo–5V–3Cr–1Zr alloy. *Acta Mater.* **2015**, *95*, 90–101. [CrossRef]

15. Barriobero-Vila, P.; Requena, G.; Warchomicka, F.; Stark, A.; Schell, N.; Buslaps, T. Phase transformation kinetics during continuous heating of a β-quenched Ti–10V–2Fe–3Al alloy. *J. Mater. Sci.* **2015**, *50*, 1412–1426. [CrossRef]

16. Zheng, Y.; Williams, R.E.A.; Sosa, J.M.; Talukder, A.; Wang, Y.; Banerjee, R.; Fraser, H.L. The indirect influence of the ω phase on the degree of refinement of distributions of the α phase in metastable β-Titanium alloys. *Acta Mater.* **2016**, *103*, 165–173. [CrossRef]

17. Zheng, Y.; Williams, R.E.A.; Wang, D.; Shi, R.; Nag, S.; Kami, P.; Banerjee, R.; Wang, Y.; Fraser, H.L. Role of ω phase in the formation of extremely refined intragranular α precipitates in metastable β-titanium alloys. *Acta Mater.* **2016**, *103*, 850–858. [CrossRef]

18. Li, T.; Kent, D.; Sha, G.; Liu, H.; Fries, S.G.; Ceguerra, A.V.; Dargusch, M.S.; Cairney, J.M. Nucleation driving force for ω-assisted formation of α and associated ω morphology in β-Ti alloys. *Scripta Mater.* **2018**, *155*, 149–154. [CrossRef]

19. Zháňal, P.; Harcuba, P.; Hájek, M.; Smola, B.; Stráský, J.; Šmilauerová, J.; Veselý, J.; Janeček, M. Evolution of ω phase during heating of metastable β titanium alloy Ti–15Mo. *J. Mater. Sci.* **2018**, *53*, 837–845. [CrossRef]

20. Nag, S.; Banerjee, R.; Srinivasan, R.; Hwang, J.Y.; Harper, M.; Fraser, H.L. ω-Assisted nucleation and growth of α precipitates in the Ti–5Al–5Mo–5V–3Cr–0.5Fe β titanium alloy. *Acta Mater.* **2009**, *57*, 2136–2147. [CrossRef]

21. Šmilauerová, J.; Harcuba, P.; Kriegner, D.; Holý, V. On the completeness of the $\beta \rightarrow \omega$ transformation in metastable β titanium alloys. *J. Appl. Crystallogr.* **2017**, *50*, 283–287. [CrossRef]

22. Kuan, T.S.; Ahrens, R.R.; Sass, S.L. The stress-induced omega phase transformation in Ti-V alloys. *Metall. Trans. A* **1975**, *6*, 1767–1774. [CrossRef]

23. Zháňal, P.; Harcuba, P.; Šmilauerová, J.; Stráský, J.; Janeček, M.; Smola, B.; Hájek, M. Phase Transformations in Ti-15Mo Investigated by in situ Electrical Resistance. *Acta Phys. Pol. A* **2015**, *128*, 779–783. [CrossRef]

24. Devaraj, A.; Nag, S.; Srinivasan, R.; Williams, R.E.A.; Banerjee, S.; Banerjee, R.; Fraser, H.L. Experimental evidence of concurrent compositional and structural instabilities leading to ω precipitation in titanium–molybdenum alloy. *Acta Mater.* **2012**, *60*, 596–609. [CrossRef]

25. Valiev, R.Z.; Ivanisenko, Y.V.; Rauch, E.F.; Baudelet, B. Structure and deformaton behaviour of Armco iron subjected to severe plastic deformation. *Acta Mater.* **1996**, *44*, 4705–4712. [CrossRef]

26. Václavová, K.; Stráský, J.; Zháňal, P.; Veselý, J.; Polyakova, V.; Semenova, I.; Janeček, M. Ultra-fine grained microstructure of metastable beta Ti-15Mo alloy and its effects on the phase transformations. *IOP Conf. Ser. Mater. Sci. Eng.* **2017**, *194*, 012021. [CrossRef]

27. Schell, N.; King, A.; Beckmann, F.; Fischer, T.; Müller, M.; Schreyer, A. The High Energy Materials Science Beamline (HEMS) at PETRA III. *Mater. Sci. Forum* **2014**, *772*, 57–61. [CrossRef]

28. Dollase, W.A. Correction of intensities for preferred orientation in powder diffractometry: Application of the March model. *J. Appl. Crystallogr.* **1986**, *19*, 267–272. [CrossRef]

29. Bartha, K. Phase transformation in ultra-fine grained titnaium alloys. Ph.D. Thesis, Charles University, Prague, Czech Republic, 19 June 2019.

30. Hauk, V. *Structural and Residual Stress Analysis by Nondestructive Methods*, 1st ed.; Elsevier: Amsterdam, Netherlands, 1997.

31. Šmilauerová, J.; Janeček, M.; Harcuba, P.; Stráský, J.; Veselý, J.; Kužel, R.; Rack, H.J. Ageing response of sub-transus heat treated Ti–6.8Mo–4.5Fe–1.5Al alloy. *J. Alloy. Compd.* **2017**, *724*, 373–380. [CrossRef]

32. Furuhara, T.; Nakamori, H.; Maki, T. Crystallography of α Phase Precipitated on Dislocations and Deformation Twin Boundaries in a β Titanium Alloy. *Mater. Trans. JIM* **1992**, *33*, 585–595. [CrossRef]

33. Zhang, B.; Yang, T.; Huang, M.; Wang, D.; Sun, Q.; Wang, Y.; Sun, J. Design of uniform nano α precipitates in a pre-deformed β-Ti alloy with high mechanical performance. *J. Mater. Res. Technol.* **2019**, *8*, 777–787. [CrossRef]

34. Semiatin, S.L.; Knisley, S.L.; Fagin, P.N.; Barker, D.R.; Zhang, F. Microstructure evolution during alpha-beta heat treatment of Ti-6Al-4V. *Metall. Mater. Trans. A* **2003**, *34*, 2377–2386. [CrossRef]

35. Legros, M.; Dehm, G. Obsevation of Giant diffusivitiy along dislocation core. *Science* **2008**, *319*, 1646–1649. [CrossRef] [PubMed]

36. Jiang, B.; Tsuchiya, K.; Emura, S.; Min, X. Effect of High-Pressure Torsion Process on Precipitation Behavior of α Phase in β-Type Ti–15Mo Alloy. *Mater. Trans.* **2014**, *55*, 877–884. [CrossRef]

37. Zafari, A.; Xia, K. Formation of equiaxed α during ageing in a severely deformed metastable β Ti alloy. *Scr. Mater.* **2016**, *124*, 151–154. [CrossRef]

38. Furuhara, T.; Makino, T.; Idei, Y.; Ishigaki, H.; Takada, A.; Maki, T. Morphology and Crystallography of α Precipitates in β Ti–Mo Binary Alloys. *Mater. Trans. JIM* **1998**, *39*, 31–39. [CrossRef]

39. Xu, W.; Wu, X.; Stoica, M.; Calin, M.; Kühn, U.; Eckert, J.; Xia, K. On the formation of an ultrafine-duplex structure facilitated by severe shear deformation in a Ti–20Mo β-type titanium alloy. *Acta Mater.* **2012**, *60*, 5067–5078. [CrossRef]

40. Xu, W.; Edwards, D.P.; Wu, X.; Stoica, M.; Calin, M.; Kühn, U.; Eckert, J.; Xia, K. Promoting nano/ultrafine-duplex structure via accelerated α precipitation in a β-type titanium alloy severely deformed by high-pressure torsion. *Scr. Mater.* **2013**, *68*, 67–70. [CrossRef]

41. Guo, S.; Meng, Q.; Hu, L.; Liao, G.; Zhao, X.; Xu, H. Suppression of isothermal ω phase by dislocation tangles and grain boundaries in metastable β-type titanium alloys. *J. Alloy. Compd.* **2013**, *550*, 35–38. [CrossRef]

42. Friedel, J. *Dislocations*, 1st ed.; Smoluchovski, R., Kurti, N., Eds.; Pergamon Press: Oxford, UK, 1964.

43. Labusch, R. A Statistical Theory of Solid Solution Hardening. *Phys. Status Solidi (b)* **1970**, *41*, 659–669. [CrossRef]

44. Zherebtsov, S.; Ozerov, M.; Klimova, M.; Stepanov, N.; Vershinina, T.; Ivanisenko, Y.; Salishchev, G. Effect of High-Pressure Torsion on Structure and Properties of Ti-15Mo/TiB Metal-Matrix Composite. *Materials* **2018**, *11*, 2426. [CrossRef] [PubMed]

Article

Effect of Severe Plastic Deformation on the Conductivity and Strength of Copper-Clad Aluminium Conductors

Rimma Lapovok [1,2,*], Michael Dubrovsky [1], Anna Kosinova [1] and Georgy Raab [3]

[1] Department of Materials Science and Engineering, Technion-Israel Institute of Technology, Haifa 3200003, Israel
[2] Institute for Frontier Materials, Deakin University, Waurn Ponds, Victoria 3216, Australia
[3] Institute of Physics of Advanced Materials, Ufa State Aviation Technical University, Ufa 450000, Russia
* Correspondence: r.lapovok@deakin.edu.au; Tel.: +61-419897830

Received: 4 August 2019; Accepted: 27 August 2019; Published: 1 September 2019

Abstract: Aluminium rods with different copper sheath thicknesses were processed by severe plastic deformation at room temperature and then annealed, to join the constituent metals and produce a nanocrystalline microstructure. A study of the effects of the deformation parameters, copper cladding thickness and annealing temperature on the electrical conductivity and hardness of the conductors is reported. It is shown that an interface forms between constituents because of intermixing in the course of severe shear deformation under high hydrostatic pressure and diffusion during the subsequent annealing. The effective conductivity of the aluminium copper-clad conductor dropped after deformation, but was recovered during annealing, especially during short annealing at 200 °C, to a level exceeding the theoretically predicted one. In addition, the annealing resulted in increased hardness at the interface and copper sheath.

Keywords: aluminium copper-clad rod; hardness; effective electrical conductivity; severe plastic deformation

1. Introduction

Copper-clad aluminium (CCA) wire is a well-known conductor produced by extrusion of an aluminium rod in a copper can. Al-Cu hybrid materials combine the high conductivity of copper with the lightweight of aluminium. These wires, used in high frequency applications, offer the advantages of better conductivity due to the skin effect, and higher strength and corrosion resistance compared to all-aluminium wire, while costing less than all-copper wire [1]. The well-known study of copper-clad wires [2] shows that at very high frequencies, these wires' electrical properties approach those of a solid copper conductor.

Extrusion of aluminium–copper bi-metallic rods is performed in several reduction steps that are combined with annealing to restore the materials' ductility. The annealing; however, leads to enhanced diffusion between constituents and the formation of brittle aluminium–copper intermetallic compounds at the interface [3,4], which affects the composite wire's conductivity [5,6]. The interface formed between aluminium and copper consists of a number of intermetallic compounds (up to 13 stable intermetallic phases according to Murray [7]), the formation of which depends on the annealing temperature and duration. The conductivity of these phases is shown to be many times lower than that of the constituent materials [5,6].

A different approach, considered here, suggests replacing the extrusion process with a severe plastic deformation (SPD) process [8,9]. We have already used a similar method for aluminium–steel conductors in [10]. Since SPD introduces a gigantic shear strain into the processed material, especially in the vicinity of the interface [11,12], it is expected that the bonding between aluminium and copper will appear at low temperatures due to intermixing. During SPD processing, this brittle intermetallic layer will either not form or be destroyed and dissolved, which would improve the strength and conductivity of the resulting wire. Low-temperature annealing was applied to assist the diffusion and strengthening of the bond region.

In this work, aluminium rods with different copper sheath thicknesses were processed using SDP at room temperature and then annealed. A study of the effect of SPD parameters, copper cladding thickness, and annealing temperature on the electrical conductivity and strength of CCA conductors is reported below.

2. Material and Experimental Techniques

Commercially-pure Al 1050 rods (Fe content of 0.352 ± 0.015 wt%) were inserted into 99.9% pure copper tubes with an outer diameter of 10 mm and wall thicknesses of 0.4 and 0.7 mm, in such a manner that compression was exercised on the aluminium rod surface, Figure 1. The copper tube was heated to 150 °C to extend its diameter sufficiently to insert the cold aluminium rod. After the tube cooled down, it shrank and created the compressive stresses at the surface of the rod. The initial average grain sizes of the aluminium and copper were around 58 ± 25 μm and 96 ± 37 μm, respectively.

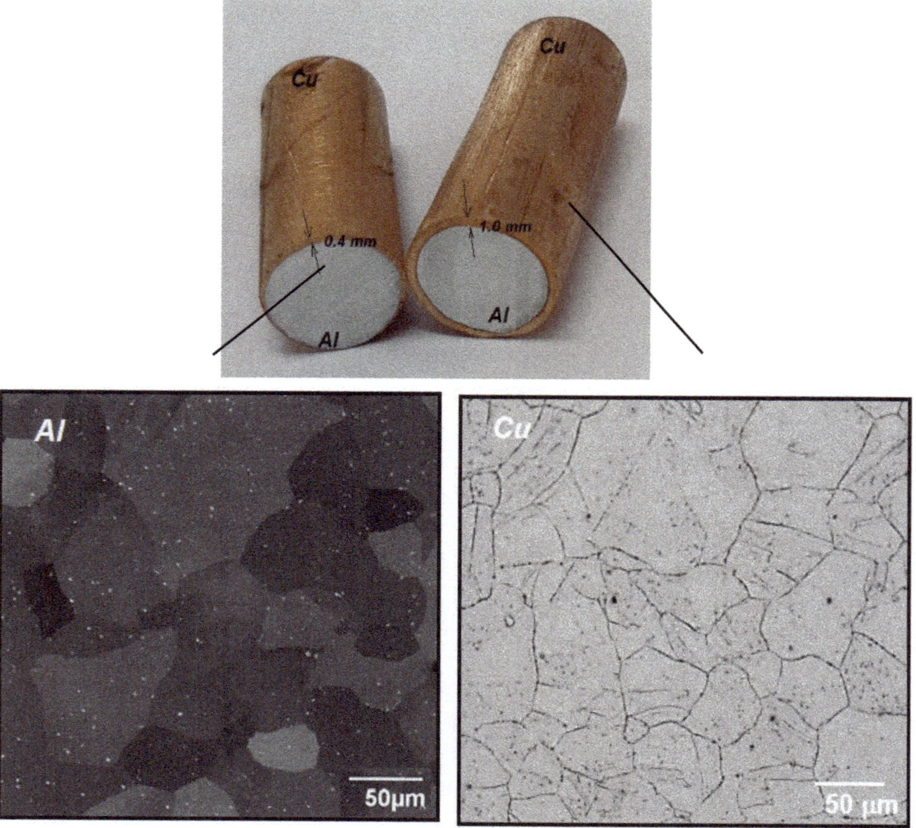

Figure 1. Aluminium rods cladded by copper with different thicknesses before deformation. (The left image is a backscattered SEM image and the right one is an optical microscope image.).

Samples were subjected to one and two passes of equal channel angular pressing (ECAP) using Route A (no rotation between passes) and Route B_C (90° rotation between passes). This was followed by annealing according to two different schedules—namely, 200 °C for 5 min and 120 °C for 2 h.

The alloy's microstructure in all conditions was characterised by high resolution scanning electron microscope (HRSEM) Zeiss Ultra Plus, Jena, Germany). The interface zone, which was formed by intermixing and diffusion, was characterised by energy-dispersive X-ray spectroscopy (EDX) (Oxford Instruments, UK) in HRSEM.

A Vickers hardness test was performed using a Buehler MMT-7 micro-hardness tester (Lake Bluff, IL, USA). The samples were polished with 500 grit sandpaper and 10 hardness measurements were performed for each sample using a load of 200 g. The standard deviations from the mean value were calculated as between 2.8 and 3.6.

The electrical resistivity of all these samples was measured at room temperature using the four-point constant-current (DC) method. The resistivity was calculated as:

$$\rho = \frac{V}{I} \cdot \frac{S}{l'},$$ (1)

where V is the voltage change between two points on the side of the tubular samples at a distance of $l' = 5$ mm from each other, measured by a Keithley 2700 multi-meter. In addition, $I = 10$ A was the constant current applied by a TDK Lambda current source through the top and bottom sides of the samples over an area S of the tube cross-section, which was in contact with copper plates under ~250 Pa pressure.

3. Results and Discussion

3.1. The Grain Refinement during SPD and Annealing

Extreme grain refinement was observed in the deformed hybrid samples, especially in the outer copper cladding where friction added to the severity of the shear strain. The microstructure, after two ECAP passes (Route B_C), followed by annealing, is shown in Figure 2. Due to SPD, the grain size in the aluminium rod was reduced to 410 ± 150 nm and in the copper sheath to 180 ± 90 nm. The annealing did not cause any grain growth in the copper as the temperature–time schedules were chosen to be within the region of thermal stability suggested for high purity copper in [13]. The presence of Fe precipitates decorating the grain boundaries in the aluminium also prevented grain growth in the aluminium rod. Therefore, the annealing resulted in recovery of the dislocations within the ultrafine grains of both constituents of the hybrid material and formation of a clean (free of dislocation within the grain interior) ultrafine microstructure with sharp boundaries.

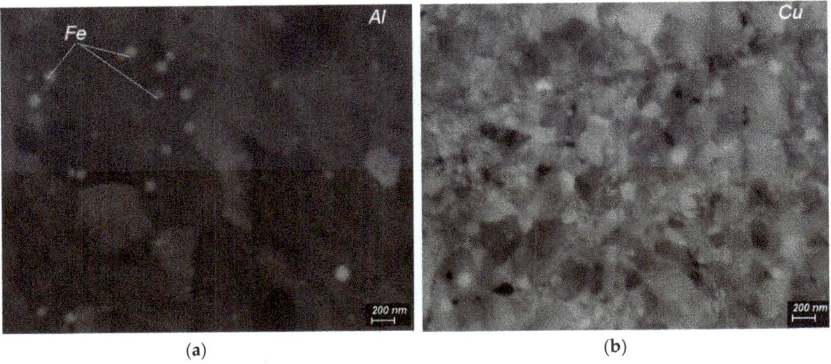

(a)   (b)

Figure 2. SEM images (back-scattered electrons) of the microstructure after two ECAP passes (Route B_C) and annealing. (a) Aluminium core; (b) copper sheath.

3.2. Interface Formation during Room Temperature Deformation

Interface formation is typically governed by three mechanisms, namely, intermixing, inter-diffusion and phase formation [14,15]. Room temperature deformation leads to intermixing when the surface asperities are crushed during contact under high hydrostatic pressure and severe shear deformation [14], mixing the constituents in a swirl flow (Figure 3). As a result of SPD; however, grain refinement by rearrangement of accumulated dislocations takes place and many new defects are introduced (for example, solid solutions of constituent atoms).

Figure 3. The inter-mixing zone resulting from intensive shear of constituents under high hydrostatic pressure (black lines show the boundary of the intermixing zone where random islands of copper within the aluminium can be seen).

Subsequent static annealing results in accelerated inter-diffusion, which adds to the formation of the interface, Figure 4. SEM-EDX analysis of selected points in the vicinity of interface points 1–5 shows the diffusion of copper atoms into the aluminium matrix to a depth of about 750 nm while aluminium atoms diffused into the copper matrix to a depth of around 250 nm, Figure 4a. The concentration of aluminium at a distance of ~200 nm from the interface within the copper matrix is around 12 at%, Figure 4b, which is below the concentration required to form any intermetallic compound [7]. These results are similar to previously reported observations of interface formation in aluminium/copper bimetallic tubes made by high-pressure tube shearing [11], and the formation of intermetallic compounds detrimental to conductivity was not observed.

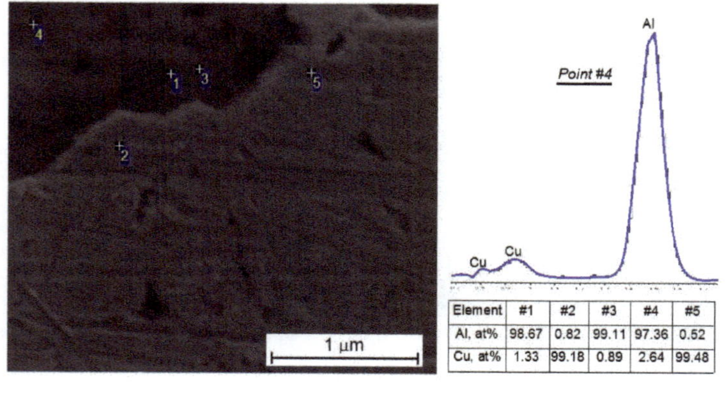

Element	#1	#2	#3	#4	#5
Al, at%	98.67	0.82	99.11	97.36	0.52
Cu, at%	1.33	99.18	0.89	2.64	99.48

(**a**)

Figure 4. *Cont.*

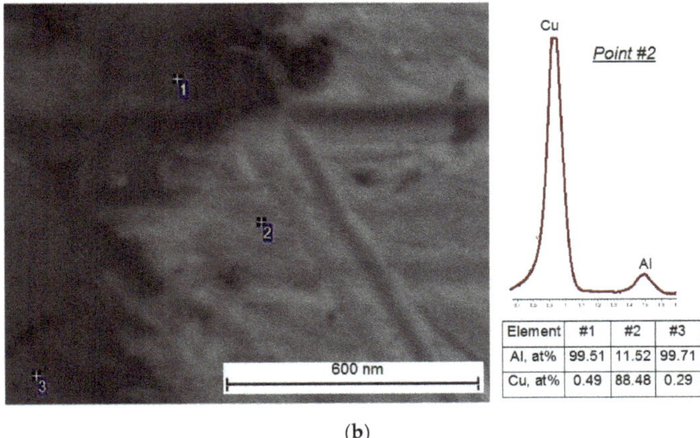

Element	#1	#2	#3
Al, at%	99.51	11.52	99.71
Cu, at%	0.49	88.48	0.29

(b)

Figure 4. SEM-EDX analysis of selected points in the vicinity of the interface at two locations: (**a**) and (**b**).

3.3. Hardness

It can be seen, Figure 5, that the aluminium hardness raises gradually as the number of passes increases. Due to the small number of ECAP passes; however, this increase is insignificant, from ~36 HV at the initial annealed conditions to 40–42 HV after one ECAP pass and to 45–48 HV after two ECAP passes. In contrast, the copper cladding hardness rises significantly from its initial annealed value of 58 HV to 85–105 HV after deformation. The Cu-Al interface hardness displays a gradual increase compared to the aluminium hardness due to non-homogeneous intermixing of constituents.

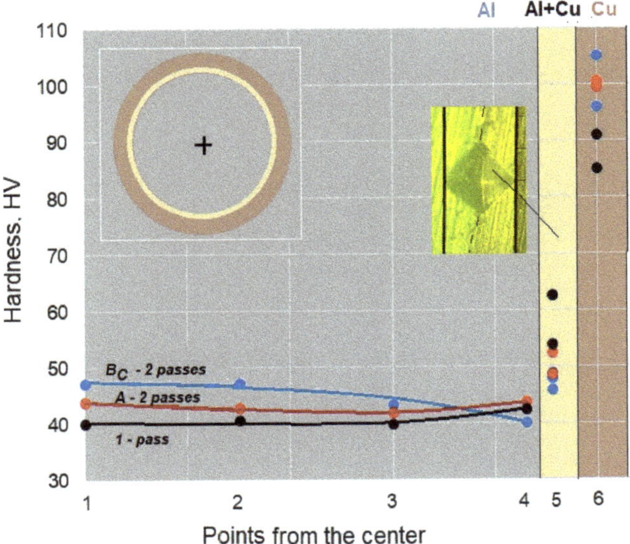

Figure 5. Hardness after deformation versus distance from the centre measured at six points in the aluminium part (points 1–4), at the interface (point 5) and in the copper cladding (point 6). (The two similarly coloured points at the interface and at the copper cladding represent measurements for thick and thin sheaths).

It should be noted that annealing results in a decrease in the aluminium hardness and an increase in the interface and copper sheath hardness (Figure 6). The effect of increased strength in ultrafine-grained (UFG) materials due to annealing has been observed and discussed quite recently in several publications [16,17]. The reason for this phenomenon was believed to be the segregated impurities at the grain boundaries. Nevertheless, the issue has still not been elucidated as high purity metals in a nano-crystalline state demonstrated a similar effect. Further, several UFG materials show hardening by annealing while others do not. For example, copper annealed after high-pressure torsion at temperatures in the range of 0.25 to 0.3 Tm (depending on the copper purity) demonstrated similar hardening behaviour, which was explained by agglomeration and annihilation of deformation-induced vacancies [18].

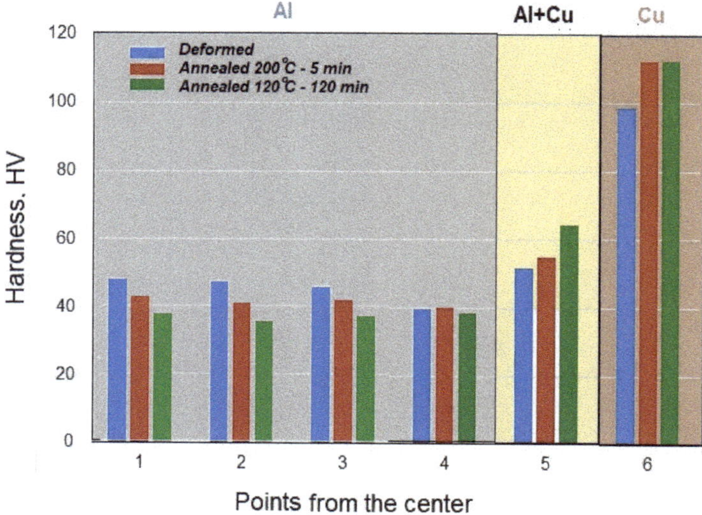

Figure 6. Comparison of hardness after deformation (two passes Route B$_C$) and annealing measured at six points located in the aluminium part (points 1–4); at the interface (point 5); and in the copper part (point 6) (measured for sample with thick sheath).

In our case, the temperature of annealing was chosen within the range defined for copper in [17]. Subsequently, hardening of the copper and the interface zone was observed. The chosen temperature; however, was high enough to change the crystallite size and dislocation density in aluminium, causing softening of the aluminium core.

3.4. Electrical Conductivity

The electrical conductivity for parallel connection of conductors can be calculated using the rule of mixture:

$$\sigma_{ef} = \sigma_{Al} f_{Al} + \sigma_{Cu} f_{Cu} = \sigma_{Al}(1 - f_{Cu}) + \sigma_{Cu} f_{Cu}$$

The effective conductivity of aluminium–copper-clad conductors, calculated using theoretical conductivities for aluminium and copper constituents, is represented by the dashed line in Figure 7. In reality; however, the effective conductivity was about 10% lower than the ideal theoretical value, and was in the range of 55% to 60% IACS. The low initial conductivity could be explained by the Fe high content in the solid solution and is in line with results published in [18].

The electrical conductivity measured after deformation shows strong dependence on the severity of the deformation. The conductivity after one pass dropped from the initial value to 36.1% ± 3.6% IACS for a thin sheath and to 60.5% ± 2.4% IACS for a thick sheath. The conductivity decreased further after two deformation passes, more when using Route B$_C$ than when using Route A.

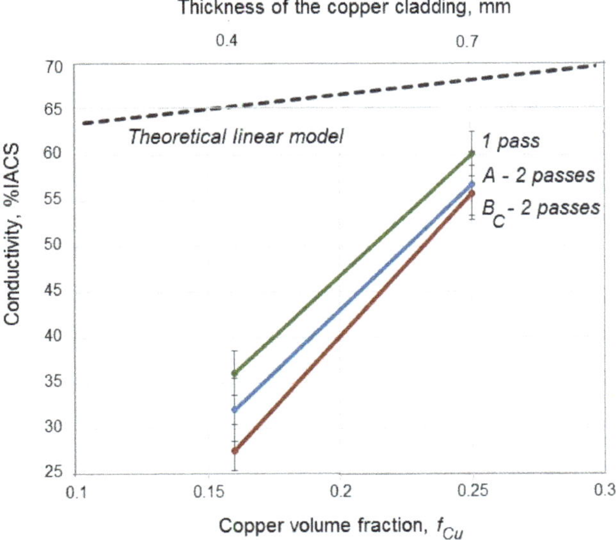

Figure 7. Effective conductivity versus thickness of copper cladding after deformation.

Short annealing at 200 °C restored conductivity to a greater extent than prolonged annealing at 120 °C, especially for samples deformed by one or two passes along Route B$_C$, as seen in Figure 8. It should be noted that conductivity after annealing of samples with the thick sheath exceeded the theoretical values (~68%IACS), represented by the dashed line in Figure 7 and reached ~75% IACS. Connectivity of samples with thin sheath was worse due to defects introduced by SPD and leads to the conclusion that the cladding thickness has the lower limit on the benefits offered by this cladding technique.

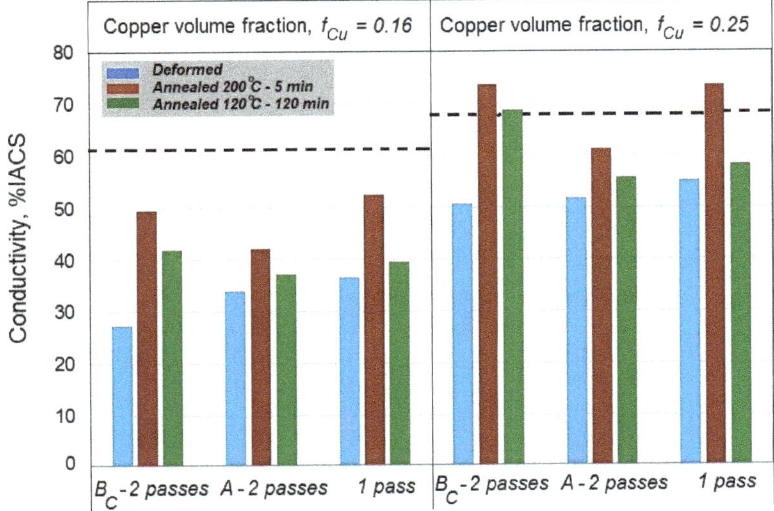

Figure 8. Effective conductivity after deformation and two annealing schedules for different thicknesses of copper cladding (the dashed line shows the theoretically predicted level of conductivity).

4. Conclusions

Copper-clad aluminium rods with different copper sheath thicknesses were produced by SPD at room temperature, followed by annealing rather than by conventional extrusion at elevated temperatures. Detrimental formation of intermetallic phases with low conductivity was not observed. An interface between constituents formed by their severe intermixing due to large shear, high hydrostatic pressure and possibly diffusion during annealing.

Severe plastic deformation was performed using ECAP for one or two passes along Routes A and B_C. Extreme grain refinement down to the nano-range size as a result of SPD processing was shown. The UFG microstructure remained stable during annealing. It is demonstrated that to achieve an interface without defects, ECAP must be performed on copper-clad aluminium rods with thickness of copper cladding at least 0.7 mm or higher.

The effective conductivity of aluminium copper-clad conductors dropped after deformation in proportion to the number of ECAP passes (more when using Route B_C than A). Annealing, especially short annealing at 200 °C; however, resulted not only in recovery of conductivity but in conductivity values exceeding the theoretically predicted ideal ones.

In addition, annealing resulted in an increase of hardness at the interface and in the copper sheath, which confirms the possibility of hardening UFG metals by annealing at temperatures in the range of 0.25 to 0.3 Tm. Thus, SPD and short annealing at the temperature within this range could be used to produce lightweight conductors with high conductivity and strength.

Author Contributions: R.L.—conceptualization, funding acquisition, supervision, writing—original draft, analysis of results; M.D.—properties measurements; A.K.—microstructure characterization, analysis of results, writing—review & editing; G.R.—samples preparation and processing, analysis of results.

Funding: This research was funded by the EU Framework Program for Research and Innovation 'HORIZON 2020' (Grant-742098).

Acknowledgments: R. Lapovok acknowledges the Marie Curie Fellowship within the EU Framework Program for Research and Innovation 'HORIZON 2020'.

Conflicts of Interest: The authors declare no conflicts of interest.

References

1. Read, D. Copper Clad Aluminum in Electrical Application. *SAE Trans.* **1968**, *77*, 1920–1931. [CrossRef]
2. Miller, J.M. *Effective Resistance and Inductance of Iron and Bimetallic Wires*; Classic Reprint Series; Forgotten Books Publisher: London, UK, 2019; p. 76.
3. Hug, E.; Bellido, N. Brittleness study of intermetallic (Cu, Al) layers in copper-clad aluminium thin wires. *Mater. Sci. Eng. A* **2011**, *528*, 7103–7106. [CrossRef]
4. Braunovic, M.; Rodrigue, L.; Gagnon, D. Nanoindentation Study of Intermetallic Phases in Al-Cu Bimetallic System. In Proceedings of the 2008 IEEE 54th Holm Conference on Electrical Contacts, Orlando, FL, USA, 27–29 October 2008. [CrossRef]
5. Pfeifer, S.; Großmann, S.; Freudenberger, R.; Willing, H.; Kappl, H. Characterization of Intermetallic Compounds in Al-Cu-Bimetallic Interfaces. In Proceedings of the 2012 IEEE 58th Holm Conference on Electrical Contacts, Portland, OR, USA, 23–26 September 2012. [CrossRef]
6. D'Heurle, F.; Alliota, C.; Angilello, J.; Brusic, V.; Dempsey, J.; Irmischer, D. The deposition by evaporation of Cu-Al alloy films. *Vacuum* **1997**, *27*, 321–327. [CrossRef]
7. Murray, J.L. The Aluminium-Copper System. *Int. Met. Rev.* **1985**, *30*, 211–233. [CrossRef]
8. Valiev, R.Z.; Estrin, Y.; Horita, Z.; Langdon, T.G.; Zehetbauer, M.J.; Zhu, Y.T. Producing bulk ultrafine-grained materials by severe plastic deformation. *JOM* **2006**, *58*, 33–39. [CrossRef]
9. Valiev, R.Z.; Estrin, Y.; Horita, Z.; Langdon, T.G.; Zehetbauer, M.J.; Zhu, Y.T. Producing Bulk Ultrafine-Grained Materials by Severe Plastic Deformation: Ten Years Later. *JOM* **2016**, *68*, 1216–1226. [CrossRef]
10. Qi, Y.; Lapovok, R.; Estrin, Y. Microstructure and electrical conductivity of aluminium/steel bimetallic rods processed by severe plastic deformation. *J. Mater. Sci.* **2016**, *51*, 6860–6875. [CrossRef]

11. Lapovok, R.; Ng, H.P.; Thomus, D.; Estrin, Y. Bimetallic Copper-Aluminium Tube by Severe Plastic Deformation. *Scr. Mater.* **2012**, *66*, 1081–1084. [CrossRef]
12. Lapovok, R.; Qi, Y.; Ng, H.P.; Toth, L.S.; Estrin, Y. Gradient structures in thin-walled metallic tubes produced by continuous high pressure tube shearing process. *Adv. Eng. Mater.* **2017**, *19*, 1700345. [CrossRef]
13. Wang, Y.L.; Lapovok, R.; Wang, J.T.; Qi, Y.S.; Estrin, Y. Thermal behavior of copper processed by ECAP with and without back pressure. *Mater. Sci. Eng. A* **2015**, *628*, 21–29. [CrossRef]
14. Medvedev, A.E.; Lapovok, R.; Koch, E.; Höppel, H.W.; Göken, M. Optimisation of Interface Formation by Shear Inclination: Example of Aluminium-Cooper Hybrid Produced by ECAP with Back-Pressure. *Mater. Des.* **2018**, *146*, 142–151. [CrossRef]
15. Mendes, A.; Timokhina, I.; Molotnikov, A.; Hodgson, P.; Lapovok, R. Role of Shear in Interface Formation of Aluminium-Steel Multilayered Composite Sheets. *Mater. Sci. Eng. A* **2017**, *705*, 142–152. [CrossRef]
16. Renk, O.; Hohenwarter, A.; Schuh, B.; Li, J.H.; Pippan, R. Hardening by annealing: Insights from different alloys. In Proceedings of the 36th Risø International Symposium on Materials Science, Risø, Denmark, 7–11 September 2015; IOP Publishing: Bristol, UK, 2015; Volume 89. [CrossRef]
17. Cengeri, P.; Kerber, M.B.; Schafler, E.; Zehetbauer, M.J.; Setman, D. Strengthening during heat treatment of HPT processed copper and nickel. *Mater. Sci. Eng. A* **2019**, *742*, 124–131. [CrossRef]
18. Medvedev, A.; Murashkin, M.; Enikeev, N.; Valiev, R.Z.; Hodgson, P.D.; Lapovok, R. Optimization of strength-electrical conductivity properties in Al-2Fe alloy by severe plastic deformation and heat treatment. *Adv. Eng. Mater.* **2018**, *20*, 1700867. [CrossRef]

 *metals*

Article

Tuneable Magneto-Resistance by Severe Plastic Deformation

Stefan Wurster [1,*], Lukas Weissitsch [1], Martin Stückler [1], Peter Knoll [2], Heinz Krenn [2], Reinhard Pippan [1] and Andrea Bachmaier [1]

[1] Erich Schmid Institute of Materials Science of the Austrian Academy of Sciences, Jahnstrasse 12, 8700 Leoben, Austria; Lukas.Weissitsch@oeaw.ac.at (L.W.); Martin.Stueckler@oeaw.ac.at (M.S.); reinhard.pippan@oeaw.ac.at (R.P.); Andrea.Bachmaier@oeaw.ac.at (A.B.)

[2] Institute of Physics, University of Graz, Universitätsplatz 5, 8010 Graz, Austria; peter.knoll@uni-graz.at (P.K.); heinz.krenn@uni-graz.at (H.K.)

*   Correspondence: stefan.wurster@oeaw.ac.at

Received: 9 October 2019; Accepted: 31 October 2019; Published: 5 November 2019

Abstract: Bulk metallic samples were synthesized from different binary powder mixtures consisting of elemental Cu, Co, and Fe using severe plastic deformation. Small particles of the ferromagnetic phase originate in the conductive Cu phase, either by incomplete dissolution or by segregation phenomena during the deformation process. These small particles are known to give rise to granular giant magneto-resistance. Taking advantage of the simple production process, it is possible to perform a systematic study on the influence of processing parameters and material compositions on the magneto-resistance. Furthermore, it is feasible to tune the magneto-resistive behavior as a function of the specimens' chemical composition. It was found that specimens of low ferromagnetic content show an almost isotropic drop in resistance in a magnetic field. With increasing ferromagnetic content, percolating ferromagnetic phases cause an anisotropy of the magneto-resistance. By changing the parameters of the high pressure torsion process, i.e., sample size, deformation temperature, and strain rate, it is possible to tailor the magnitude of giant magneto-resistance. A decrease in room temperature resistivity of ~3.5% was found for a bulk specimen containing an approximately equiatomic fraction of Co and Cu.

Keywords: severe plastic deformation; high pressure torsion; microstructural characterization; magnetic properties; hysteresis; magneto-resistance

---

1. Introduction

The giant magneto-resistance (GMR), independently discovered by the two groups of Fert and Grünberg [1,2] at the end of the 1980s, was first observed for stacks of very thin multilayers of alternating ferromagnetic/antiferromagnetic Fe and Cr. These layers couple magnetically, resulting in a giant decrease of resistivity with an increasingly applied magnetic field. Some years after the discovery of GMR, it was found that this phenomenon is not only restricted to layered systems but can also be found for materials containing dispersed ferromagnetic particles (granules) [3,4]; thus it was labeled as granular GMR.

If grains are not subjected to an external magnetic field, a random orientation of magnetic moments or domains prevails. With an increasing magnetic field, the magnetic domains gradually align by rotating the magnetization, and become aligned parallel to the magnetic field. This results in an overall decrease of resistance. It was shown that the increase in resistance for randomly oriented magnetic particles originates from spin-dependent scattering of conduction electrons at the magnetic-nonmagnetic interfaces [5,6]. Rabedeau et al. [5] found this fact by using small angle X-ray scattering measurements on thin films, making the particle sizes accessible. If the GMR originates from scattering within the ferromagnetic particles, the GMR would weakly depend on the particle size (providing that all the ferromagnetic atoms can be found in the particles). However, as GMR scaled with the inverse of the cluster size (interfaces per volume, r^2/r^3) instead, an interfacial spin-dependent scattering was proposed.

Upon applying a magnetic field, the gradual change in the magnetization direction of single domain particles leads to a gradual change of the resistivity and this property can be directly linked to the hysteresis loop of the material. One method to characterize the relationship between the specimens' resistance and the magnetic field is the squared global relative magnetization $\mu\,(H) = (M\,(H)/M_S)^2$ [4], which is the ratio of the magnetization $M\,(H)$ at a certain applied field H and the saturation magnetization M_S. The GMR-effect is described in the following way:

$$GMR = \frac{\Delta R}{R} = \frac{R(H) - R(H=0)}{R(H=0)} = A\,\mu(H)^2, \tag{1}$$

where A determines the effect amplitude and is different for each experimental setup. In some cases, the GMR of Equation (1) is expressed by $R\,(H)$ in the denominator instead of $R\,(H=0)$, or $R\,(H=0)$ is replaced by $R\,(H=H_C)$. It is stated [4] that the change in resistance, which is a measure for the GMR-effect, is proportional to $A^*\,(M\,(H)/M_s)^2$; thus, the strength of the GMR-effect can be quantified by the proportionality factor A. For a magnetron sputtered thin film specimen consisting of 84 at% Cu and 16 at% of Co ($Cu_{84}Co_{16}$) and a temperature of 5 K, this proportionality factor A was found to be -0.065 [4]. To allow comparison, all compositions in this work will be given in at%, except stated otherwise.

Research on granular GMR first started with thin films. They were produced using a variety of different techniques, such as magnetron sputtering [3,4,7–11], molecular beam epitaxy [5], ion beam co-sputtering [12], cluster beam deposition [13], thermal evaporation [14], or by electrochemical deposition [15–17]. Later, research was extended to bulk materials and mixed granular materials were produced using different techniques such as mechanical alloying/ball milling [18–22] or melt spinning [9,23–25]. Research focused on a small number of binary, sometimes ternary systems such as CuCo [3–5,9,14,16,17,19,21,23,25,26], AgCo [4,6,7,9,11,12,18], CuFe [13,19], CoFe-Cu [10,20,22], CrFe [8], and AuCo [9,24]. A common feature of these systems is the small mutual solubility of ferromagnetic and non-magnetic elements, as well as the nonmagnetic phase representing the major phase. Reduced solubility promotes the production of small, finely dispersed ferromagnetic particles within a nonmagnetic metal matrix; either directly during production or after adequate annealing treatments. With increasing ferromagnetic content, a transition towards anisotropic magneto-resistance (AMR) is found [10]. Thomson discovered this anisotropy of ferromagnetic materials in magnetic fields [27], where the resistance for currents parallel and perpendicular to the magnetic field is different. For parallel alignment, the magneto-resistance increases with an increasing field and for the perpendicular alignment, the magneto-resistance decreases. The effect of AMR is typically in the size of the GMR or about one magnitude smaller and with a change of the sign of A (A > 0), depending on the investigated material.

The restricted mutual solubility is known for Cu and Co. and the granular GMR was discovered on magnetron sputtered CuCo [3,4]. They found a strong dependence of the resistance with the applied magnetic field, and the resistivity being highest in the initial non-magnetized state decreases with increasing applied field. The resistivity increases again upon decreasing the field and reaches a local maximum at the coercive field. However, resistivity is slightly lower than in the initial, non-magnetized state. To give a first idea on the amount of GMR present, some results of original works [3,4] are presented: Berkowitz et al. [3] investigated Cu containing 12, 19, and 28 at% Co and found for $Cu_{81}Co_{19}$ (as an example) a GMR of 10% at 10 K at the highest applied magnetic fields of 20 kOe. Negligible GMR was found at room temperature. Xiao et al. [4] found a GMR of 16.5% at 5 K for magnetron sputtered $Cu_{80}Co_{20}$, annealed for 10 min at 500 °C. A review on GMR, including granular GMR, is provided in reference [28].

The goal of this work was to produce bulk materials of different amounts of ferromagnetic and diamagnetic components. This enables the investigation of the influence of composition and processing parameters on the evolving microstructure, on the development of small, ferromagnetic particles and thus on the GMR. The chosen method is high pressure torsion (HPT) [29], a special method of severe plastic deformation (SPD), as it provides the opportunity to easily produce bulk samples from elemental powder mixtures [30]. The principle idea of HPT is based on the work of Bridgman [31], where material is confined under high hydrostatic pressure between two anvils. One anvil is rotated against the other and the material is severely deformed by shear deformation and the microstructure gets refined. This refinement saturates at a certain grain size—mostly depending on the amount of alloying elements, impurities, and deformation temperature [32].

Regarding the investigation of different magnetic properties of materials deformed by HPT, numerous studies focusing on the magnetic properties of HPT-processed materials are available [33–42]. Other techniques to apply severe deformation onto materials include ball milling, mechanical alloying, and equal channel angular pressing (ECAP), with some studies focusing on the magnetic properties of these alternative processing routes [18–21,26,43,44]. However, to the best knowledge of the authors, there are only two studies on HPT-deformed materials, which also contain information on magneto-resistive properties [34,37]. In references [34,37], the authors used arc-melted Cu-10 wt % Co for HPT deformation. The magneto-resistive drop was ~0.25% at room temperature and ~ 2.5% at 77 K, with both measured in fields of 6 kOe.

In summary, a detailed GMR—Study of the influences of HPT process parameters including deformation temperature and composition on GMR is lacking, which is the motivation for and aim of the presented work. Within this study, the Cu-Co-system was thoroughly investigated to understand the dependency of composition on the GMR, and to demonstrate the applicability of HPT throughout the whole compositional range.

2. Materials and Methods

Commercially available pure elemental powders were used as a starting material: Fe (MaTeck, Jülich, Germany, 99.9%, −100 + 200 mesh), Co (Goodfellow, Hamburg, Deutschland, 99.9% 50–150 μm), high purity Co (Alfa Aesar—Puratronic, Ward Hill, MA, USA, 99.998%, −22 mesh), and Cu (Alfa Aesar, Ward Hill, MA, USA, 99.9%, −170 + 400 mesh). The reason for using two grades of Co powder was the following: although the diameter of the particles of the less pure powder is rather large, the grains are small, meaning HPT deformation of high Co containing materials became difficult. For comparison, scanning electron micrographs of both types of Co powders are presented in Figure 1. Description of the other powders used is given in reference [40].

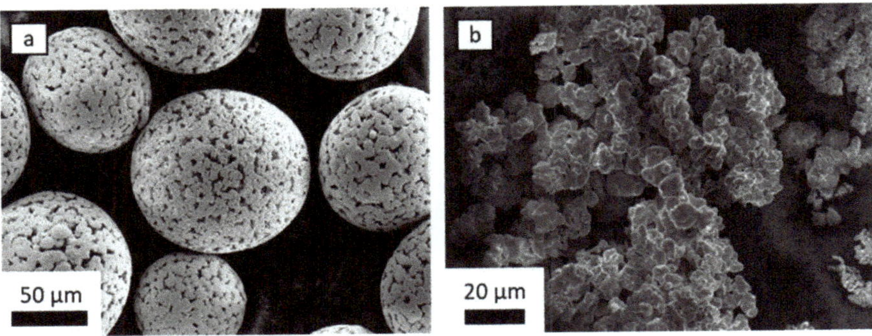

Figure 1. scanning electron microscopy (SEM) micrographs of Co particles. The main difference is the particle shape, the purity, and the resulting grain size. (**a**) Purity 99.9%. (**b**) Purity 99.998%.

For preventing oxidation, the powders are stored and prepared for further processing in an argon (Ar)-filled glove box. For the first processing step (pre-compaction of mixed powders in the HPT-tool), the powders are contained in a specially-made, Ar-filled compaction device [40]. After pre-compaction at a pressure of 5 GPa, the samples are deformed in air, at room temperature (RT) or at elevated temperatures, as described below (Table 1). Samples of 8 mm diameter and approximately 0.5 mm thickness were produced. To investigate the effects of varying process conditions on microstructure and the GMR, several samples of about 50/50 at% were made. They vary in sample size, processing temperature, and applied strain rate. To prove the feasibility of upscaling the HPT-process, one sample with approximately equiatomic composition was made with a different, larger HPT-tool. It is capable of applying a force of 4 MN; thus, samples of diameters of 30 mm (p_{max} = 5.6 GPa) and thicknesses of several millimeters can be deformed. The strain rate which was applied to the small specimens was ~0.6 s^{-1}, whereas it was approximately 0.05 s^{-1} for the large sample. In the latter case, the powder mixture was simply poured in the anvil's cavity and deformed under atmospheric conditions. According to Equation (2)

$$\varepsilon_{v.M.} = 2\,\pi\,n\,r\,/\,(t\,\sqrt{3}), \tag{2}$$

the equivalent von-Mises-strain $\varepsilon_{v.M.}$ applied to the material upon HPT is given by the radius r, the number of applied rotations n, and the thickness t. For samples of 8 mm diameter, this yields an applied strain of 2200 at a radius of 3 mm after 100 rotations, assuming a thickness of 0.5 mm. For the large sample (diameter 30 mm), a strain of 1300 was applied at a radius of 10 mm (rotations: 250, thickness: ~7 mm). Although the applied strain is smaller and the strain rate is considerably lower, the large sample heats up during HPT due to plastic deformation and a lack of heat dissipation [45]. Therefore, it was deformed at elevated temperature although no external heating device was used.

Table 1. Summary of investigated samples, varying in composition (energy dispersive X-ray -measurement) and processing parameters including HPT-tool size, processing temperature, number of rotations, and purity of the used Co-powder (where applicable).

Composition [at%]	HPT-Tool/Sample Diameter [mm]	Processing Temperature	No. of Rotations	Purity of Co-Powder [%]
Cu	Small/8	RT	>25	n.a.
$Cu_{81}Co_{19}$	Small/8	RT	100	99.9
$Cu_{64}Co_{36}$	Small/8	RT	100	99.9
$Cu_{55}Co_{45}$	Large/30	Elevated	250	99.9
$Cu_{52}Co_{48}$	Small/8	RT	100	99.9
$Cu_{49}Co_{51}$	Small/8	200 °C	100	99.9
$Cu_{43}Co_{57}$	Small/8	RT	100	99.9
$Cu_{22}Co_{78}$	Small/8	300 °C	100	99.998
Co	Small/8	RT	50	99.998
$Cu_{85}Fe_{15}$	Small/8	RT	100	n.a.

To enable a thorough microstructural and magneto-resistive investigation, at least two small HPT samples were made out of each powder mixture, but only one large sample was made since several GMR specimens and specimens for microstructural analysis could be made from one sample. All relevant sample and process parameters are summarized in Table 1.

For detailed investigations of the microstructure scanning electron microscopy (SEM) and transmission electron microscopy (TEM) studies as well as synchrotron measurements were performed. Vickers hardness was measured to ensure there was a microstructurally saturated state of the sample. GMR-specimens were taken from this microstructurally saturated region. Hardness measurements were made along the radius of the HPT-disc in a tangential direction (Micromet 5104, Buehler, Lake Bluff, IL, USA). SEM micrographs using a backscattered electron detection mode (BSE) were acquired with a LEO 1525 (Zeiss, Oberkochen, Germany) that was further equipped with an electron backscatter diffraction (EBSD) system (e⁻ Flash^{FS}, Bruker, Berlin, Germany) and an energy dispersive X-ray (EDX) system (XFlash 6-60, Bruker, Berlin, Germany). The conventional EBSD can easily be changed to a transmission EBSD geometry (Transmission Kikuchi Diffraction, TKD) [46,47]; thus, EBSD-TKD was performed for a chosen TEM specimen ($Cu_{55}Co_{45}$). The TEM specimen was made using conventional metallographic thinning methods and a final ion polishing step with grazing incidence. For analysing the results of EBSD-TKD and EDX measurements, the software package Esprit 2.1 from Bruker (Billerica, MA, USA) was used. The TEM/TKD—Specimen was taken at a radius of ~10 mm from the large HPT-disc, where the incident electron beam was parallel to the axial HPT-direction. Using an acceleration voltage of 30 kV and a step size of 5 nm to 9 nm (four scans in total), the orientation data was recorded. After scanning, a clean-up of the data was performed, where certain zero solutions are absorbed by the surrounding well-defined matrix. Grain boundaries were defined by having more than 15° misorientation between the two grains, and grains below the size of five pixels were rejected.

Using the same specimen, TEM micrographs were made with a JEOL-TEM (JEM2200FS, Tokyo, Japan). For phase analysis, transmissive synchrotron X-ray diffraction measurements were made at PETRA III, P07 at DESY, Hamburg, Germany (Deutsches Elektronen Synchrotron). A beam energy of 98.25 keV was used and recorded spectra were analyzed with the software PyFAI and compared with CeO_2–standards. Magnetic properties were measured with a superconducting quantum interference device (SQUID, Quantum Design MPMS-XL-7, Darmstadt, Germany). For small HPT-discs, the specimens for SEM and synchrotron investigations were taken out in accordance with reference [40]. For the sample made with the large HPT-tool, the TEM study was performed in an axial view. For SQUID measurements, the applied field was parallel to the axial HPT-direction.

For GMR-measurements, long and thin specimens were cut out of the HPT-disc along a secant line. According to Equation (2), the center of the HPT-disc does not receive a lot of deformation; thus, it was avoided for resistivity measurements. The closest distance of the GMR-specimen to the former center of the HPT-disc was at least 1.5 to 2 mm. The length of the GMR-specimens was several millimeters; their thickness was reduced by grinding and polishing to approximately 200 μm in order to increase the voltage drop. We sought to make sure that the impact of heat during specimen preparation was kept as low as reasonably possible; no microstructural changes are expected to occur upon sample preparation. In literature, it was found that for measurements on thin films, the applied current and magnetic field are often within the film plane [48]. Following this, the radial-tangential-plane was identified as the "film plane" for small HPT-samples. In this case, the applied current has components in the radial and shear direction. For the large HPT-sample, the current was applied in an axial direction. The specimen was placed in a four-point resistance setup, which again is placed within the air gap of an electromagnet (Type B-E 30, Bruker, Karlsruhe, Germany). The air gap was fixed with 50 mm and the diameter of the conical pole pieces was 176 mm (max. field = 22.5 kOe) providing enough space within the homogeneous magnetic field for the four-point resistance setup. The voltage was measured using a multimeter (model 2000, Keithley, Cleveland, OH, USA), the current source was a sourcemeter (model 2400, Keithley, Cleveland, OH, USA), and the applied current was 500 mA to 800 mA. The strength and stability of the magnetic field was measured using a Gaussmeter (Model 475 DSP, Lakeshore, Westerville, OH, USA). The electromagnet's power supply, the current source, and the voltage and field measurements were connected for communication and data collection.

3. Results

3.1. Microstructural Characterization

A typical feature of HPT-deformed materials is an increasing hardness and decreasing microstructural size with an increasing distance from the sample's center. As soon as a microstructural saturation is reached, the hardness does not increase further [32]. For a valid comparison of samples of varying Co-content (Figure 2), Vickers hardness values were taken from the saturation region of constant hardness. Measurements within these regions show a clear trend of increasing hardness with increasing Co-content. Figure 2 was drawn using samples from this work and data from samples presented in reference [40]. The trend breaks down for pure Co of highest purity, whose hardness was found to be 368 HV. In the latter case, neither impurities (as would be the case for the less pure Co-powder) nor another element obstruct grain boundary motion; the grains are larger and the material becomes softer.

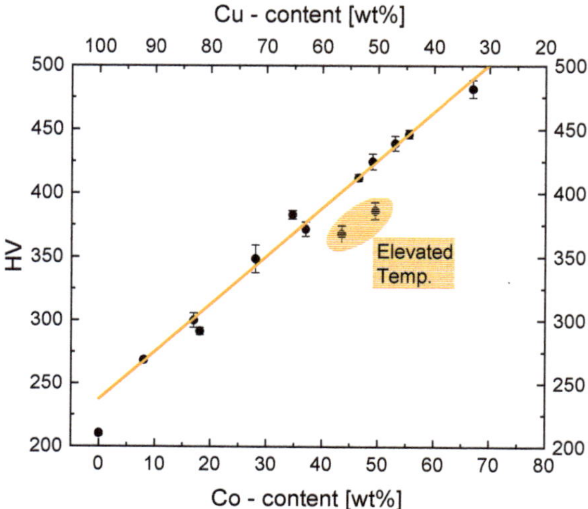

Figure 2. Hardness HV as a function of Cu-Co-content, showing a linear trend for room temperature deformed samples. The highlighted measurements were made on samples $Cu_{55}Co_{45}$ (large HPT, elevated temperature) and $Cu_{49}Co_{51}$ (small HPT, 200 °C) Hardness values from samples presented in reference [40] are included. The linear fit between 10 wt % and 70 wt % Co, representing a simple rule of mixture, is a guide for the eye.

SE micrographs already give a first indication of the enhanced, mutual solubility of Cu- and Co-phases upon HPT. Figure 3 shows micrographs taken in tangential direction from the saturation region for binary CuCo composites. They were made at a radius of 3 mm ($\varepsilon_{v.M.} \sim 2200$) with the exception of r = 10 mm ($\varepsilon_{v.M.} \sim 1300$) for the large sample (Figure 3c).

At room temperature and for 200 °C deformed samples, the microstructure appears to be homogeneous on a scale of ~1 µm, while the microstructure for the samples deformed at elevated temperature (Figure 3c) and 300 °C (Figure 3f) appears to be granular. EDX measurements of the sample presented in Figure 3c) using low energy electrons (5 kV, smaller excitation volume) show changes in the chemical composition at the same length scale (Figure 4). The small sample $Cu_{49}Co_{51}$ (200 °C deformation temperature) does not show this segregation of Co. Therefore, the microstructure of the large sample resembles more like the one of the Co-rich $Cu_{22}Co_{78}$, which experienced a deformation temperature of 300 °C.

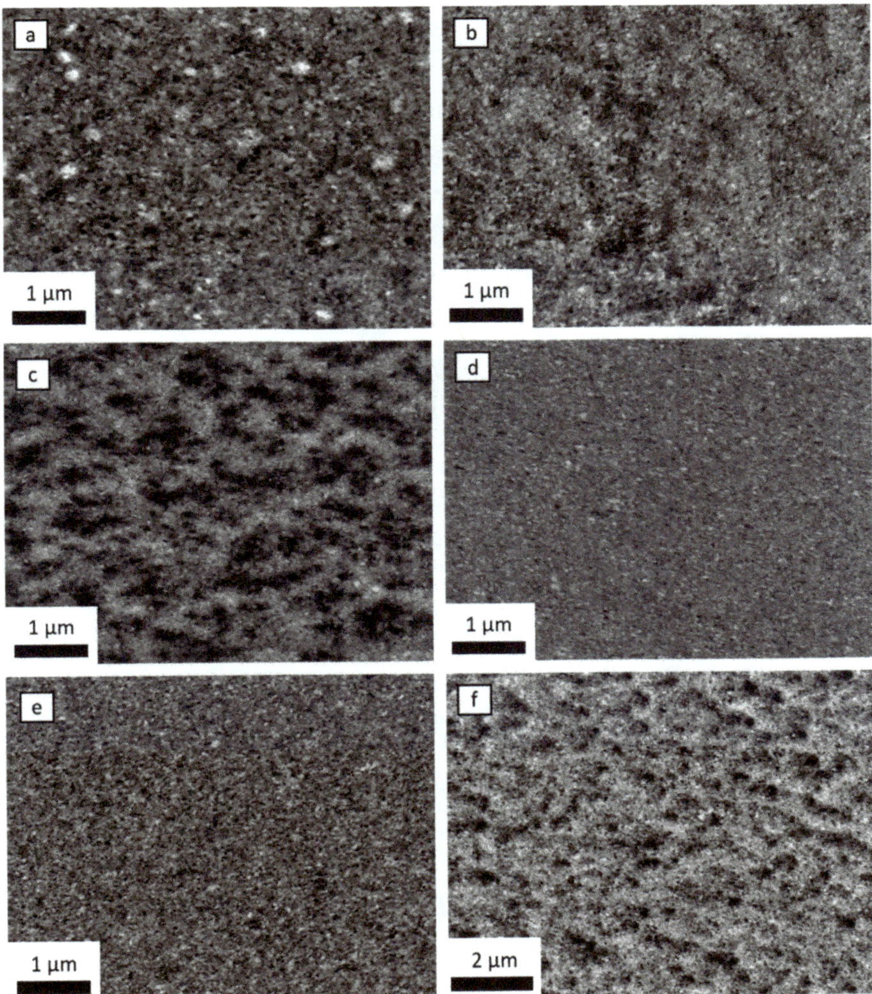

Figure 3. SEM micrographs of HPT-deformed samples, taken at r = 3 mm (with the exception of (**c**)) in tangential direction. (**a**) $Cu_{81}Co_{19}$; (**b**) $Cu_{64}Co_{36}$; (**c**) $Cu_{55}Co_{45}$ (r = 10 mm, large HPT-tool, deformed at elevated temperature); (**d**) $Cu_{52}Co_{48}$; (**e**) $Cu_{49}Co_{51}$ deformed at 200 °C; (**f**) $Cu_{22}Co_{78}$ deformed at 300 °C.

For the determination of the actual chemical composition and elemental distribution of the deformed samples and to detect possible deviations from the nominal composition of the powder mixture, several EDX spectra were recorded at large radii, and their mean values are presented for the chemical composition in Table 1. To demonstrate the good co-deformation and the increasing intermixing with high applied strains, 60 EDX spectra were recorded right at the center and at radii of 1, 2, and 3 mm. As an example, the sample $Cu_{85}Fe_{15}$ was used. In Figure 5a, it can be seen that still some Fe-particles (dark particles) can be found for small radii (r = 0 mm, r = 1 mm) and the distribution of results of EDX-analysis (Figure 5d) is widespread (blue lines), due to the existence of Cu-rich and Fe-rich regions. For r = 2 mm and r = 3 mm, the distribution become narrower, demonstrating improved intermixing of Cu and Fe on the scale of the EDX-analyzed volume. This is also confirmed by micrographs in Figure 5b,c.

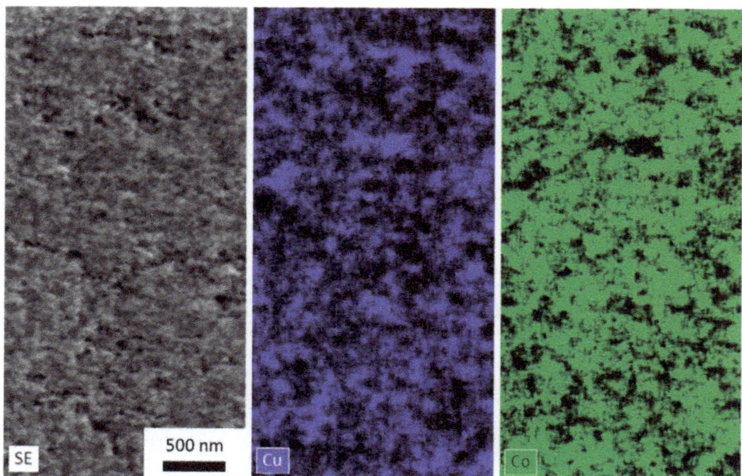

Figure 4. SEM micrograph of the large HPT sample Cu$_{55}$Co$_{45}$, taken at r = 10 mm, with EDX maps of the same region showing the distribution of Cu (**center**, blue) and Co (**right**, green).

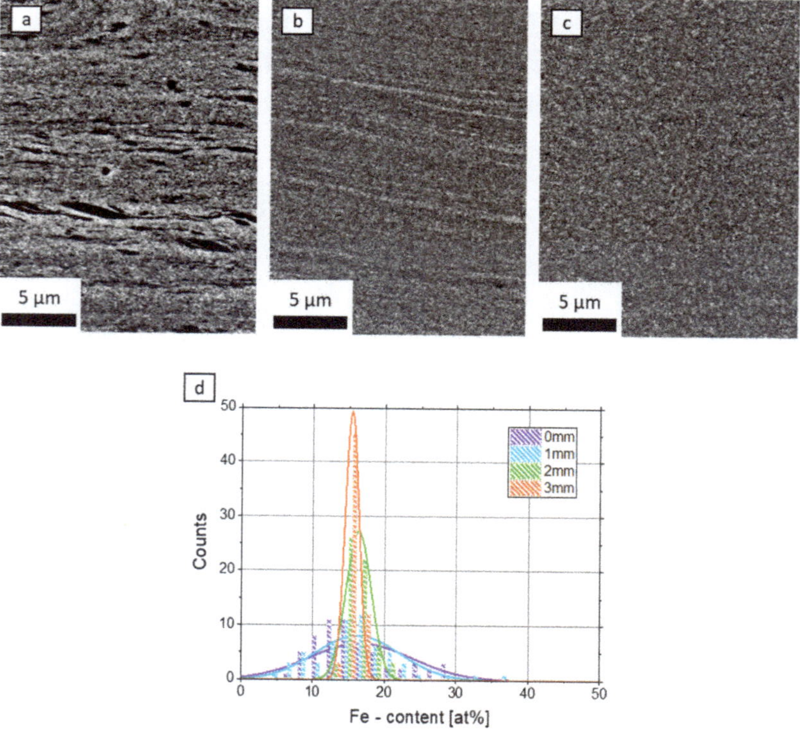

Figure 5. SEM micrographs of the sample containing Cu$_{85}$Fe$_{15}$ at (**a**) r ~ 0 mm, (**b**) r = 1 mm, (**c**) r = 2 mm, from left to right (No. rotations = 100). The progressing dissolution of Fe into the Cu matrix can be seen. (**d**) Distribution of EDX-results for the same sample and for different radii. 60 spectra were recorded at each position. As a guide for the eye, a normal distribution was fitted to the data.

As typical examples taken from different ranges of the CuCo system, the following three specimens were subjected to synchrotron diffraction phase analysis: $Cu_{81}Co_{19}$, $Cu_{55}Co_{45}$, $Cu_{22}Co_{78}$. The results, logarithmic intensity as a function of scattering vector q are presented in Figure 6 for different radii. To compare with, the peaks of fcc-Cu (lattice constant d = 3.615 Å [49]), fcc-Co (d = 3.554 Å [50]) and of hcp-Co, data from reference [51], are included. For the low Co-containing material, small peaks of hcp-Co can be found for r = 1 mm; however, these peaks vanish for larger radii. The Cu-peak slightly deviates from the position of pure Cu, due to supersaturating the crystal with Co. This effect is stronger for the higher-Co containing material (Figure 6b); however, traces of hcp-Co can be found for all radii. No hcp-Co can be found in the large sample $Cu_{55}Co_{45}$ (Figure 6c,d). The plateau-like shape of the peaks – especially at high q – is explained most likely due to the occurrence of two fcc-phases for Co and Cu.

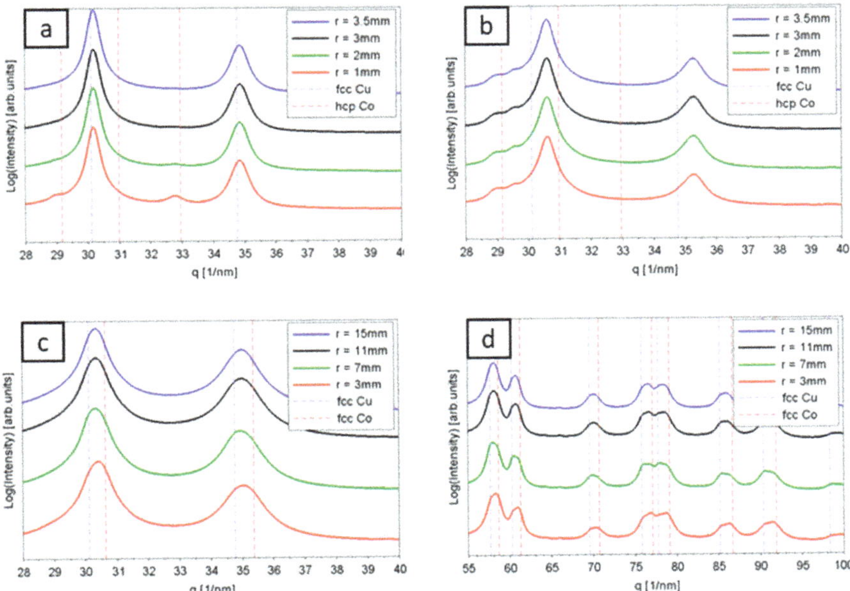

Figure 6. Synchrotron data for three CuCo specimens (**a**) $Cu_{81}Co_{19}$ deformed at room temperature, (**b**) $Cu_{22}Co_{78}$ deformed at 300 °C, (**c**) $Cu_{55}Co_{45}$ deformed at elevated temperature and lower applied strain rate. Note the larger radius of the specimen, (**d**) like (**c**) but showing a different regime of scattering vector q.

Further microstructural investigation, using TKD, was made for $Cu_{55}Co_{45}$ to obtained deeper insights into the as-deformed microstructure. As an example for TKD results, Figure 7a is shown. Although the hcp-Co phase was searched for, there are no clear indications of hcp-Co grains of detectable size. Distinction between grains rich in fcc-Co and rich in fcc-Cu cannot be made since the difference in lattice parameter is too small. Identified grains are in the 100 nm regime and a texture analysis, which is not shown here, yielded the shear texture typical of fcc metals [52]. In total, four scans were made, and their aggregated area weighted grain size distribution $f(d)$ was fitted by a lognormal distribution.

$$f(d) = \frac{1}{\sqrt{2\pi}\sigma d}e^{-\frac{(\ln(d)-\mu)^2}{2\sigma^2}} \tag{3}$$

With the mean μ and the standard deviation σ, the median value of the grain size e^{μ} is found to be 79 nm. Stückler et al. [40] found similar results for other CuCo materials, which were processed the same way: A decreasing grain size of 100 nm, 78 nm, and 77 nm for increasing Co-content of Cu-Co28 wt %, Cu-Co49 wt %, and Cu-Co67 wt %, respectively.

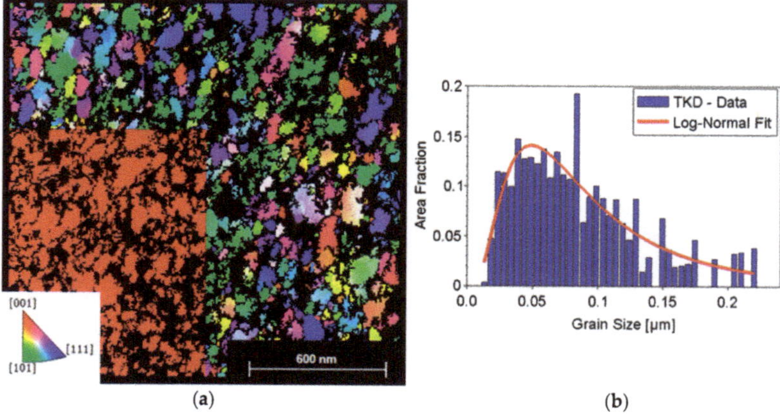

(a)   (b)

Figure 7. (**a**) Inverse pole figure map of $Cu_{55}Co_{45}$, taken at a radius of ~10 mm in an axial direction. As an overlay in the lower left part, the Cu-phase map (fcc–phase map, respectively) of the identical area is shown. After the clean-up, all identified grains are found to be fcc. No hcp-particles can be detected. (**b**) Aggregated grain size distribution from four TKD scans. In total, 1648 grains were taken into account.

Regarding spatial resolution, TKD-EBSD is better than conventional EBSD in many cases. Ge et al. [17] made a related analysis on Co-particle sizes in electrodeposited $Cu_{84}Co_{16}$, using TEM, before and after annealing for 30 min at 695 K and found mean values of 10 nm and 12 nm respectively, being even below the resolution limit of TKD. Thus, a high angle annular dark field (HAADF) image was recorded in scanning TEM, Figure 8. HAADF is sensitive to the atomic number, while Co-enriched (dark) and Cu-enriched (bright) regions can be identified within the matrix.

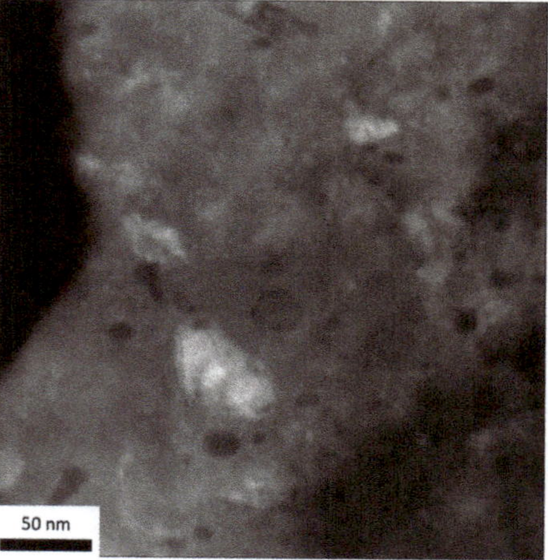

Figure 8. High angle annular dark field (HAADF) image of $Cu_{55}Co_{45}$, using the same specimen, which was used for deriving the results shown in Figure 7. The Co-particles (dark regions) with a size of several tens of nanometers become visible.

3.2. Magnetic Properties—Magneto-Resistivity

To investigate the change in magneto-resistive behavior for all produced samples, four-point resistance measurements were performed at room temperature in magnetic fields up to 22.5 kOe. The results of the drop in resistance with applied magnetic field, displayed according to Equation (1), are summarized in Figures 9 and 10. Two different testing directions were taken into account. First, applied current and applied magnetic field are parallel; second, current and magnetic field are perpendicular to each other. Probe current flows within the shear-tangential HPT-plane, except for $Cu_{55}Co_{45}$. Depending on the composition and processing parameters, different types of resistive behavior within magnetic fields can be identified.

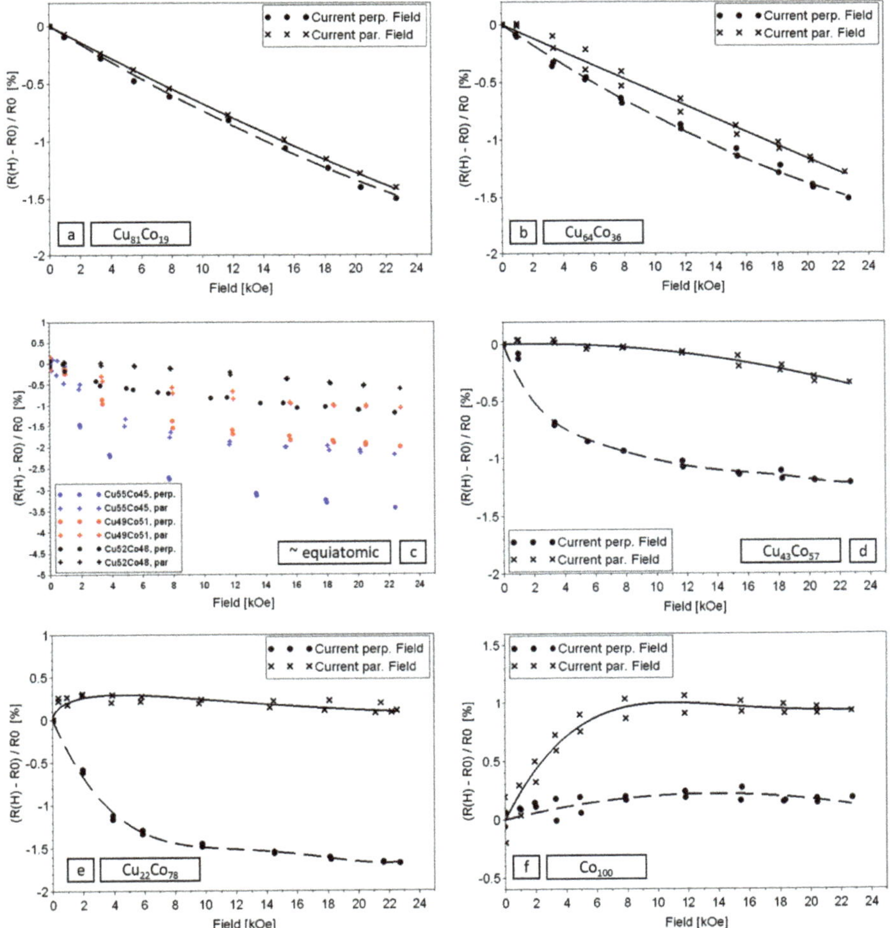

Figure 9. Drop in room-temperature resistance as a function of magnetic fields for Co-based HPT-deformed samples. (**a**) $Cu_{81}Co_{19}$. (**b**) $Cu_{64}Co_{36}$. (**c**) black: $Cu_{52}Co_{48}$ deformed at room temperature; red: $Cu_{49}Co_{51}$ deformed at 200°C; blue: $Cu_{55}Co_{45}$ deformed at elevated temperature with the large HPT-tool. (**d**) $Cu_{43}Co_{57}$. (**e**) $Cu_{22}Co_{78}$, deformed at 300 °C. (**f**) Pure Co. The continuous lines (parallel alignment) and dashed lines (perpendicular alignment) are derived from polynomial fits and serve only as a guide for the eye.

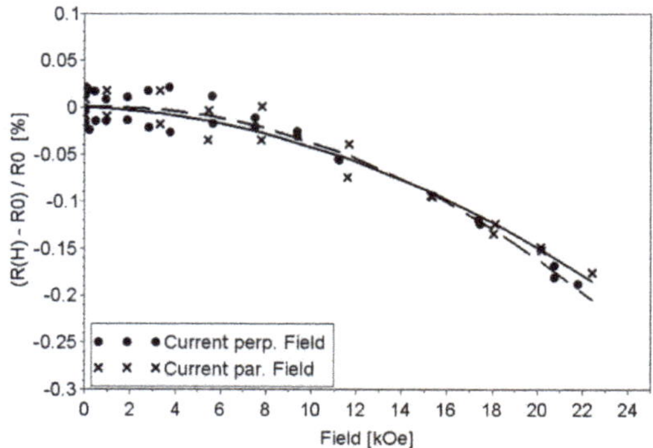

Figure 10. Drop in room-temperature resistance as a function of applied magnetic fields for HPT-deformed $Cu_{85}Fe_{15}$. The continuous line (parallel alignment) and dashed line (perpendicular alignment) are derived from polynomial fits and serve only as a guide for the eye.

For small Co-content, an almost isotropic dependence of resistance with magnetic field can be found (Figure 9a,b). With increasing ferromagnetic content, there is a gradual development of an anisotropic component, which is due to larger, percolating ferromagnetic regions. Thus, for medium and high ferromagnetic content, the resistance curve can be seen as a superposition of GMR and AMR. Pure Co (Figure 9f) is the best example for AMR with the resistance being higher for parallel current and magnetic field. For pure Co, the difference in resistivity between parallel and perpendicular orientation increases quickly for small fields; for higher fields, the curves are about parallel.

In case for replacing Co with Fe, the effect for low ferromagnetic content is identical for Fe and Co (Figure 10). The decrease in resistance with applied field is immediate at low fields. However, the effect is much smaller for Fe by around a factor of ten for fields as high as 22.5 kOe. A detailed discussion of the results of Figures 9 and 10 is provided in Section 4.2.

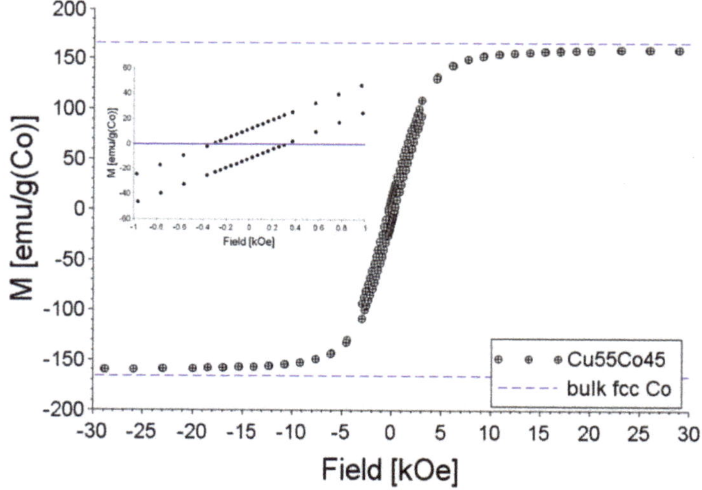

Figure 11. Hysteresis loop (specific magnetic moment versus applied field) of $Cu_{55}Co_{45}$, measured at 300 K, with the inset showing a detailed view for lower applied fields. For comparison, the magnetic moment of fcc-Co (164.8 emu/g [53]), is indicated as a dashed line.

The large sample deformed at elevated temperature ($Cu_{55}Co_{45}$) shows the largest reduction in resistance (~3.5% in perpendicular orientation) including a saturation of the decrease in resistance for high fields. For smaller Co content, an even larger GMR effect is expected—e.g., according to reference [17], the largest effect can be expected for a Co-concentration of 16%. Thin films, which were electrodeposited and annealed, were used for this work. However, as HPT-samples of these compositions saturate at very high fields due to paramagnetic contributions [40], the applied magnetic field is not large enough to saturate them. A sample that shows a satisfactory GMR as well as saturation in GMR at lower fields is $Cu_{55}Co_{45}$. It was chosen to be thoroughly investigated using SQUID. Figure 11 shows the result of hysteresis measurements at 300 K. The hysteresis was measured with the specimen's axial direction being parallel to the applied field. The coercivity was determined by linear interpolating between fields of +/− 1000 Oe and was found to be 316 Oe. For high applied fields, the saturation magnetization of bulk fcc-Co (164.8 emu/g [53]) is almost reached.

4. Discussion

4.1. Influence of Processing Conditions on Microstructural Evolution

Severe plastic deformation, especially at low temperatures, will promote the formation of supersaturated solid solutions. Diffusion, also taking place during the deformation process, will lead to segregation and phase formation, and as a result the resulting microstructure and elemental distribution depends on the strength of the one or the other process. As already displayed in Figure 2, the microstructure of the large sample made of $Cu_{55}Co_{45}$, resembles the one of high Co-content, deformed at high temperatures. For equiatomic composition however, there is quite a difference regarding Co agglomeration when comparing the sample processed with the large HPT and small HPT-samples. The following two paragraphs will discuss the influence of sample heating and strain rate on the evolving microstructure:

When deforming a sample with a large HPT tool, more volume is deformed and more strain energy ($\int \tau \gamma \, dV$) is transferred to heat during the whole experiment. Applied shear γ is proportional to r and integrating the infinitesimal volume $V = 2\pi r h \, dr$ yields a r^3-dependency of strain energy. Using typical values: the hardness value (~350 HV) gives an approximate shear strength of 500 MPa and 150 s per rotation result in a power of ~150 W for large HPT-samples, whereas one rotation taking 45 s for the small HPT-equipment results in a heating power of approximately 10 W. The increased energy released upon deforming a sample with larger HPT-equipment is counterbalanced by the decreased strain rate and by the increased diameter of the sample and the anvils, representing the samples "cooling finger". The latter point loses importance for long-lasting experiments (~10 h in total for the $Cu_{55}Co_{45}$ sample) as the anvils heat up and their cooling capability decreases. As a result, a large sample, which is nominally deformed at room temperature, heats up more easily, leading to accelerated diffusion and segregation processes.

The second important point that has to be taken into account regarding microstructural evolution is the lower applied strain rate for the large HPT-tool. The strain rate is proportional to angular velocity divided by thickness times the radius and for the conditions described above, smaller samples were subjected to an approximately 12 times higher strain rate compared to the large sample. During HPT, the tendency of Cu and Co to segregate by diffusional processes is counter-acted by severe shear deformation. For the large HPT-sample segregation processes are more pronounced and as a result the microstructure evolves, as can be seen in Figure 12. The microstructure is shown in tangential view for radii r = 5 mm, 10 mm, 15 mm and a clear increase in the size of dark regions is visible. Thus, segregation is promoted for large radii and an increased local temperature could therefore be deduced for regions of higher strain rate. Edalati et al. determined a temperature difference of several degrees between the center and the rim of the HPT-disc. This was done by FEM simulations for a sample of 10 mm diameter [45].

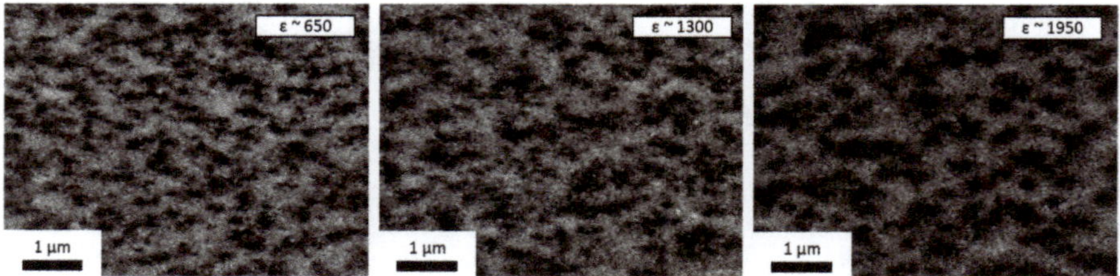

Figure 12. SEM—backscattered electron detection mode (BSE) micrograph of $Cu_{55}Co_{45}$, deformed with the large HPT tool at elevated temperatures. Micrographs were taken in tangential direction and for increasing radius (r = 5 mm, 10 mm, 15 mm) from left to right.

Regarding the results from synchrotron X-ray diffraction measurements (Figure 6), the low Co-containing sample ($Cu_{81}Co_{19}$) shows pure fcc phases for large radii. Diffraction lines from hcp-Co, still visible e.g., at a radius of 1 mm, vanish for higher applied strains. For high Co-contents ($Cu_{22}Co_{78}$), where deformation was possible to be performed at 300 °C, there are still hcp-Co particles remaining for large radii, as can be seen e.g., by a shoulder in the diffraction peak at q ~29 nm^{-1}. Nevertheless, a solid solution of the two elements can be seen for both samples. The lines are not at their nominal position for pure elements, but slightly shifted to the right (from Cu towards fcc-Co, for low-Co concentration) or to the left (from Co towards Cu, for high Co concentration). The situation is different for the $Cu_{55}Co_{45}$ sample. Here, the diffractogram is dominated by very broad peaks—especially for higher q-values, which can be easily explained by overlapping peaks at (or close to) the original position of Cu and fcc-Co.

4.2. Dependence of Magneto-Resistive Properties on Processing and Composition

The result of coercive field measurement from SQUID magnetometry shows a slightly increased coercivity for CuCo of about equiatomic composition, when compared to reference [40]. Therein, the powder compacted and room temperature HPT-deformed samples were measured with SQUID at 300 K and for the variety of compositions between Cu-28 wt % Co and Cu-67 wt % Co, coercive fields of ~70 Oe ranging down to ~0 Oe, below the resolution limit of SQUID magnetometry, were determined.

As described earlier, there is a variety of methods to produce thin film or bulk materials, showing a GMR. Depending on the production route, yielding different microstructures and particle size distribution—GMR is thus not only depending on the material's composition. In case the material is fully supersatured (e.g., a minor ferromagnetic element is fully dissolved in the matrix) and no ferromagnetic particles can be found, GMR is inexistent. For precipitating ferromagnetic particles, GMR will start to rise up to a maximum value. For further growing ferromagnetic particles, GMR will decrease as the particle size are too large to form single magnetic domain particles and the particle number becomes too small to be efficient scattering sites. The dependence of the strength of GMR on the size distribution is described in reference [54].

Depending on the production route, the material might not be in the state yielding the highest GMR-effect right after processing. Very often, proper thermal treatments lead to an increase of the effect. After the production of bulk materials showing granular GMR, Ikeda et al. [20] found the maximum GMR ratio (6.4%, room temperature) for ball milled $Cu_{80}Co_{20}$ annealed at 450 °C for 1 h in vacuum. Nagamine et al. [22] report 4% (room temperature) for ball milled $(Co_{0.7}Fe_{0.3})_{20}Cu_{80}$, which was annealed for 15 min at 500 °C. For the as-milled powder they found a negligible effect (<0.2%), due to the almost perfect supersaturation of Fe and Co in Cu. Champion et al. [21] found a value of ~4% at 4.2 K for $Cu_{60}Co_{40}$ and $Cu_{50}Co_{50}$ (both in vol%) for ball milled materials. For an as-deposited, magnetron sputtered CuCo thin film, a negligible GMR at room temperature, most likely as a result of an almost perfect supersaturated state was found [3]. Comparing with our results, it is evident that the incomplete supersaturation of Cu with Co improves the GMR effect at room temperature – considering as-prepared samples. With a short annealing of the $Cu_{81}Co_{19}$ thin film [3] (10 min at 484 °C) the GMR substantially increased. Upon annealing, an increase in GMR from negligibility at room temperature to ~22% at 10 K was found.

For a granular system, the GMR might also get as large as ~50% as reported for multilayer FeCr systems at 4.2 K [1]: For another type of binary alloys, $Ag_{80}Co_{20}$ thin films, Xiong et al. [7] report on values as large as 84% (They normalize the change in resistivity to the resistance at high fields, which is different to Equation (1).) at 4.2 K for sputtered and subsequently annealed (330 °C for 10 min) specimens. Values for GMR of granular systems presented in studies on CuCo-systems are in close agreement with the values reported here. Improved thermo-mechanical treatment during HPT-processing or subsequent annealing of the as-deformed sample should lead to a closure of this small gap. Here, only HPT-processed states have been investigated and will be discussed.

In the following, the different behaviors of magneto-resistive curves shall be discussed with respect to the Co-content of the samples. Coming back to Figure 9: Starting with low ferromagnetic contents ($Cu_{81}Co_{19}$, $Cu_{64}Co_{36}$, $Cu_{85}Fe_{15}$) the dependence of the resistance with applied magnetic field is almost linear and isotropic. The effect is high for $Cu_{81}Co_{19}$ with an GMR of close to 2% for the highest applied field. No saturation in resistance drop could have been achieved even for the highest magnetic fields of 22.5 kOe. Replacing Co by Fe leads to a decrease in GMR. Through the use of SQUID magnetometry, it was shown in reference [40] that low-Fe containing CuFe composites, processed the same way as described in this work, do not saturate in magnetization even in very high fields of 70 kOe. The CuFe samples with a low Fe-content show a very pronounced paramagnetic behavior. This is explainable by a better dilution of Fe in Cu. Using GMR data from Figures 9 and 10 and comparing the GMR curves for $Cu_{85}Fe_{15}$ and $Cu_{81}Co_{19}$, which are almost identical in ferromagnetic composition, the same conclusion of an increased dilution of Fe in Cu compared to Co in Cu can be drawn.

For increasing Co content, approaching about equiatomic composition, the GMR curves for perpendicular and parallel current alignment start to differ from each other, with the curve for parallel alignment being higher. One reason could be a non-perfect globular shape of the particles. Scattering occurs at the nonmagnetic—Ferromagnetic interfaces; thus, a change in the cross-sectional area for different current flow directions could lead to a change in scattering behavior for cigar- or pancake-shaped particles. However, the specimen for probing the GMR for the large $Cu_{55}Co_{45}$ sample was not taken out the same way as for the smaller HPT-samples. At the same time, the current flow was in the tangential-radial plane (along a secant) for all small specimens, the current flow was in axial direction for the $Cu_{55}Co_{45}$ specimen. The same qualitative behavior of GMR in differently orientated specimens seems to rule out the particle shape anisotropy as an explanation for the non-perfect isotropic GMR. Another explanation for this might be the existence of large Co-particle or large percolating domains of a Co-rich phase, as these are likely to contain multiple domains. As stated in reference [10], multi-domain particles do not contribute to the GMR and as a result, a superposition of GMR and AMR occurs.

For the equiatomic composition, it was shown that deforming the sample at 200 °C leads to increased GMR compared to room temperature processing. The drop in resistivity is even higher for slightly increased process temperature and strongly reduced strain rate. Using the large HPT-tool provides temperature and time and a lower strain rate, for small Co particles to develop. According to Equation (1), an increase in GMR may as well originate from a decreased overall resistivity of the specimen—as can be expected from subsequent annealing processes but also from higher processing temperatures. The measured resistances of the specimen bear a large source for errors as the geometry is of high importance and the production of perfectly prismatic specimens is difficult. Thus, just rough values but more importantly the sequence of measured resistivities shall be given: The room temperature deformed $Cu_{52}Co_{48}$ has the highest specific resistivity (~0.52 Ω mm^2 m^{-1}), followed by $Cu_{55}Co_{45}$ (~0.18 Ω mm^2 m^{-1}) and finally $Cu_{49}Co_{51}$ (~0.13 Ω mm^2 m^{-1}). The specific resistivities were calculated, taking into account the individual specimen sizes of ~ 5 mm × 1 mm × 0.2 mm. It can be stated that the higher GMR effect for $Cu_{55}Co_{45}$ is not due to the same amount of GMR sitting atop of a smaller quantity (residual resistivity) but is truly due to a change in GMR (i.e., Co-segregation behavior), which is a consequence of changing process parameters.

SQUID measurements (Figure 11) show that the magnetization is almost fully saturated in fields of about 20 kOe and thus the approach of drawing the change in resistivity versus relative magnetization according to reference [4] can be followed. A quadratic fit of relative resistivity drop as a function of M/M_s (Figure 13) yields a proportional constant A of −0.031. This value is lower than the one stated in reference [4], who used magnetron sputtered $Cu_{84}Co_{16}$, resulting in a value of −0.065 at a temperature of 5 K. The shape of the two MR-curves for this composition (Figure 9c) can be explained by a parallel connection of varistor-like components as shown in Figure 13b). On the one hand, there is conduction in the Cu phase, where small Co particles can be found. On the other hand, there is conduction in the partially percolating Co-phase, which gives rise to an anisotropic behavior. Both pathways have individual dependencies on the direction of applied magnetic field. The strength of both resistive branches determines the sign and magnitude of the proportional constant A.

With further increasing Co content, the fraction of AMR becomes more pronounced but still a markedly high drop in parallel alignment can be seen. For pure Co, the difference between both types of about 1% matches the value given by McGuire and Potter [55] of 1.9%. The difference might be explained by the overall higher resistivity of HPT-deformed, nanocrystalline Co, reducing the relative fraction of the AMR regarding total resistivity.

For pure Cu, no change in resistivity with applied magnetic field has been found within the accuracy of the used measurement setup.

The number of parameters influencing the microstructure and – as a consequence – the magneto-resistive properties is very large. Although this study is very detailed, it is not complete and leaves plenty of ideas for further investigations. In future, further interesting tasks shall be tackled: (i) Larger GMR effects will be investigated by measuring the most promising samples at cryogenic temperatures. (ii) The influence of subsequent thermal treatments of the as-HPT-deformed materials on the microstructure as well as on magneto-resistivity will be investigated.

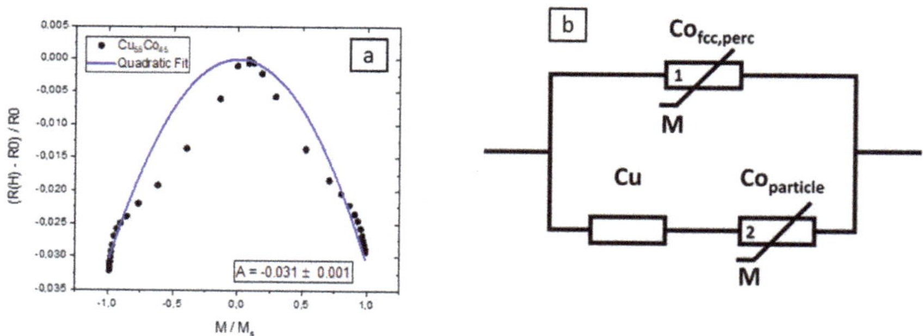

Figure 13. (**a**) Quadratic fit of room temperature GMR data for $Cu_{55}Co_{45}$ in accordance with reference [4]. (**b**) Schematic representation of electron conduction channels in case of $Cu_{55}Co_{45}$, consisting of Cu, containing Co-particles on the one side and Co enriched, percolating areas on the other side.

5. Conclusions

The influence of the SPD conditions such as deformation temperature and strain rate on the microstructural evolution and the magneto-resistive response of different combinations of ferromagnetic and diamagnetic elements has been investigated and the following conclusions can be drawn:

It has been shown that not only the processing temperature but also the strain rate is a very important parameter regarding the deformation behavior and microstructural evolution of composite materials. The strain rate influences, in combination with the applied (or naturally evolving) temperature, the diffusion, segregation, and dissolution mechanisms taking place during severe plastic deformation. As a result, SPD by HPT is a versatile tool for achieving different microstructural states and particle sizes, respectively, when the process parameters are chosen wisely.

Depending on the ferromagnetic content of the HPT-deformed materials, different behaviors regarding magneto-resistivity at room temperature develop. When there is a small ferromagnetic content, isotropic magneto-resistive behavior (GMR) can be found. The highest drop in resistivity that could be measured within the available magnetic field was found for an approximately equiatomic composition of Cu and Co. This sample was deformed at elevated temperatures and—in respect to typical HPT-deformation processes—at a small strain rate. For medium and high Co-content, the characteristics of magneto-resistance show the occurrence of both GMR and AMR.

The investigations of GMR in connection with HPT-deformed materials are interlinked: On the one side, it is possible to first adjust the occurrence of particles and then adjust the particle size distribution. This can be done by first changing all of the material's composition and then by changing the HPT-process parameters such as deformation temperature and strain rate. On the other side, GMR-measurements are a versatile tool to study the as-deformed (and annealed) microstructures regarding the distribution of ferromagnetic particles, thereby gaining deeper insights on the deformation and segregation mechanisms acting during high pressure torsion.

Author Contributions: Conceptualization, S.W., R.P. and A.B.; methodology, S.W. and A.B.; software, S.W.; validation, S.W.; formal analysis, S.W.; investigation, S.W., M.S., L.W., P.K., H.K.; resources, P.K., H.K., R.P., A.B.; data curation, S.W.; writing—original draft preparation, S.W.; writing—review and editing, S.W., L.W., M.S., H.K., P.K, R.P., A.B.; visualization, S.W., M.S.; supervision, A.B.; project administration, A.B.; funding acquisition, A.B.

Funding: This project has received funding from the European Research Council (ERC) under the European Union's Horizon 2020 research and innovation programme (Grant No. 757333).

Acknowledgments: The authors are in deep gratitude to Roland Grössinger, who passed away in 2018. He always was a source of fruitful and supportive discussion and he arranged the transfer of parts of the used equipment from the Technical University of Vienna to the Erich Schmid Insitute of Materials Science. Without him, these investigations of severely deformed materials would not have been possible. The measurements leading to some of the results presented here, have been performed at PETRA III P07 at DESY Hamburg (Germany), a member of the Helmholtz Association. The authors thank of the assistance of Norbert Schell and Karoline Kormout, Stefan Zeiler, Florian Spieckermann, Pradipta Gosh and Niraj Chawake for helping with synchrotron measurements and analysis. S.W. deeply appreciates the help of Christoph Gammer for performing the TEM analysis, Mirjam Spuller and Alexander Paulischin for performing the GMR specimen preparation and assistance in resistance measurements.

References

1. Baibich, M.N.; Broto, J.M.; Fert, A.; Van Dau, F.N.; Petroff, F.; Etienne, P.; Creuzet, G.; Friederich, A.; Chazelas, J. Giant Magnetoresistance of (001)Fe/(001)Cr Magnetic Superlattices. *Phys. Rev. Lett.* **1988**, *61*, 2472–2475. [CrossRef] [PubMed]

2. Binasch, G.; Grünberg, P.; Saurenbach, F.; Zinn, W. Enhanced magnetoresistance in layered magnetic structures with antiferromagnetic interlayer exchange. *Phys. Rev. B* **1989**, *39*, 4828–4830. [CrossRef] [PubMed]

3. Berkowitz, A.E.; Mitchell, J.R.; Carey, M.J.; Young, A.P.; Zhang, S.; Spada, F.E.; Parker, F.T.; Hutten, A.; Thomas, G. Giant magnetoresistance in heterogeneous Cu-Co alloys. *Phys. Rev. Lett.* **1992**, *68*, 3745. [CrossRef] [PubMed]

4. Xiao, J.Q.; Jiang, J.S.; Chien, C.L. Giant magnetoresistance in nonmultilayer magnetic systems. *Phys. Rev. Lett.* **1992**, *68*, 3749. [CrossRef] [PubMed]

5. Rabedeau, T.A.; Toney, M.F.; Marks, R.F.; Parkin, S.S.P.; Farrow, R.F.C.; Harp, G.R. Giant magnetoresistance and Co-cluster structure in phase-separated Co-Cu granular alloys. *Phys. Rev. B* **1993**, *48*, 16810–16813. [CrossRef] [PubMed]

6. Rubin, S.; Holdenried, M.; Micklitz, H. A model system for the GMR in granular systems: Well-defined Co clusters embedded in an Ag matrix. *J. Magn. Magn. Mater.* **1999**, *203*, 97–99. [CrossRef]

7. Xiong, P.; Xiao, G.; Wang, J.Q.; Xiao, J.Q.; Jiang, J.S.; Chien, C.L. Extraordinary Hall effect and giant magnetoresistance in the granular Co-Ag system. *Phys. Rev. Lett.* **1992**, *69*, 3220–3223. [CrossRef]

8. Sugawara, T.; Takanashi, K.; Hono, K.; Fujimori, H. Study of giant magnetoresistance behavior in sputter-deposited Cr Fe alloy films. *J. Magn. Magn. Mater.* **1996**, *159*, 95–102. [CrossRef]

9. Bernardi, J.; Hütten, A.; Thomas, G. GMR behavior of nanostructured heterogeneous M-Co (M = Cu, Ag, Au) alloys. *Nanostructured Mater.* **1996**, *7*, 205–220. [CrossRef]

10. Franco, V.; Batlle, X.; Labarta, A. CoFe–Cu granular alloys: From noninteracting particles to magnetic percolation. *J. Appl. Phys.* **1999**, *85*, 7328–7335. [CrossRef]

11. Chen, Y.J.; Ding, J.; Si, L.; Cheung, W.Y.; Wong, S.P.; Wilson, I.H.; Suzuki, T. Magnetic domain structures and magnetotransport properties in Co-Ag granular thin films. *Appl. Phys. A* **2001**, *73*, 103–106. [CrossRef]

12. Sang, H.; Jiang, Z.S.; Guo, G.; Ji, J.T.; Zhang, S.Y.; Du, Y.W. Study on GMR in Co Ag thin granular films. *J. Magn. Magn. Mater.* **1995**, *140–144*, 589–590. [CrossRef]

13. Hihara, T.; Sumiyama, K.; Onodera, H.; Wakoh, K.; Suzuki, K. Characteristic giant magnetoresistance of Fe/Cu granular films produced by cluster beam deposition and subsequent annealing. *Mater. Sci. Eng. A* **1996**, *217–218*, 322–325. [CrossRef]

14. Mebed, A.M.; Howe, J.M. Spinodal induced homogeneous nanostructures in magnetoresistive CoCu granular thin films. *J. Appl. Phys.* **2006**, *100*, 074310. [CrossRef]

15. Zaman, H. Magnetoresistance Effect in Co-Ag and Co-Cu Alloy Films Prepared by Electrodeposition. *J. Electrochem. Soc.* **1998**, *145*, 565. [CrossRef]

16. Ueda, Y.; Houga, T.; Zaman, H.; Yamada, A. Magnetoresistance Effect of Co–Cu Nanostructure Prepared by Electrodeposition Method. *J. Solid State Chem.* **1999**, *147*, 274–280. [CrossRef]

17. Ge, S.; Li, H.; Li, C.; Xi, L.; Li, W.; Chi, J. Giant magnetoresistance in electro-deposited Co-Cu granular film. *J. Phys. Condens. Matter* **2000**, *12*, 5905–5916. [CrossRef]

18. Coey, J.M.D.; Fagan, A.J.; Skomski, R.; Gregg, J.; Ounadjela, K.; Thompson, S.M. Magnetoresistance in nanostructured Co-Ag prepared by mechanical-alloying. *IEEE Trans. Magn.* **1994**, *30*, 666–668. [CrossRef]

19. Yermakov, A.Y.; Uimin, M.A.; Shangurov, A.V.; Zarubin, A.V.; Chechetkin, Y.V.; Shtolz, A.K.; Kondratyev, V.V.; Konygin, G.N.; Yelsukov, Y.P.; Enzo, S.; et al. Magnetoresistance and Structural State of Cu-Co, Cu-Fe Compounds Obtained by Mechanical Alloying. *MSF* **1996**, *225–227*, 147–156. [CrossRef]

20. Ikeda, S.; Houga, T.; Takakura, W.; Ueda, Y. Magnetoresistance in (CoxFe1—x)20Cu80 granular alloys produced by mechanical alloying. *Mater. Sci. Eng. A* **1996**, *217–218*, 376–380. [CrossRef]

21. Champion, Y.; Meyer, H.; Bonnentien, J.-L.; Chassaing, E. Fabrication of Cu–Co nanogranular bulk materials by mixing of nanocrystalline powders. *J. Magn. Magn. Mater.* **2002**, *241*, 357–363. [CrossRef]

22. Nagamine, L.C.C.M.; Chamberod, A.; Auric, P.; Auffret, S.; Chaffron, L. Giant magnetoresistance and magnetic properties of (Co0.7Fe0.3)20Cu80 prepared by mechanical alloying. *J. Magn. Magn. Mater.* **1997**, *174*, 309–315. [CrossRef]

23. Wecker, J.; von Helmolt, R.; Schultz, L.; Samwer, K. Giant magnetoresistance in melt spun Cu-Co alloys. *Appl. Phys. Lett.* **1993**, *62*, 1985–1987. [CrossRef]

24. Hütten, A.; Bernardi, J.; Friedrichs, S.; Thomas, G.; Balcells, L. Microstructural influence on magnetic properties and giant magnetoresistance of melt-spun gold-cobalt. *Scr. Metall. Mater.* **1995**, *33*, 1647–1666. [CrossRef]

25. Allia, P.; Knobel, M.; Tiberto, P.; Vinai, F. Magnetic properties and giant magnetoresistance of melt-spun granular Cu 1 0 0- x- Co x alloys. *Phys. Rev. B* **1995**, *52*, 15398. [CrossRef]

26. Aizawa, T.; Zhou, C. Nanogranulation process into magneto-resistant Co–Cu alloy on the route of bulk mechanical alloying. *Mater. Sci. Eng. A* **2000**, *285*, 1–7. [CrossRef]

27. Thomson, W., XIX. On the electro-dynamic qualities of metals:—Effects of magnetization on the electric conductivity of nickel and of iron. *Proc. R. Soc. Lond.* **1857**, *8*, 546–550.

28. Ennen, I.; Kappe, D.; Rempel, T.; Glenske, C.; Hütten, A. Giant Magnetoresistance: Basic Concepts, Microstructure, Magnetic Interactions and Applications. *Sensors* **2016**, *16*, 904. [CrossRef]

29. Valiev, R.; Islamgaliev, R.; Alexandrov, I. Bulk nanostructured materials from severe plastic deformation. *Prog. Mater. Sci.* **2000**, *45*, 103–189. [CrossRef]

30. Bachmaier, A.; Hohenwarter, A.; Pippan, R. New procedure to generate stable nanocrystallites by severe plastic deformation. *Scr. Mater.* **2009**, *61*, 1016–1019. [CrossRef]

31. Bridgman, P.W. Effects of High Shearing Stress Combined with High Hydrostatic Pressure. *Phys. Rev.* **1935**, *48*, 825–847. [CrossRef]

32. Pippan, R.; Scheriau, S.; Taylor, A.; Hafok, M.; Hohenwarter, A.; Bachmaier, A. Saturation of Fragmentation during Severe Plastic Deformation. *Annu. Rev. Mater. Res.* **2010**, *40*, 319–343. [CrossRef]

33. Scheriau, S.; Rumpf, K.; Kleber, S.; Pippan, R. Tailoring the Magnetic Properties of Ferritic Alloys by HPT. In Proceedings of the Materials Science Forum to the 4th International Conference on Nanomaterials by Severe Plastic Deformation, Goslar, Germany, 18–22 August 2008; Trans Tech Publ.: Bäch, Switzerland, 2008; Volume 584, pp. 923–928.

34. Suehiro, K.; Nishimura, S.; Horita, Z.; Mitani, S.; Takanashi, K.; Fujimori, H. High-pressure torsion for production of magnetoresistance in Cu–Co alloy. *J. Mater. Sci.* **2008**, *43*, 7349–7353. [CrossRef]

35. Straumal, B.B.; Protasova, S.G.; Mazilkin, A.A.; Baretzky, B.; Goll, D.; Gunderov, D.V.; Valiev, R.Z. Effect of severe plastic deformation on the coercivity of Co–Cu alloys. *Philos. Mag. Lett.* **2009**, *89*, 649–654. [CrossRef]

36. Scheriau, S.; Kriegisch, M.; Kleber, S.; Mehboob, N.; Grössinger, R.; Pippan, R. Magnetic characteristics of HPT deformed soft-magnetic materials. *J. Magn. Magn. Mater.* **2010**, *322*, 2984–2988. [CrossRef]

37. Nishihata, S.; Suehiro, K.; Arita, M.; Masuda, M.; Horita, Z. High-Pressure Torsion for Giant Magnetoresistance and Better Magnetic Properties. *Adv. Eng. Mater.* **2010**, *12*, 793–797. [CrossRef]

38. Straumal, B.B.; Protasova, S.G.; Mazilkin, A.A.; Kogtenkova, O.A.; Kurmanaeva, L.; Baretzky, B.; Schütz, G.; Korneva, A.; Zięba, P. SPD-induced changes of structure and magnetic properties in the Cu–Co alloys. *Mater. Lett.* **2013**, *98*, 217–221. [CrossRef]

39. Bachmaier, A.; Krenn, H.; Knoll, P.; Aboulfadl, H.; Pippan, R. Tailoring the magnetic properties of nanocrystalline Cu-Co alloys prepared by high-pressure torsion and isothermal annealing. *J. Alloys Compd.* **2017**, *725*, 744–749. [CrossRef]

40. Stückler, M.; Krenn, H.; Pippan, R.; Weissitsch, L.; Wurster, S.; Bachmaier, A. Magnetic Binary Supersaturated Solid Solutions Processed by Severe Plastic Deformation. *Nanomaterials* **2019**, *9*, 6. [CrossRef]

41. Stückler, M.; Weissitsch, L.; Wurster, S.; Felfer, P.; Krenn, H.; Pippan, R.; Bachmaier, A. Magnetic Dilution by Severe Plastic Deformation. *AIP Adv.* Submitted.

42. Stückler, M.; Krenn, H.; Kürnsteiner, P.; Felfer, P.; Weissitsch, L.; Wurster, S.; Pippan, R.; Bachmaier, A. Intermixing of Iron and Copper on the atomic scale via High-Pressure Torsion: SQUID AC-Susceptometry and Atom Probe Tomography. In preparation.

43. Rattanasakulthong, W.; Sirisathitkul, C. Large negative magnetoresistance in capsulated Co–Cu powder prepared by mechanical alloying. *Phys. B Condens. Matter* **2005**, *369*, 160–167. [CrossRef]

44. Suehiro, K.; Nishimura, S.; Horita, Z. Change in Magnetic Property of Cu-6.5 mass% Co Alloy through Processing by ECAP. *Mater. Trans.* **2008**, *49*, 102–106. [CrossRef]

45. Edalati, K.; Miresmaeili, R.; Horita, Z.; Kanayama, H.; Pippan, R. Significance of temperature increase in processing by high-pressure torsion. *Mater. Sci. Eng. A* **2011**, *528*, 7301–7305. [CrossRef]

46. Trimby, P.W. Orientation mapping of nanostructured materials using transmission Kikuchi diffraction in the scanning electron microscope. *Ultramicroscopy* **2012**, *120*, 16–24. [CrossRef] [PubMed]

47. Brodusch, N.; Demers, H.; Gauvin, R. Nanometres-resolution Kikuchi patterns from materials science specimens with transmission electron forward scatter diffraction in the scanning electron microscope. *J. Microsc.* **2013**, *250*, 1–14. [CrossRef]

48. Chien, C.L.; Xiao, J.Q.; Jiang, J.S. Giant negative magnetoresistance in granular ferromagnetic systems (invited). *J. Appl. Phys.* **1993**, *73*, 5309–5314. [CrossRef]

49. Cu Crystal Structure-Springer Materials. Available online: https://materials.springer.com/isp/crystallographic/docs/sd_0250160 (accessed on 26 September 2019).

50. Co alpha (Co ht) Crystal Structure-Springer Materials. Available online: https://materials.springer.com/isp/crystallographic/docs/sd_1324295 (accessed on 26 September 2019).

51. α-Co (Co rt) Crystal Structure-Springer Materials. Available online: https://materials.springer.com/isp/crystallographic/docs/sd_0531743 (accessed on 26 September 2019).

52. Hafok, M.; Pippan, R. High-pressure torsion applied to nickel single crystals. *Philos. Mag.* **2008**, *88*, 1857–1877. [CrossRef]

53. Chen, Y.T.; Jen, S.U.; Yao, Y.D.; Wu, J.M.; Lee, C.C.; Sun, A.C. Magnetic properties of face-centered cubic Co films. *IEEE Trans. Magn.* **2006**, *42*, 278–282. [CrossRef]

54. Wang, C.; Guo, Z.; Rong, Y.; Hsu (Xu Zuyao), T.Y. A phenomenological theory of the granular size effect on the giant magnetoresistance of granular films. *J. Magn. Magn. Mater.* **2004**, *277*, 273–280. [CrossRef]
55. McGuire, T.; Potter, R. Anisotropic magnetoresistance in ferromagnetic 3d alloys. *IEEE Trans. Magn.* **1975**, *11*, 1018–1038. [CrossRef]

Article

On the Use of Functionally Graded Materials to Differentiate the Effects of Surface Severe Plastic Deformation, Roughness and Chemical Composition on Cell Proliferation

Laurent Weiss [1,2,*], Yaël Nessler [2], Marc Novelli [1], Pascal Laheurte [1,2] and Thierry Grosdidier [1,2]

[1] Laboratoire d'Etude des Microstructures et de Mécanique des Matériaux, Université de Lorraine, 57073 Metz, France; Marc.novelli@univ-lorraine.fr (M.N.); pascal.laheurte@univ-lorraine.fr (P.L.); thierry.grosdidier@univ-lorraine.fr (T.G.)

[2] Laboratory of Excellence for Design of Alloy Metals for Low-Mass Structures ('DAMAS' Labex), Université de Lorraine, 57070 Metz, France; nessler.yael@hotmail.fr

* Correspondence: Laurent.weiss@univ-lorraine.fr; Tel.: +33-3-72-74-77-87

Received: 20 November 2019; Accepted: 6 December 2019; Published: 13 December 2019

Abstract: Additive manufacturing allows the manufacture of parts made of functionally graded materials (FGM) with a chemical gradient. This research work underlines that the use of FGM makes it possible to study mechanical, microstructural or biological characteristics while minimizing the number of required samples. The application of severe plastic deformation (SPD) by surface mechanical attrition treatment (SMAT) on FGM brings new insights on a major question in this field: which is the most important parameter between roughness, chemistry and microstructure modification on biocompatibility? Our study demonstrates that roughness has a large impact on adhesion while microstructure refinement plays a key role during the early stage of proliferation. After several days, chemistry is the main parameter that holds sway in the proliferation stage. With this respect, we also show that niobium has a much better biocompatibility than molybdenum when alloyed with titanium.

Keywords: surface mechanical attrition treatment (SMAT); ultrasonic shot peening (USP); functionally graded materials (FGM); titanium niobium alloys; titanium molybdenum alloys; human mesenchymal stem cells culture; cell adhesion; cell proliferation

1. Introduction

Improving the biocompatibility of titanium and titanium alloy prostheses is a major challenge for the biomedical industry and has been the subject of much scientific interest in recent years [1,2]. Among all the techniques allowing the improvement of the cells adhesion and proliferation, severe plastic deformation (SPD) of the material has been studied for about one decade and, since then, been the subject of several investigations. SPD makes it possible to refine the microstructure to obtain, under some specific conditions, a nanostructure. The first biocompatibility study on pure titanium deformed by SPD was carried out by Kim et al. in 2007 [3]. The results have shown that the refinement of the microstructure by equal-channel angular pressing (ECAP) leads to a better wettability and therefore a better adhesion and proliferation of fibroblasts after two and five days. In a second study, the same research group has shown that the cytotoxicity remained unchanged after SPD [4]. Other teams have shown that ECAP also leads to a better proliferation of mouse fibroblasts after 72 h [5] and a better adhesion of human mesenchymal stem cells (MSCs) [6]. Further, other investigations have shown that ECAP can be followed by chemical etching or sanding to create surface pores that further increase the bio-integration [7,8]. The high pressure torsion (HPT) [9] is another technique of grain refinement by core SPD that can improve the biocompatibility as illustrated on NiTi [10] alloys as well as on pure titanium by modification of its protective oxide layer [11]. This modification of the oxide layer on Ti-based materials has recently been demonstrated by other teams using other SPD techniques [12–14]. These positive results on the effect of core SPD have nonetheless been counterbalanced by some studies which have shown no beneficial effect of SPD or equivalent results for coarse-grain and nano-grain microstructures [15,16].

These core SPD techniques impart the plastic deformation to the overall material while, in many cases, only the surface is an important issue in terms of bio-integration. Thus, an interesting prospect for bio-integration is to modify the surface properties while preserving at core the mechanical properties of the raw material. With this respect, techniques have been applied to create surface graded microstructures by surface SPD such as, for example, sliding friction treatment (SFT) [17] as well as surface mechanical attrition treatment (SMAT) [18]. The first one has shown promising results in terms of biocompatibility [13] but remains very difficult to implement industrially. The SMAT technique, also called ultrasonic shot peening (USSP), is already used in industry to increase the fatigue resistance of prosthesis [19] and has already proven its ability to increase the biocompatibility of titanium alloys. SMAT consists of plastically deforming the sample surface by moving shots set in motion by an ultrasonic vibrating part called sonotrode. The main difference compared to conventional directional peening is that, in the case of SMAT, the shots are moving within a sealed chamber giving them random impact trajectories on the surface, enhancing the superficial structural refinement [18,20]. A complete description of the treatment can be found in [21]. This treatment results in nanostructuring of the material but also in an increase in the surface roughness [18,20,21] and, apparently, an increase in surface wettability [22]. Results on pure titanium have revealed an increase in adhesion and viability of MSCs [23]. In-vivo tests on rabbits have been done to compare SMATed and raw Ti6Al4V and the study showed better bio-integration of the surface SPD materials within the bones after four, eight and 12 weeks [24].

The results on bio-integration by SPD nevertheless raise an important question: to greatly improve the biocompatibility, is it better to change the roughness, the size of the microstructure or the chemistry of the material? To answer this question, the latest advances in additive manufacturing that allow to manufacture parts made of functionally graded material (FGM) are here tested in combination with SMAT to ensure that all preparations have been made., Following this, the cell cultures are made under exactly the same atmosphere and in the same environment (temperature, time).

These types of materials, born in the 1980s in Japan [25], can be classified into three distinct groups depending on the nature of the gradients: gradient in microstructure, gradient in porosity and gradient in composition. In the present work, gradients in composition (Ti-Nb and Ti-Mo) will be tested. The functionally graded composition is defined by Pei et al. as "a change in composition across the bulk volume of a material aimed to dynamically mix and vary the ratios of materials within a three dimensional volume to produce a seamless integration of monolithic functional structures with varied properties" [26]. FGM are already used to provide an enhanced substitute for the coating in orthopedic implants, thus avoiding the sudden change in chemical composition and the "peeling-off" effect of the coated [27]. Several advantages can be achieved, such as improving the fixation of implant to bone, enhancing the stress shielding phenomena, hardening the articulating surface, and removing interfacial stresses between the implant and bone.

Bogdanski et al. were the first to use a FGM metal (Ni-Ti) ranging from 0 to 100% to find the composition with the best biocompatibility [28]. Unfortunately, the combination of cold isostating pressing (CIP) + hot isostating pressure (HIP) + vacuum welding processes used at this time was both time-consuming and costly.

Nowadays, direct energy deposition (DED) makes it possible to manufacture FGMs more rapidly and at a lower cost with two or more materials. This additive manufacturing technique, that uses a powder spray, has been successfully tested on FGM titanium-based alloys [29,30]. Since it can be used to manufacture prostheses with surface properties that are different from those of the core, these

FGMs have a strong potential for the future in the biomedical industry [31]. Among all material couples, titanium-niobium alloys and titanium-molybdenum alloys are excellent candidates for new prostheses generation [32,33]. In this interest, the aim of this study is thus to investigate the effect of chemistry, roughness and SMATed microstructure on cell adhesion and proliferation. To this end, FGMs have been made based on the Ti-Nb and Ti6Al4V-Mo constituents and then treated by SMAT. The FGMs compositions range from 100% Ti to 100% Nb as well as 100% Ti6Al4V to 100% Mo and were manufactured with a DED technique called CLAD® [34,35]. The FGMs were investigated in their polished (P), SMATed (S) and SMATed + polished (S+P) conditions then human mesenchymal stem cells were deposited and counted as function of time from three up to 21 days.

2. Experimental Procedure

2.1. Sample Preparation

The samples were manufactured at Irepa Laser using a CLAD®machine (Magic 800, BEAM company, Strasbourg, France) under a neutral argon atmosphere using a Ti6Al4V support plate. In this technology, the powder is injected into a laser beam and melts before falling onto the substrate. The machine can be fed simultaneously by different powder feeders, so it becomes possible to real-time control the chemical composition of the powder mixture. The size distribution of the powder particles measured by laser diffraction analysis is given in Table 1. As sketched in Figure 1, the manufactured gradient materials contain respectively 0, 25, 50, 75 and 100% of Nb or Mo and, conversely, 100, 75, 50, 25 and 0% of Ti or Ti6Al4V. The size of the parts is also given in Figure 1a. The powders and manufacturing data (power, speed, pitch, etc.) of the samples have been detailed in the works of Schneider et al. [34,35].

Table 1. Size distribution of the different powders in µm [34,35].

Material	D_{10}	D_{50}	D_{90}
Ti	54	73	100
Nb	41	56	76
Ti6Al4V	58	80	109
Mo	45	61	82

The FGM walls were cut into 1.5 mm thick slices. To take into account the effects of the surface roughness, the samples have been divided into three categories: polished (P), SMATed (S) and SMATed then polished (S + P). The P samples were polished using OPS (Oxide Polishing Suspension, Struers, Champigny-sur-Marne, France) to obtain a mirror-like surface. Samples S and S + P were SMATed for 5 min (frequency 20 kHz, amplitude 60 microns) with 2 mm spherical shots made of 100C6 steel. The shot blasting time was deliberately chosen to be short enough to minimize any possible transfer of chemical species coming from the shot-balls or tooling that are known to pollute the surface under SMAT [21,36–40]. This type of pollution, which is known to develop for long term peening, has been demonstrated to affect the corrosion properties of some parts [39,40] can indeed also have an effect on biological tests. After SMAT, the S + P samples were polished with OP-S over 2 or 3 µm to modify the roughness issued from the SMAT process while remaining in the area where the microstructure was modified. This experimental procedure using FGMs coupled with surface SPD as well as polishing has several advantages. While the comparison between the S and S + P samples makes it possible to study the effect of the roughness created by the SMAT for a given type of refined structure, the comparison between the P and S + P samples authorizes to analyze the effect of the microstructure modification by SPD for a mirror-like surface. In addition, the comparison of the response along the same sample can be used to depict the effect of the chemistry as well as the comparison between Ti-Nb and Ti6Al4V-Mo can be used to find the most biocompatible alloying element.

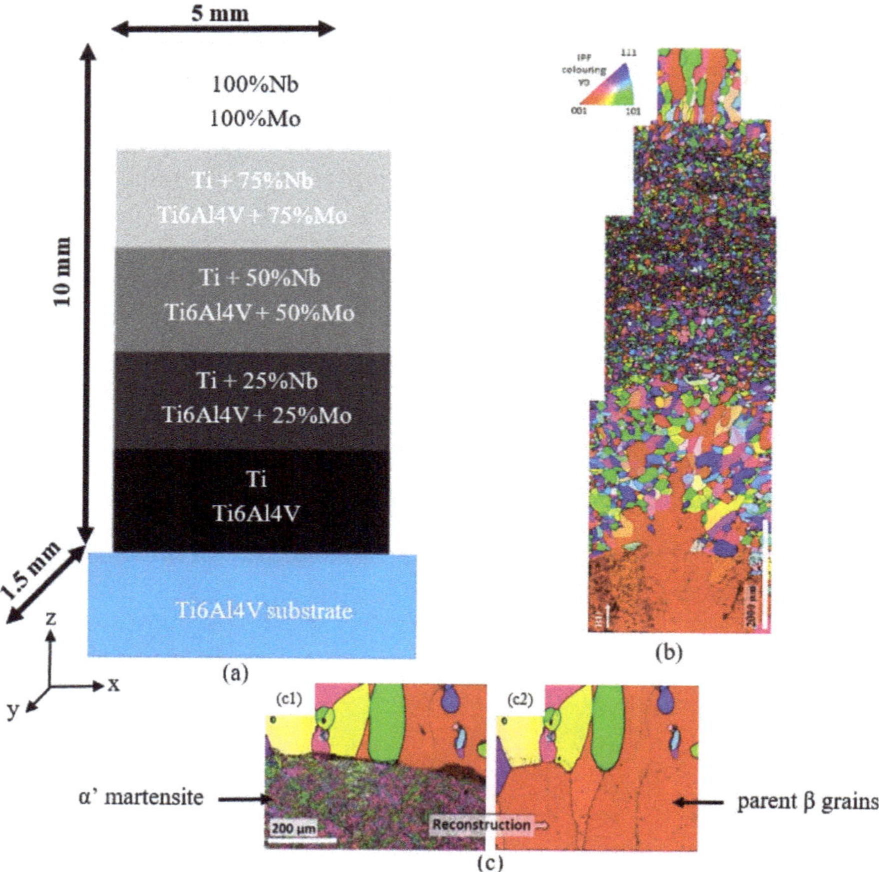

Figure 1. (**a**) Size and chemical composition of the samples for the Ti/Nb and Ti6Al4V/Mo couples; (**b**) complete map of the Ti6Al4V-Mo FGM after reconstruction from Schneider et al. [35]; (**c**) example of the interface Ti6Al4V/Ti6Al4V + 25% Mo before (**c1**) and after beta reconstruction (**c2**).

2.2. Roughness and Microscopy

The roughness measurements were carried out with a focus-variation microscopy using an Alicona Infinite Focus apparatus. The values presented are the average of 200 lines of 400 μm long. The microstructure of the samples was analyzed on the cross-section after OPS polishing with a Supra40 Scanning Electron Microscope (Carl Zeiss, Oberkochen, Germany).

2.3. Sterilization and Cell Culture

The samples were first placed in 12-well plates and then immersed in 1 mL of ethanol for 30 min. The plates were then placed in a sterile hood where they were exposed to UV for 30 min. To ensure perfect sterilization, the samples were then cleaned with bleach and then autoclaved for 4 h at 120 °C.

For each sample (P, S and S + P), the tests were performed three times with MSCs from three different donors. The cells were thawed one week before seeding and all manipulations were performed under hood using sterile equipment. The inoculated samples were placed in an incubator at 37 °C, 5% CO_2 and 90% humidity. After 45 min, 1 mL of α MEM (Minimum Essential Medium) was added per well and the plates were put back into the incubator. The culture medium was changed twice a week.

The viability of the cells was checked after 3, 7, 14 and 21 days. For this, an Alamar Blue test (a resazurin blue colored indicator which transforms into pink resorufin thanks to the reducing power of mitochondria) was used. The percent difference in reduction is proportional to the metabolic activity of the cells, and, thus, to their viability. The tests were done in a dark sterile hood and the color change was detected using a spectrometer.

2.4. Cell Adhesion and Cell Counting

Paraformaldehyde was intended to freeze the cell structures in a state as close as possible to their living state. For each configuration and donor, the cells of three samples of each pair of material were fixed at 3, 7, 14 and 21 days after the start of the cell culture. The samples were placed in a new 12-well plate and then 1 mL of 4% paraformaldehyde (PAF) was added in each. The PAF acted for 10 min, then it was removed and the samples were washed with phosphate buffered saline (PBS).

A 0.1% solution of 4′, 6-diamidino-2-phenylindole (DAPI), i.e., 1 µL of DAPI per 999 µL of PBS, was deposited in each well. The DAPI inserted into the DNA of the nucleus cells and emitted a 455 nm blue light when excited by 345 nm light. This allowed us to observe the cells under a fluorescence optical microscope and to count them on each section of the chemical gradient.

3. Results and Discussion

3.1. Evolution of the Microstructure Along the Gradient

The microstructure features of the different graded materials in their as-deposited states has been detailed previously [34,35]. The main features are just recalled in Figure 1b,c for the sake of clarity. The evolution of the microstructure, and in particular the parent grain morphologies after a reconstruction from the martensitic ′ variants, is illustrated in Figure 1b,c. The evolution of the microstructure for the Ti-Nb and TA64-Mo walls shared a number of aspects and the features shown in Figure 1b,c for TA64-Mo pertain for the other FGM sample. Because of the fast cooling rate [34,35], the initial layers of Ti or Ti6Al4V were characterized respectively by a Widmanstätten or a martensitic α' microstructure, issued from the high temperature parent phase nucleated from the melt (Figure 1c). With the addition of Nb or Mo, the microstructure became beta equiaxed with a decreasing grain size as the alloying element amount increased (Figure 1b). In the pure Niobium (Nb) or pure Molybdenum (Mo) layers, the beta grains were columnar parallel to the z axis (deposition axis).

Figures 2 and 3 give SEM images showing details of the microstructures after the application of the SMAT process. In Ti and Ti6Al4V as well as at the lowest amount of additions (25% of Mo), the top surface microstructure was refined over a depth of the order of 5 to 15 µm. This is particularly visible in Figure 3a,b where the grains are refined and equiaxed near the surface while elongated grains are visible at the sub-surface. In addition, for the initial structure, kink bands appeared in the subsurface over significant thicknesses (>200 µm). This type of kink band has already been observed in highly alloyed beta titanium alloys during deformation at high speed rate [41]. The width of the zone where these kink bands were present was thicker for the case of the Nb addition than for the Mo one (Figures 2b and 3b). Indeed, the kink bands reached the surface for the Ti-25Nb alloy, whereas for the Ti6Al4V-25Mo alloy, the nanostructured layer was present from the surface towards 15 µm below the surface (Figure 3b). From 50% of alloying elements, the kink bands disappeared and the contrast visible within the beta grains at the SMATed surface witnessed strong internal misorientations. Figures 2c–e and 3c–e show that the thickness of this layer depends on the amount in alloying element. A comparison of the blue and red arrows in Figures 2c–e and 3c–e indicates, however, that its thickness is equivalent for both the Ti-Nb and Ti6Al4V-Mo alloys at a given amount of alloying element.

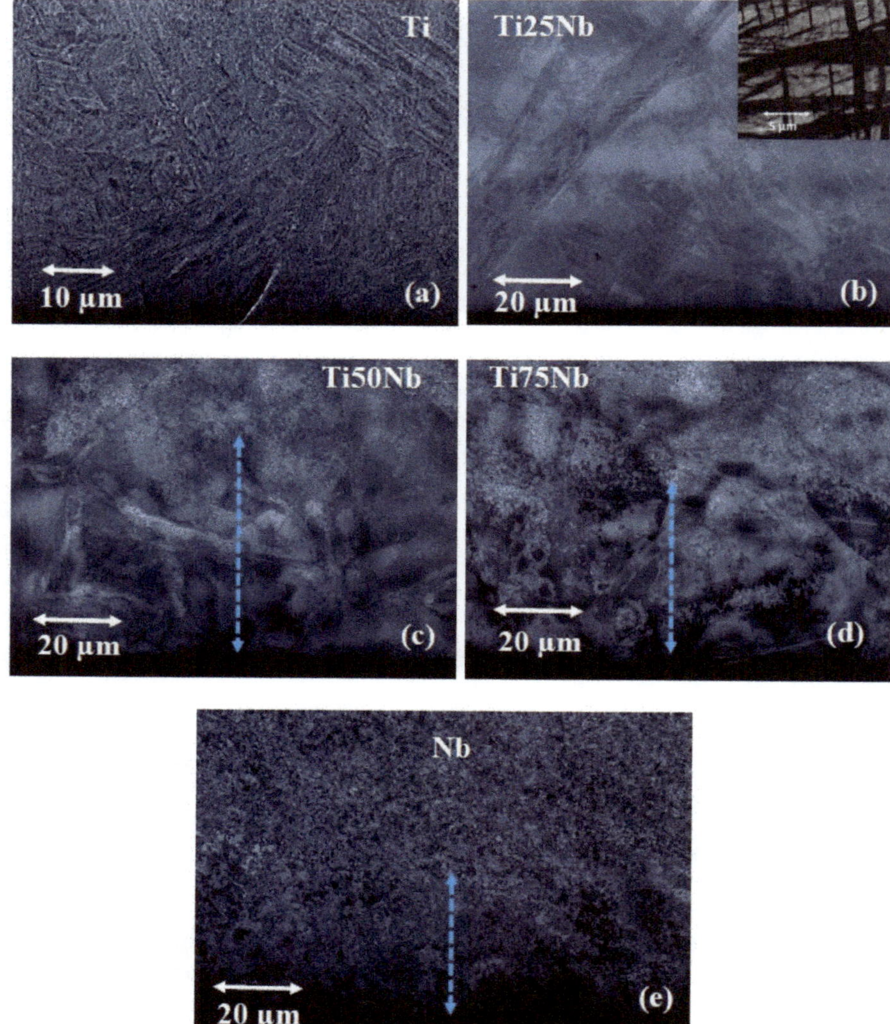

Figure 2. (**a–e**): cross section pictures of the surface mechanical attrition treatment (SMAT)ed TiNb sample focused on the severe plastic deformed region (the SMATed edges are always at the bottom of the pictures); (**b**): in the top right corner, focus on the kink bands at 20 µm from the surface in the Ti-25Nb; (**c–e**): the blue arrow represents the affected depth.

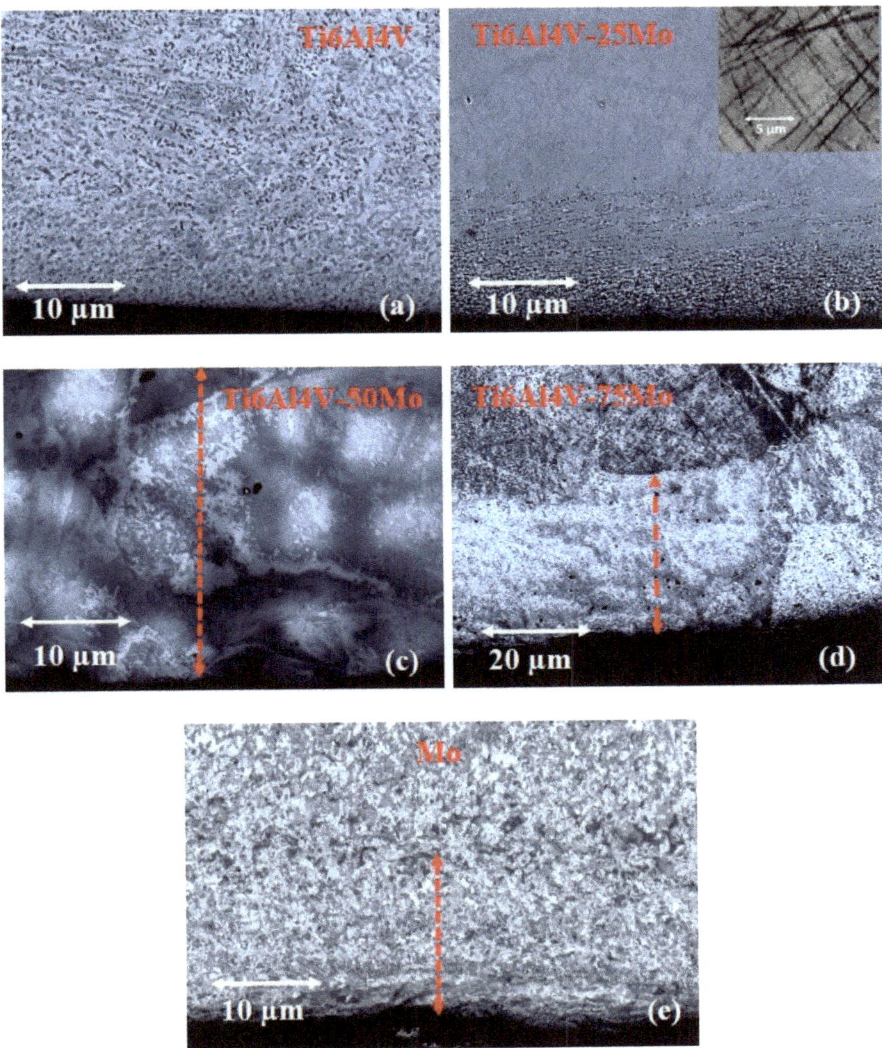

Figure 3. (**a–e**): Cross section pictures of the SMATed TiMo sample focused on the severe plastic deformed region (the SMATed edges are always at the bottom of the pictures); (**b**): in the top right corner, focus on the kink bands at 20 μm from the surface in the Ti6Al4V-25Mo; (**c–e**): the red arrow represents the affected depth.

3.2. Hardness and Roughness Evolutions

Figure 4 shows the evolution of the hardness and roughness before and after the different SMAT and SMAT + polishing treatments for the Ti-Nb wall. For the polished sample, the maximum hardness was reached for 25% of Nb (Figure 4a). According to the literature [42,43], this increase is due to the presence of a small fraction of ω-phase, especially when the composition is around 25% Nb. The amount of omega phase was increased by the successive remelting and tempering from the successive layer by layer deposition generated by the DED-CLAD process. This tempering promoted the formation of the ω-phase, which explains the increase in the microhardness, as was also observed for TiMo alloys [44].

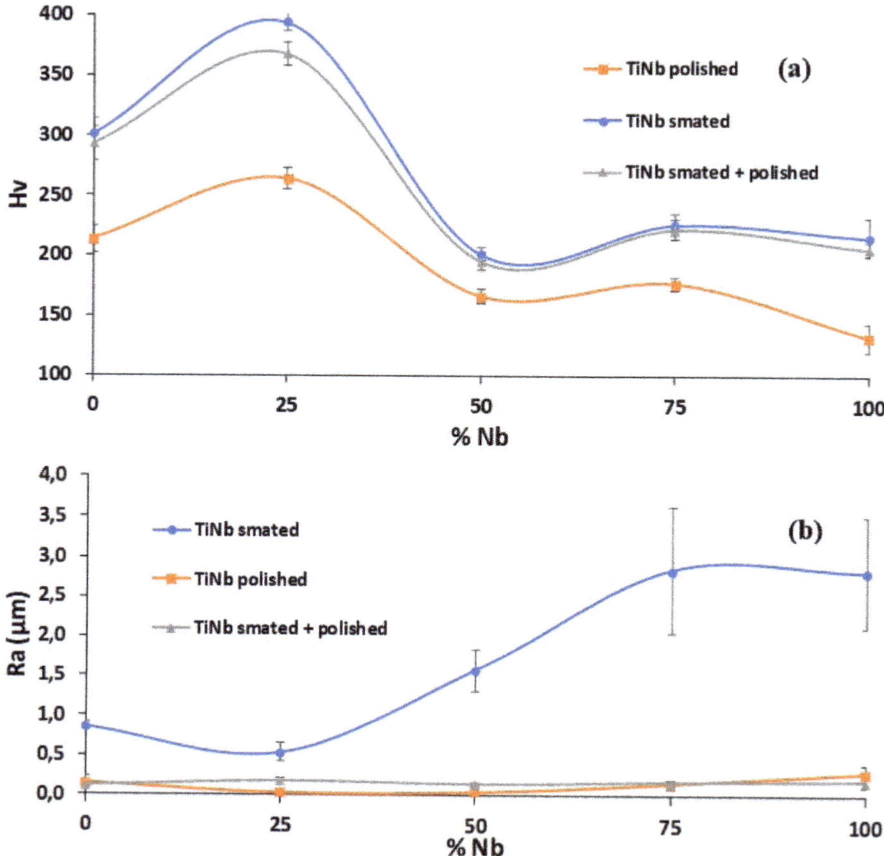

Figure 4. Evolution of (**a**) the hardness (HV$_{0.3}$) and (**b**) the roughness (Ra) as a function of the Nb alloying amount for the P, S and P + S conditions of the TiNb walls.

After the SMAT modifications, it was first interesting to notice that the hardness evolutions for the S and S + P samples are almost equal (Figure 4a). This meant that the polishing has allowed to remove a few microns to smooth the roughness (Figure 4b) while remaining in the hyperdeformed thickness. The range of hardness increase by shot peening depended slightly on the alloying amount. The maximum increase was obtained due to the appearance of the kink bands in the Ti25Nb, which made it possible to change hardness from 250 to 400 Hv (60% increase). Sadeghpour et al. [45] have recently shown that there is a relation between the presence of omega phase and the formation of kink bands due to the accumulation of simple dislocation and dislocation channels, which correlates with our study. The strong internal misorientation of the 50 to 100% Nb containing samples also increased the hardness but to a lesser extent (15 to 40% increase).

The evolutions of roughness presented in Figure 4b shows that the roughness after SMAT was inversely related to the hardness. When the material was hard, the roughness increase remained moderate. This is because—for a given impact energy—the balls created deeper craters when the material was softer (Figure 4a). Because of the polishing procedure, the P and S + P samples have the same low roughness, independently of the initial hardness. These modifications of hardness versus roughness made it possible to dissociate the effect of the roughness from the other effects (chemistry and SPD induced by SMAT). Indeed, several papers have suggested that the surface roughness plays a major role in cell biology: the higher the roughness, the higher the cell adhesion [46–48].

3.3. Cell Adhesion and Cell Proliferation

3.3.1. Comparison of the Effect of Nb and Mo on Titanium Biocompatibility

Figure 5 shows the number of cells on each gradient after three and seven days for the same donor (other donors generating the same trends). This allowed us to rapidly compare the effects of chemistry, roughness and hyperdeformation. Figure 5a shows that pure titanium and Ti6Al4V have the same trends on adhesion and proliferation after three and seven days. Ti6Al4V appeared to have a slightly more beneficial effect than pure titanium on adhesion. The number of cells on polished samples after seven days remained almost unchanged which confirms the results obtained by Johansson et al. [49]. The S + P samples for both materials had the same behavior as the polished ones. The Ti SMATed samples were unlike the Ti6Al4V SMATed samples which had a sharp decrease in the number of cells after seven days. Nevertheless, whatever the material, the SMAT had a positive effect on the adhesion because of the increase of the roughness, which is consistent with previous studies from the literature [46–48]. Figure 5b shows that the biocompatibility of these two materials does not seem to be affected by the presence of low amount of alloying elements (≤25%).

When the percentage of alloying element is increased (50 and 75%), the behavior of the cells was strongly modified. Niobium was found to have an excellent biocompatibility, which greatly increased the proliferation of the cells (Figure 5c). This effect of the chemistry was more important after seven days if the samples were SMATed. This is because of the increase in the number of adhering cells, which further increased subsequently after seven days, due to the cell division process in an excellent biocompatible environment. As shown in Figure 5d, the harmfulness of molybdenum began to have a very important effect from 75%. Even for the polished sample where the adhesion was best after three days, the cells then died quickly avec seven days. The toxicity of this alloy was even greatly increased by the microstructure modification induced by the SMAT. Indeed, on the samples, S and S + P, no cell survived after seven days. These results validated the differences in biocompatibility between Nb and Mo observed by Eisenbarth et al. [50].

In the case of pure elements, as no cells adhered or survived for all conditions of pure Mo, Figure 5e demonstrates the toxicity of pure Mo and the better biocompatibility of Nb.

All these results mainly demonstrated that the best biocompatibility was obtained for the SMATed TiNb alloys when the concentration in niobium was between 50 and 75%. In addition, it was clear that SMAT has a beneficial effect only during the adhesion stage by increasing the roughness and that the toxicity of molybdenum was increased by the microstructure modification induced by SMAT.

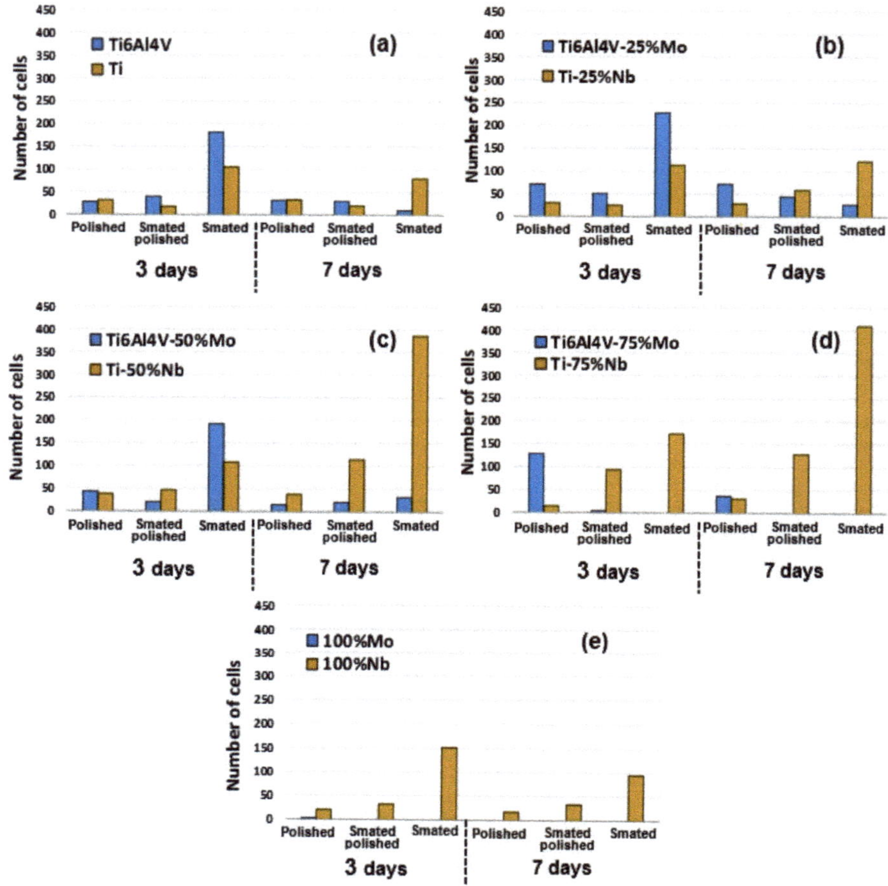

Figure 5. Comparison of the effect of Nb and Mo on cells adhesion and subsequent proliferation for different amount of alloying element. (**a**) 0%; (**b**) 25%; (**c**) 50%; (**d**) 75%; (**e**) 100%.

As the molybdenum has a poor biocompatibility compare to niobium, only the results on TiNb are showed in the rest of this paper.

3.3.2. Cell Adhesion

Figure 6 shows the average cell distributions of three donors after three and seven days. Because of the differences on the cells quality and viability between donors, the total number of healthy cells varied and it was better to calculate, for each gradient, the distribution in terms of percentages.

The comparison between Figures 4b and 6 demonstrates that, for the SMATed samples, the adhesion was preferentially done in the gradients where the roughness was important (75 and 100% of Nb). For the same roughness (P and S + P samples), the cells seemed to be randomly distributed. After seven days, the distribution trended to be the same for all samples but the nanostructured microstructure decreased the time needed to reach the distribution equilibrium. This indicated that the microstructure modification (dislocations cells, finer grains, etc.) induced by severe plastic deformation only played a role just after adhesion, at the early stage of proliferation.

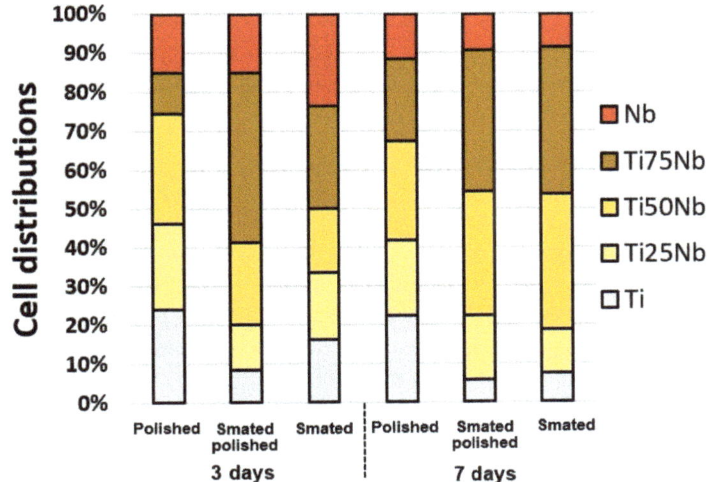

Figure 6. Cells distribution on TiNb after three and seven days.

3.3.3. Long Term Cell Proliferation on FGM

As demonstrated in Figure 5, the cells could not proliferate for the Mo containing sample and even not adhere at high Mo concentration, so only the Nb containing sample were investigated in term of cell proliferation. Figure 7 presents an example of the number of cells after three, seven, 14 and 21 days in the polished TiNb sample for the same donor. Figure 7 shows that if the total number of cells increased during the 21 days, the proliferation depended on the composition. Niobium had the fewest cells in the first week but then became the location in the gradient where proliferation was the most important. Thus, after three weeks, it became the best material with the highest cell number. Despite the fact that niobium was not used as such in prostheses because of its mechanical properties that were too far from those of the bones (higher Young modulus) [51], this test demonstrated the perfect biocompatibility of pure niobium [50]. In addition, this type of test validated the fact that FGM can be used successfully in biocompatible experiments to decrease the number of needed samples for long term proliferation test.

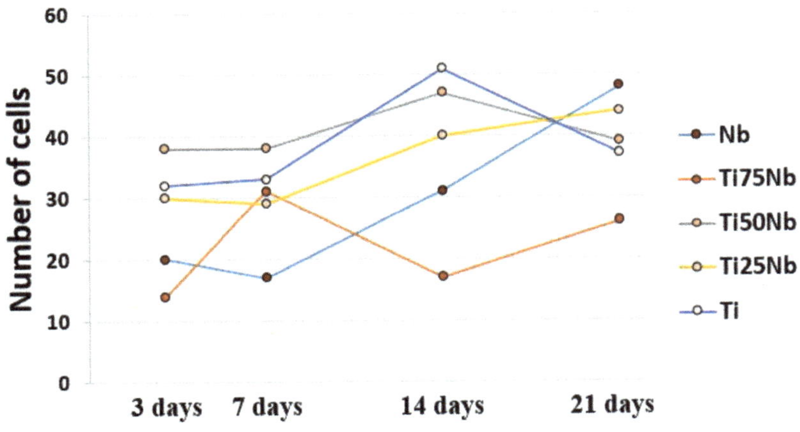

Figure 7. Cells proliferation: example of polished TiNb.

4. Conclusions

The aim of this study was to describe separately, with a single test, the effect of chemistry, roughness and SPD microstructure on MSC cell adhesion and proliferation using SMAT on two binary FGMs (Ti-Nb and TA64-Mo).

- The major findings can be summarized has follows:
- The use of FGMs can greatly reduce the number of samples needed in biocompatibility studies by ensuring at the same time that all tests are done under the same conditions.
- Increase in roughness was induced by SMAT trends to improve the cellular adhesion.
- Comparatively, this increase in roughness dud not modify the proliferation capability of the cells.
- The microstructure refinement and the presence of structural defects induced by severe plastic deformation have an effect on cell distribution during the first stages of proliferation.
- Chemistry modifications are the most important factor to ensure long-term cell proliferation.
- Niobium has better long-term biocompatibility than molybdenum when it is pure or when it is alloyed with titanium.

Author Contributions: Results analysis and paper writing; L.W. and T.G.; samples preparation and biological tests, Y.N.; microscopy, M.N.; FGMs analysis, P.L.

Funding: This study was supported by the French State through the program "Investment in the Future" operated by the National Research Agency (ANR), referenced by ANR-11-LABX-0008-01 (Labex DAMAS).

Acknowledgments: The authors gratefully acknowledge the Labex Damas for its financial support, Catherine Schneider-Maunoury from Irepa Laser for the production of FGM and Natalia De Isla from Imopa laboratory in Nancy (France) for the supervision of biological tests.

Conflicts of Interest: The authors declare no conflicts of interest.

References

1. Asri, R.I.M.; Harun, W.S.W.; Samykano, M.; Lah, N.A.C.; Ghani, S.A.C.; Tarlochan, F.; Raza, M.R. Corrosion and surface modification on biocompatible metals: A review. *Mater. Sci. Eng. C* **2017**, *77*, 1261–1274. [CrossRef] [PubMed]

2. Bagherifard, S.; Ghelichi, R.; Khademhosseini, A.; Guagliano, M. Cell Response to Nanocrystallized Metallic Substrates Obtained through Severe Plastic Deformation. *ACS Appl. Mater. Interface* **2014**, *6*, 7963–7985. [CrossRef] [PubMed]

3. Kim, T.N.; Balakrishnan, A.; Lee, B.C.; Kim, W.S.; Smetana, K.; Park, J.K.; Panigrahi, B.B. In vitro biocompatibility of equal channel angular processed (ECAP) titanium. *Biomed. Mater.* **2007**, *2*, S117–S120. [CrossRef] [PubMed]

4. Kim, T.N.; Balakrishnan, A.; Lee, B.C.; Kim, W.S.; Dvorankova, B.; Smetana, K.; Park, J.K.; Panigrahi, B.B. In vitro fibroblast response to ultra fine grained titanium produced by a severe plastic deformation process. *J. Mater. Sci. Mater. Med.* **2008**, *19*, 553–557. [CrossRef] [PubMed]

5. Valiev, R.; Semenova, I.P.; Jakushina, E.; Latysh, V.V.; Rack, H.J.; Lowe, T.C.; Petruželka, J.; Dluhoš, L.; Hrušák, D.; Sochová, J. Nanostructured SPD Processed Titanium for Medical Implants. *Mater. Sci. Forum* **2008**, *584*, 49–54. [CrossRef]

6. Estrin, Y.; Ivanova, E.P.; Michalska, A.; Truong, V.K.; Lapovok, R.; Boyd, R. Accelerated stem cell attachment to ultrafine grained titanium. *Acta Biomater.* **2011**, *7*, 900–906. [CrossRef]

7. Zheng, C.Y.; Nie, F.L.; Zheng, Y.F.; Cheng, Y.; Wei, S.C.; Valiev, R.Z. Enhanced in vitro biocompatibility of ultrafine-grained titanium with hierarchical porous surface. *Appl. Surf. Sci.* **2011**, *257*, 5634–5640. [CrossRef]

8. Park, J.W.; Kim, Y.J.; Park, C.H.; Lee, D.H.; Ko, Y.G.; Jang, J.H.; Lee, C.S. Enhanced osteoblast response to an equal channel angular pressing-processed pure titanium substrate with microrough surface topography. *Acta Biomater.* **2009**, *5*, 3272–3280. [CrossRef]

9. Edalati, K.; Horita, Z. A review on high-pressure torsion (HPT) from 1935 to 1988. *Mater. Sci. Eng. A* **2016**, *652*, 325–352. [CrossRef]

10. Nie, F.L.; Zheng, Y.F.; Cheng, Y.; Wei, S.C.; Valiev, R.Z. In vitro corrosion and cytotoxicity on microcrystalline, nanocrystalline and amorphous NiTi alloy fabricated by high pressure torsion. *Mater. Lett.* **2010**, *64*, 983–986. [CrossRef]

11. Faghihi, S.; Azari, F.; Zhilyaev, A.; Szpunar, J.; Vali, H.; Tabrizian, M. Cellular and molecular interactions between MC3T3-E1 pre-osteoblasts and nanostructured titanium produced by high-pressure torsion. *Biomaterials* **2007**, *28*, 3887–3895. [CrossRef] [PubMed]

12. Korotin, D.M.; Bartkowski, S.; Kurmaev, E.Z.; Neumann, M.; Yakushina, E.B.; Valiev, R.Z.; Cholakh, S.O. Surface Studies of Coarse-Grained and Nanostructured Titanium Implants. *J. Nanosci. Nanotechnol.* **2012**, *12*, 8567–8572. [CrossRef] [PubMed]

13. Lu, J.; Zhang, Y.; Huo, W.; Zhang, W.; Zhao, Y.; Zhang, Y. Electrochemical corrosion characteristics and biocompatibility of nanostructured titanium for implants. *Appl. Surf. Sci.* **2018**, *434*, 63–72. [CrossRef]

14. Bahl, S.; Aleti, B.T.; Suwas, S.; Chatterjee, K. Surface nanostructuring of titanium imparts multifunctional properties for orthopedic and cardiovascular applications. *Mater. Des.* **2018**, *144*, 169–181. [CrossRef]

15. Nie, F.L.; Wang, S.G.; Wang, Y.B.; Wei, S.C.; Zheng, Y.F. Comparative study on corrosion resistance and in vitro biocompatibility of bulk nanocrystalline and microcrystalline biomedical 304 stainless steel. *Dent. Mater.* **2011**, *27*, 677–683. [CrossRef]

16. Zhao, M.; Wang, Q.; Lai, W.; Zhao, X.; Shen, H.; Nie, F.; Zheng, Y.; Wei, S.; Ji, J. In vitro bioactivity and biocompatibility evaluation of bulk nanostructured titanium in osteoblast-like cells by quantitative proteomic analysis. *J. Mater. Chem. B* **2013**, *1*, 1926. [CrossRef]

17. Zhang, Y.S.; Zhang, L.C.; Niu, H.Z.; Bai, X.F.; Yu, S.; Ma, X.Q.; Yu, Z.T. Deformation twinning and localized amorphization in nanocrystalline tantalum induced by sliding friction. *Mater. Lett.* **2014**, *127*, 4–7. [CrossRef]

18. Lu, K.; Lu, J. Nanostructured surface layer on metallic materials induced by surface mechanical attrition treatment. *Mater. Sci. Eng. A* **2004**, *375–377*, 38–45. [CrossRef]

19. Shot Peening for Medical Indusrty. Available online: https://sonats-et.com/en/our-markets/shot-peening-medical-industry/ (accessed on 12 December 2019).

20. Bagheri, S.; Guagliano, M. Review of shot peening processes to obtain nanocrystalline surfaces in metal alloys. *Surf. Eng.* **2009**, *25*, 3–14. [CrossRef]

21. Grosdidier, T.; Novelli, M. Recent Developments in the Application of Surface Mechanical Attrition Treatments for Improved Gradient Structures: Processing Parameters and Surface Reactivity. *Mater. Trans.* **2019**, *60*, 1344–1355. [CrossRef]

22. Arifvianto, B.; Suyitno; Mahardika, M.; Dewo, P.; Iswanto, P.T.; Salim, U.A. Effect of surface mechanical attrition treatment (SMAT) on microhardness, surface roughness and wettability of AISI 316L. *Mater. Chem. Phys.* **2011**, *125*, 418–426. [CrossRef]

23. Lai, M.; Cai, K.; Hu, Y.; Yang, X.; Liu, Q. Regulation of the behaviors of mesenchymal stem cells by surface nanostructured titanium. *Colloids Surf. B Biointerfaces* **2012**, *97*, 211–220. [CrossRef] [PubMed]

24. Zhao, C.; Ji, W.; Han, P.; Zhang, J.; Jiang, Y.; Zhang, X. In vitro and in vivo mineralization and osseointegration of nanostructured Ti6Al4V. *J. Nanoparticle Res.* **2011**, *13*, 645–654. [CrossRef]

25. Koizumi, M. FGM activities in Japan. *Compos. Part. B Eng.* **1997**, *28*, 1–4. [CrossRef]

26. Pei, E.; Loh, G.H.; Harrison, D.; de Amorim, H.A.; Monzón Verona, M.D.; Paz, R. A study of 4D printing and functionally graded additive manufacturing. *Assem. Autom.* **2017**, *37*, 147–153. [CrossRef]

27. Roop Kumar, R.; Wang, M. Functionally graded bioactive coatings of hydroxyapatite/titanium oxide composite system. *Mater. Lett.* **2002**, *55*, 133–137. [CrossRef]

28. Bogdanski, D.; Köller, M.; Müller, D.; Muhr, G.; Bram, M.; Buchkremer, H.P.; Stöver, D.; Choi, J.; Epple, M. Easy assessment of the biocompatibility of Ni–Ti alloys by in vitro cell culture experiments on a functionally graded Ni–NiTi–Ti material. *Biomaterials* **2002**, *23*, 4549–4555. [CrossRef]

29. Qian, T.; Liu, D.; Tian, X.; Liu, C.; Wang, H. Microstructure of TA2/TA15 graded structural material by laser additive manufacturing process. *Trans. Nonferrous Met. Soc. China* **2014**, *24*, 2729–2736. [CrossRef]

30. Ren, H.S.; Liu, D.; Tang, H.B.; Tian, X.J.; Zhu, Y.Y.; Wang, H.M. Microstructure and mechanical properties of a graded structural material. *Mater. Sci. Eng. A* **2014**, *611*, 362–369. [CrossRef]

31. Naebe, M.; Shirvanimoghaddam, K. Functionally graded materials: A review of fabrication and properties. *Appl. Mater. Today* **2016**, *5*, 223–245. [CrossRef]

32. Bartakova, S.; Prachar, P.; Kudrman, J.; Bresina, V.; Podhorna, B.; Cernochova, P.; Vanek, J.; Strecha, J. New titanuim β-alloys for dental implantology and their laboratory-based assays of biocompatibility. *Scr. Med.* **2009**, *82*, 76–82.

33. Cremasco, A.; Messias, A.D.; Esposito, A.R.; de Rezende Duek, E.A.; Caram, R. Effects of alloying elements on the cytotoxic response of titanium alloys. *Mater. Sci. Eng. C* **2011**, *31*, 833–839. [CrossRef]

34. Schneider-Maunoury, C.; Weiss, L.; Perroud, O.; Joguet, D.; Boisselier, D.; Laheurte, P. An application of differential injection to fabricate functionally graded Ti-Nb alloys using DED-CLAD®process. *J. Mater. Process. Technol.* **2019**, *268*, 171–180. [CrossRef]

35. Schneider-Maunoury, C.; Weiss, L.; Acquier, P.; Boisselier, D.; Laheurte, P. Functionally graded Ti6Al4V-Mo alloy manufactured with DED-CLAD®process. *Addit. Manuf.* **2017**, *17*, 55–66. [CrossRef]

36. Novelli, M.; Bocher, P.; Grosdidier, T. Effect of cryogenic temperatures and processing parameters on gradient-structure of a stainless steel treated by ultrasonic surface mechanical attrition treatment. *Mater. Charact.* **2018**, *139*, 197–207. [CrossRef]

37. Samih, Y.; Novelli, M.; Thiriet, T.; Bolle, B.; Allain, N.; Fundenberger, J.J.; Marcos, G.; Czerwiec, T.; Grosdidier, T. Plastic deformation to enhance plasma-assisted nitriding: On surface contamination induced by Surface Mechanical Attrition Treatment. *IOP Conf. Ser. Mater. Sci. Eng.* **2014**, *63*, 012020. [CrossRef]

38. Alikhani Chamgordani, S.; Miresmaeili, R.; Aliofkhazraei, M. Improvement in tribological behavior of commercial pure titanium (CP-Ti) by surface mechanical attrition treatment (SMAT). *Tribol. Int.* **2018**, *119*, 744–752. [CrossRef]

39. Wen, L.; Wang, Y.; Zhou, Y.; Guo, L.X.; Ouyang, J.H. Iron-rich layer introduced by SMAT and its effect on corrosion resistance and wear behavior of 2024 Al alloy. *Mater. Chem. Phys.* **2011**, *126*, 301–309. [CrossRef]

40. Wen, L.; Wang, Y.; Jin, Y.; Ren, X. Comparison of corrosion behaviour of nanocrystalline 2024-T4 Al alloy processed by surface mechanical attrition treatment with two different mediums. *Corros. Eng. Sci. Technol.* **2015**, *50*, 425–432. [CrossRef]

41. Zheng, Y.; Zeng, W.; Wang, Y.; Zhang, S. Kink deformation in a beta titanium alloy at high strain rate. *Mater. Sci. Eng. A* **2017**, *702*, 218–224. [CrossRef]

42. Lee, C.M.; Ju, C.P.; Chern Lin, J.H. Structure-property relationship of cast Ti-Nb alloys. *J. Oral Rehabil.* **2002**, *29*, 314–322. [CrossRef] [PubMed]

43. Thoemmes, A.; Bataev, I.A.; Belousova, N.S.; Lazurenko, D.V. Microstructure and mechanical properties of binary Ti-Nb alloys for application in medicine. In Proceedings of the 2016 11th International Forum on Strategic Technology (IFOST), Novosibirsk, Russia, 1–3 June 2016; pp. 26–29.

44. Ho, W.F. Effect of omega phase on mechanical properties of Ti-Mo alloys for biomediacal applications. *J. Med. Biol. Eng.* **2007**, *28*, 47–51.

45. Sadeghpour, S.; Abbasi, S.M.; Morakabati, M.; Karjalainen, L.P. Effect of dislocation channeling and kink band formation on enhanced tensile properties of a new beta Ti alloy. *J. Alloys Compd.* **2019**, *808*, 151741. [CrossRef]

46. Deligianni, D.D.; Katsala, N.; Ladas, S.; Sotiropoulou, D.; Amedee, J.; Missirlis, Y.F. Effect of surface roughness of the titanium alloy Ti}6Al}4V on human bone marrow cell response and on protein adsorption. *Biomaterials* **2001**, *22*, 1241–1251. [CrossRef]

47. Huang, H.H.; Ho, C.T.; Lee, T.H.; Lee, T.L.; Liao, K.K.; Chen, F.L. Effect of surface roughness of ground titanium on initial cell adhesion. *Biomol. Eng.* **2004**, *21*, 93–97. [CrossRef]

48. Rosales-Leal, J.I.; Rodríguez-Valverde, M.A.; Mazzaglia, G.; Ramón-Torregrosa, P.J.; Díaz-Rodríguez, L.; García-Martínez, O.; Vallecillo-Capilla, M.; Ruiz, C.; Cabrerizo-Vílchez, M.A. Effect of roughness, wettability and morphology of engineered titanium surfaces on osteoblast-like cell adhesion. *Colloids Surf. Physicochem. Eng. Asp.* **2010**, *365*, 222–229. [CrossRef]

49. Johansson, C.B.; Albrektsson, T.; Ericson, L.E.; Thomsen, P. A quantitative comparison of the cell response to commercially pure titanium and Ti-6Al-4V implants in the abdominal wall of rats. *J. Mater. Sci. Mater. Med.* **1992**, *3*, 126–136. [CrossRef]

50. Eisenbarth, E.; Velten, D.; Müller, M.; Thull, R.; Breme, J. Biocompatibility of β-stabilizing elements of titanium alloys. *Biomaterials* **2004**, *25*, 5705–5713. [CrossRef]
51. Ju, C.P.; Lee, C.M.; Chern Lin, J.H. Medical Implant Made of Biocompatible Low Modulus High Strength Titanium-Niobium Alloy and Method of Using the Same. Patent US20020162608A1, 7 November 2002.

MDPI

St. Alban-Anlage 66

4052 Basel

Switzerland

Tel. +41 61 683 77 34

Fax +41 61 302 89 18

www.mdpi.com

Metals Editorial Office

E-mail: metals@mdpi.com

www.mdpi.com/journal/metals